Socio-Economic Impacts of Carbon Sequestration on Livelihoods and Future Climate

Socio-Economic Impacts of Carbon Sequestration on Livelihoods and Future Climate

Editors

Marina Cabral Pinto
Amit Kumar
Munesh Kumar

MDPI • Basel • Beijing • Wuhan • Barcelona • Belgrade • Manchester • Tokyo • Cluj • Tianjin

Editors

Marina Cabral Pinto
GeoBioTec, Department of Geosciences
University of Aveiro
Aveiro
Portugal

Amit Kumar
School of Hydrology and Water Resources
Nanjing University of Information Science Technology
Ningliu Road No. 219, Nanjing
China

Munesh Kumar
Department of Forestry Natural Resources,
H.N.B. Garhwal University
Srinagar Garhwal, Uttarakhand 249161
India

Editorial Office
MDPI
St. Alban-Anlage 66
4052 Basel, Switzerland

This is a reprint of articles from the Special Issue published online in the open access journal *Land* (ISSN 2073-445X) (available at: www.mdpi.com/journal/land/special_issues/Carbon_Sequest).

For citation purposes, cite each article independently as indicated on the article page online and as indicated below:

LastName, A.A.; LastName, B.B.; LastName, C.C. Article Title. *Journal Name* **Year**, *Volume Number*, Page Range.

ISBN 978-3-0365-2991-2 (Hbk)
ISBN 978-3-0365-2990-5 (PDF)

Contents

About the Editors

Marina Cabral Pinto

Dr Marina Cabral-Pinto works in environmental geochemistry research on an array of topics such as medical geology, epidemiology, water, soil, air quality, geo-bioremediation, ecological health assessment, carbon sequestration, climate changes, etc. She completed her degree in Geological Engineering, MsD (University of Aveito and University of Coimbra), and obtained her Ph.D and Postdoc from University of Aveiro, where she is now working as an investigator and invited professor. During her academic career, she has published more than 60 research/review articles of international repute and attended several conferences and training events. In addition to this, she has handled Special Issues of (*MDPI*), *Land* (*MDPI*), *Applied Sciences* (*MDPI*), and *Mine Water Association* (*Springer*). She has supervised more than 15 PhD and MSc students. She was granted the Orlando Leit ˜ao premium by the Portuguese Society of Neurology due her research and results on links between environmental potentially toxic elements and cognitive disorders and diseases, such Alzheimer's disease. She was funded by FCT for the Covid19 Project and by CNRS for the LabEx DRIIHM research: Exposure and Health in an industrial area (Portugal); ChildIngest –Ingestion of dust by young children: physicochemical characterization, bioaccessibility and genotoxicity: https://orcid.org/0000-0002-6908-1596 Scopus Author ID: 22133337700 Ciˆencia ID: B110-D819-0C69 https://www.researchgate.net/profile/Marina-Cabral-Pinto/publications?sorting=recentlyAdded&editMode=1Deliv

Amit Kumar

Dr. Amit Kumar (Associate Professor) is an environmentalist-cum-hydrologist who has worked on wide areas such as sources, fate and transport of carbon in water and soil, greenhouse gas emissions from reservoirs/lakes/forest, water quality and ecological health assessments, etc. He completed his M.Tech and Ph.D from one of the top technical institutes of India, IIT Roorkee, under the umbrella of the MHRD fellowship, and undertook postdoctoral research at Hohai University (College of Hydrology and Water Resources, Nanjing, China) and is now working with Nanjing University of Information Science and Technology, School of Hydrology and Water Resources, Nanjing, China, as an associate professor. During his academic career, he has published more than 75 research/review articles of international repute and attended several training events and conferences with the fellowship of ICAR, ETH, CAS, MOEF, DAAD, etc. Beside received NSFC-young scientist project in the year 2021. In addition to this, he has handled Special Issues in *Land* (*MDPI*), *Applied Sciences* (*MDPI*), and Frontier in Environmental Science, and an advisory board member of Ecological Indicator (*Elsevier*), and an Associate Editor in Applied Water Science (*Springer*) and (*Taylor and Francis*).

Munesh Kumar

Dr. Munesh has worked on forestry, Himalaya's biodiversity and climate change, carbon sequestration, and carbon fluxes for the last 15 years. Currently, he is working as an Associate Professor at H.N.B Garhwal University's, a Central University, Department of Forestry & Natural Resources, Uttarakhand, India. He is involved in research and teaching work. During his academic career, he has published over 100 research and review articles. Dr Munesh is guest editor in Land, Frontier in Environmental Science and a reviewer for various prestigious journals.

land

MDPI

Editorial

Editorial for Special Issue "Socio-Economic Impacts of Carbon Sequestration on Livelihoods and Future Climate"

Amit Kumar [1,*], Munesh Kumar [2,†] and Marina Cabral Pinto [3,†]

1 School of Hydrology and Water Resources, Nanjing University of Information Science & Technology, Ningliu Road No. 219, Nanjing 210044, China
2 Department of Forestry and Natural Resources, HNB Garhwal University, Srinagar-Garhwal 249161, Uttarakhand, India; muneshmzu@yahoo.com
3 GeoBioTec, Department of Geosciences, Campus de Santiago, University of Aveiro, 3810-193 Aveiro, Portugal; mcp@uc.pt
* Correspondence: amitkdah@nuist.edu.cn
† Authors contributing an equal share.

check for updates

Citation: Kumar, A.; Kumar, M.; Cabral Pinto, M. Editorial for Special Issue "Socio-Economic Impacts of Carbon Sequestration on Livelihoods and Future Climate". *Land* **2022**, *11*, 51. https://doi.org/10.3390/land11010051

Received: 22 December 2021
Accepted: 29 December 2021
Published: 30 December 2021

In the modern era of industrial revolution, urbanization, and deforestation of forest land, carbon (C) sequestration through well-known activities called "land use, land-use change and forestry (LULUCF)" could establish a win–win situation from a climate change and sustainable development perspective [1]. Equally important are the socio-economic co-benefits of C sequestration, given their implications on properly designed policies especially on restoration and/or conservation of forests located in the tropical eco-regions. Further, the huge contribution of C sequestered in the vegetation and its underlain soil helps to protect socio-economic damages from climate change [2]. The literature suggests that a spatial relationship between types of vegetation, C sequestration, and its dynamics are urgently needed to estimate their socio-economic benefits [3]. However, it is crucial to design this activity for environmentalists, foresters, policymakers, etc., due to the inadequate availability of data, the methodologies used to predict the baseline values of C stocks, and the inadequate facilities used to assess the broader environmental and socio-economic effects. These key challenges are robust and need to be integrated with livelihood costs, benefits, and carbon trade-offs [4].

In the past two decades, the Himalayas are vulnerable to LULC change and climate change, from both ecological and socio-economic perspectives [2,5]. In the scientific fraternity, numerous efforts have been carried out to increase the C storage, particularly in the Himalayan range. However, the presence of vegetation biomass and its underlying soils is beneficial for ecosystem functions and services, assisting in the provision of adaptation and maintenance of natural resource-based livelihoods [6]. The planned adaptation can reduce the impacts of climate change and further overcome the risk of key vulnerability but is found to be scarce or costly [7]. Mitigation measures to reduce greenhouse gas (GHG; particularly CO_2, N_2O, and CH_4) emissions in advance will diminish the potential risk associated with most key vulnerabilities [8]. Usually, soil enriched with soil organic carbon (SOC) could help to improve the soil's structure, fertility and ecosystem services, ecological health, and water-holding capacity [9]. The decomposition of organic matter (OM) by microbes (aerobically and/or anaerobically) provides direct benefits through the biogeochemical cycle and maintains the fertility status of the soil [10]. Proper land-use management practices, when adopted in advance, could increase the carbon stock (vegetation and soil) and further help with climate change mitigation. However, a trade-off exists as, to realize the benefits to livelihood and ecosystem services that can be derived from SOC, it must be depleted through extensive crop production resulting from a net release of carbon dioxide [11]. Additionally, the trade-offs between these goals must be identified. These represent key challenges in the modern era [12].

Generally, carbon enters the terrestrial ecosystem by the process of photosynthesis in green plants, which assimilate carbon dioxide (CO_2) and fix into organic forms, such as SOC and inorganic carbon (SIC) pools, before subsequently returning to the atmosphere [5]. All these activities depend on the fertility status of the soil, its climate, and its management operations [2,11]. Carbon sequestered in soil systems is determined based on: (i) the input rate of OM and its decomposition rate; (ii) the physical protection of either intra-aggregate and/or organo-mineral complexes; and (iii) the soil depths at which the organic carbon (OC) is placed. The carbon of plants in the SOC pool enters either through above-ground litter or root material and is further consumed by herbivores and their excreta. It then enters into the SOC pool. Further, fragments of plants change into small particulate OM fractions in the soil. The literature revealed that a large proportion of the carbon that enters the soil through various sources returns to the atmosphere through the respiration activities [13] carried out by both microbes and root systems (CO_2 efflux falling in the range of from 16 to 95%). Trumbore [14] stated that assessments of autotrophic and/or heterotrophic respiration in the soil system are difficult. Apart from this, Ghosh and Maiti [15] carried out work on coal mining and its vicinity and suggested that open-cast coal mining usually caused a complete loss of C-sink potential due to the destruction of vegetation and soil systems, and thus increased C sequestration potential by restoring the mine spoils area. The mechanism of afforestation activities is widely accepted.

In general, global carbon (C) is categorized into five large pools, where biotic carbon (550 Pg; Pg denotes petagram = 10^{15} g) contributes the minimum share [16], followed by atmospheric (800 Pg), soil-based (950 Pg in SIC and 2,460 Pg in SOC), geologic carbon (5000 Pg), and maximum by oceanic carbon (38,000 Pg) [17]. It is estimated that annually 9 Pg of carbon is released into the atmosphere due to industrial activity and the burning of fossil fuels. Similarly, about 1.5 Pg C was released due to changes in land use and deforestation [18]. However, each year, 60 Pg C is exchanged between terrestrial ecosystems and the atmosphere, which will become important in the next few decades, allowing one to buy time while new C-saving, -capture, and -sequestration technologies are being developed [19]. In the global context, soils contain 1500 Pg of SOC at depths of up to 1 m, which is two times that of atmospheric carbon. It is also estimated that 160 Pg of OC is stored in the soils of agricultural croplands worldwide [20]. In a meta-analysis, the long-term use of native forest and pastureland for agricultural productions usually reduced SOC stocks to 42% and 59%, respectively. It is also suggested that if improved agricultural management practices are adopted, world agricultural soils have the potential to sequester C from 0.4–0.8 Pg per year. These improved management practices in agriculture, such as a conversion to permanent pasture, crop rotation, and fertilizer application, usually increase C in soil [21]. The soil organic matter (SOM) retained by grassland is usually influenced by various management practices, which further impacted the rate of carbon sequestration. Guo and Gifford [22], in their meta-analysis study, suggested that the conversion of cropland to pasture increases 30% of the soil carbon. Conversely, the conversion of arable land through grassland decreases the soil C amount by close to 60% [22]. Other management practices, such as increased forage production, irrigation, the use of fertilization, the sowing of grasses and legumes, and introducing earthworms, also drastically increase SOM and change the environmental chemistry [21].

The CO_2 concentration in the atmosphere exceeded the pre-industrial period by ~40%, and forests, agricultural crops, soil, and water bodies play an important role in reducing its level, behaving as major carbon sinks [2,8,11]. Furthermore, numerous innovative approaches have been implemented, proposed, and formulated in the recent past to mitigate the effects of excessive GHG emissions on global climate change. Reforestation through the conversion of non-forest lands to forest land is undoubtedly an important dimension of C sequestration. Recently, C sequestration in the terrestrial ecosystems is an important clean development mechanism (CDM) and a long-lasting solution to the mitigation and management of GHG emissions. The impact of socio-economic factors, such as a rapid increase in population and economic growth, changing lifestyle, a change in

LULC, and various environmental policies, are important driving forces for future climate change [23,24]. Reducing CO_2 in particular and other major GHG is considered the most pressing environmental challenge [17]. As the United States is a signatory of the United Nations Framework Convention on Climate Change (UNFCCC), it is actively involved in finding the key solutions to the problems created by climate change.

It is evident from the literature that very few studies have been conducted on the carbon dynamics, rate of C sequestration in the terrestrial ecosystem, and its medium-/long-term impact on socio-economics and livelihoods. Although filling these gaps in the research is an urgent task for sustainable environmental management, further state-of-the-art studies on C sequestration and its dynamics (sink and/or source) combining soil sciences, forestry, agriculture and their impact on socio-economics are missing. The present Special Issue explores the C sequestration of vegetation and its underlying soil, deforestation, as well as its impact on climate change, a vulnerability risk assessment for the climate, socio-economic impacts, and the mitigation of future climate impact strategies. The theme of the Special Issue extends across environmental policy (e.g., the Paris Agreement and REDD+), C sequestration in the terrestrial ecosystems, the applicability of land use, and the C credit generated at regional and global scales. This Special Issue is highly useful for environmentalists, hydrologists, soil scientists, and policymakers to understand the C sequestration potential of the vegetation and underlying soil in the terrestrial ecosystems at the regional, national and global scale to further study the long-term impacts on socio-economic development resulting from its implementation via climate change modeling.

Author Contributions: Conceptualization, A.K. and M.K.; writing—original draft preparation, A.K. and M.K.; writing—review and editing, A.K., M.K. and M.C.P. All authors have read and agreed to the published version of the manuscript.

Acknowledgments: We thank all the reviewers for their feedback on earlier versions of the manuscripts in this Special Issue.

Conflicts of Interest: The authors declare no conflict of interest.

References

1. Ji, Y.; Guo, X.; Zhong, S.; Wu, L. Financialization, Uncoordinated Development of Population Urbanization and Land Urbanization, and Economic Growth: Evidence from China. *Land* **2020**, *9*, 481. [CrossRef]
2. Kumar, A.; Kumar, M.; Pandey, R.; ZhiGuo, Y.; Cabral-Pinto, M. Forest soil nutrient stocks along altitudinal range of Uttarakhand Himalayas: An aid to Nature Based Climate Solutions. *CATENA* **2021**, *207*, 105667. [CrossRef]
3. Magnani, M.C.; Azofeifa, A.S.; Metternicht, G.; Laakso, K. Integration of remote-sensing based metrics and econometric models to assess the socio-economic contributions of carbon sequestration in unmanaged tropical dry forests. *Environ. Sustain. Indic.* **2021**, *9*, 100100. [CrossRef]
4. Stringer, L.C.; Dougill, A.J.; Thomas, A.D.; Spracklen, D.; Chesterman, S.; Speranza, C.I.; Rueff, H.; Riddell, M.; Williams, M.; Beedy, T.; et al. Challenges and opportunities in linking carbon sequestration, livelihoods and ecosystem service provision in drylands. *Environ. Sci. Policy* **2012**, *19*, 121–135. [CrossRef]
5. Tamang, M.; Chettri, R.; Vineeta; Shukla, G.; Bhat, J.A.; Kumar, A.; Kumar, M.; Suryawanshi, A.; Cabral-Pinto, M.; Chakravarty, S. Stand Structure, Biomass and Carbon Storage in *Gmelina arborea* Plantation at Agricultural Landscape in Foothills of Eastern Himalayas. *Land* **2021**, *10*, 387. [CrossRef]
6. Huo, C.; Chen, L. Research on the Impact of Land Circulation on the Income Gap of Rural Households: Evidence from CHIP. *Land* **2021**, *10*, 781. [CrossRef]
7. Nor Diana, M.I.; Muhamad, N.; Taha, M.R.; Osman, A.; Alam, M.M. Social Vulnerability Assessment for Landslide Hazards in Malaysia: A Systematic Review Study. *Land* **2021**, *10*, 315. [CrossRef]
8. Kumar, A.; Yu, Z.G.; Klemeš, J.J.; Bokhari, A. A state-of-the-art review of greenhouse gas emissions from Indian hydropower reservoirs. *J. Clean. Prod.* **2021**, *320*, 128806. [CrossRef]
9. Holm, A.M.; Watson, I.W.; Loneragan, W.A.; Adams, M.A. Loss of patch-scale heterogeneity on primary productivity and rainfall-use efficiency in Western Australia. *Basic Appl. Ecol.* **2003**, *4*, 569–578. [CrossRef]
10. Scholes, R.J.; Monteiro, P.M.S.; Sabine, C.L.; Canadell, J.G. Systematic long-term observations of the global carbon cycle. *Trends Ecol. Evol.* **2009**, *24*, 427–430. [CrossRef]
11. Kumar, A.; Bhatia, A.; Sehgal, V.K.; Tomer, R.; Jain, N.; Pathak, H. Net Ecosystem Exchange of Carbon Dioxide in Rice-Spring Wheat System of Northwestern Indo-Gangetic Plains. *Land* **2021**, *10*, 701. [CrossRef]

12. Powlson, D.S.; Whitmore, A.P.; Goulding, K.W.T. Soil carbon sequestration to mitigate climate change: A critical re-examination to identify the true and the false. *Eur. J. Soil Sci.* **2011**, *62*, 42–55. [CrossRef]
13. Darrah, P.R. Rhizodeposition under ambient and elevated CO_2 levels. *Plant Soil* **1996**, *187*, 265–275. [CrossRef]
14. Trumbore, S. Age of soil organic matter and soil respiration: Radiocarbon constraints on belowground dynamics. *Ecol. Appl.* **2000**, *10*, 399–411. [CrossRef]
15. Ghosh, D.; Maiti, S.K. Eco-Restoration of Coal Mine Spoil: Biochar Application and Carbon Sequestration for Achieving UN Sustainable Development Goals 13 and 15. *Land* **2021**, *10*, 1112. [CrossRef]
16. Houghton, R.A. Balancing the global carbon budget. *Annu. Rev. Earth Planet. Sci.* **2007**, *35*, 313–347. [CrossRef]
17. IPCC (Intergovernmental Panel on Climate Change). *Climate Change 2007: Mitigation of Climate Change. Working Group III*; Cambridge University Press: Cambridge, UK, 2007.
18. Global Carbon Project. Carbon Budget 2008: An Annual Update of the Global Carbon Budget and Trends. Available online: http://www.globalcarbonproject.org/carbonbudget/index.htm (accessed on 1 December 2021).
19. Morgan, J.A.; Follett, R.F.; Allen, L.H.; Del Grosso, S.; Derner, J.D.; Dijkstra, F.; Franzluebbers, A.; Fry, R.; Paustian, K.; Schoeneberger, M.M. Carbon sequestration in agricultural lands of the United States. *J. Soil Water Conserv.* **2010**, *65*, 6A–13A. [CrossRef]
20. Stockmann, U.; Adams, M.A.; Crawford, J.W.; Field, D.J.; Henakaarchchi, N.; Jenkins, M.; Minasny, B.; McBratney, A.B.; de Courcelles, V.d.R.; Singh, K.; et al. The knowns, known unknowns and unknowns of sequestration of soil organic carbon. *Agric. Ecosyst. Environ.* **2013**, *164*, 80–99. [CrossRef]
21. Lam, S.; Chen, D.; Mosier, A.; Roush, R. The potential for carbon sequestration in Australian agricultural soils is technically and economically limited. *Sci. Rep.* **2013**, *3*, 2179. [CrossRef]
22. Guo, L.B.; Gifford, R.M. Soil carbon stocks and land use change: A meta analysis. *Glob. Change Biol.* **2002**, *8*, 345–360. [CrossRef]
23. Chen, L.; Ye, W.; Huo, C.; James, K. Environmental Regulations, the Industrial Structure, and High-Quality Regional Economic Development: Evidence from China. *Land* **2020**, *9*, 517. [CrossRef]
24. Bhardwaj, D.R.; Tahiry, H.; Sharma, P.; Pala, N.A.; Kumar, D.; Kumar, A. Influence of Aspect and Elevational Gradient on Vegetation Pattern, Tree Characteristics and Ecosystem Carbon Density in Northwestern Himalayas. *Land* **2021**, *10*, 1109. [CrossRef]

land

MDPI

Article

Net Ecosystem Exchange of Carbon Dioxide in Rice-Spring Wheat System of Northwestern Indo-Gangetic Plains

Amit Kumar [1,2], Arti Bhatia [1,*], Vinay Kumar Sehgal [3], Ritu Tomer [1], Niveta Jain [1] and Himanshu Pathak [4]

[1] Center for Environment Science and Climate Resilient Agriculture (CESCRA), ICAR-IARI, New Delhi 110012, India; amitkumar.csb@gov.in (A.K.); ritutomer@gmail.com (R.T.); niveta.jain@icar.gov.in (N.J.)
[2] Central Muga Eri Research and Training Institute Lahdoigarh, Jorhat 785700, India
[3] Agricultural Physics, ICAR-IARI, New Delhi 110012, India; vk.sehgal@icar.gov.in
[4] IVCAR-National Institute of Abiotic Stress Management, Baramati 413115, India; himanshu.pathak@icar.gov.in
* Correspondence: arti.bhatia@icar.gov.in

Abstract: Rice growing under anaerobic conditions followed by spring wheat under an aerobic environment differentially impact the net ecosystem exchange (NEE) of carbon dioxide (CO2) in rice-wheat systems of the north-western Indo-Gangetic Plains (IGP). This is the first estimation of the NEE in a rice-spring wheat sequence via the eddy covariance technique in the north-western Indo-Gangetic Plains, which was partitioned into gross primary productivity (GPP) and ecosystem respiration (RE) and correlated with the environmental variables. Higher CO_2 uptake of -10.43 g C m^{-2} d^{-1} was observed in wheat during heading as compared to -7.12 g C m^{-2} d^{-1} in rice. The net uptake of CO_2 was 25% lower in rice. The average daily NEE over the crop season was -3.74 and -5.01 g C m^{-2} d^{-1} in rice and wheat, respectively. The RE varied from 0.07–9.00 g C m^{-2} d^{-1} in rice and from 0.05–7.09 g C m^{-2} d^{-1} in wheat. The RE was positively correlated with soil temperature at 5 cm depth (0.543, $p < 0.01$) in rice and with air temperature (0.294, $p < 0.01$) in wheat. The GPP was positively correlated with air temperature (0.129, $p < 0.05$) and negatively correlated with vapor pressure deficit (VPD) (-0.315, $p < 0.01$) in rice. In wheat, GPP was positively correlated with air temperature (0.444, $p < 0.01$) and soil moisture (0.471, $p < 0.01$). The rate of GPP over the crop duration was nearly the same in both rice and wheat, however, the RE was higher in rice as compared to wheat, thus, the ratio of cumulative RE/GPP was 0.51 in rice and much lower at 0.34 in spring wheat. Rice contributed 46% and 43% to the annual totals of RE and GPP, respectively, while spring wheat contributed 36% and 51%. The NEE of CO_2 was higher in spring wheat at -576 g C m^{-2} d^{-1} as compared to -368 g C m^{-2} in rice. Thus, while estimating the carbon sink potential in the intensively cultivated northern IGP, we need to consider that spring wheat may be a moderately stronger sink of CO_2 as compared to rice in the rice-wheat system.

Keywords: net ecosystem exchange; ecosystem respiration; gross primary productivity; eddy covariance micrometeorological technique

Citation: Kumar, A.; Bhatia, A.; Sehgal, V.K.; Tomer, R.; Jain, N.; Pathak, H. Net Ecosystem Exchange of Carbon Dioxide in Rice-Spring Wheat System of Northwestern Indo-Gangetic Plains. *Land* **2021**, *10*, 701. https://doi.org/10.3390/land10070701

Academic Editor: Marko Scholze

Received: 28 January 2021
Accepted: 16 March 2021
Published: 2 July 2021

1. Introduction

Carbon dioxide (CO_2) is an important greenhouse gas (GHG) that is emitted from agricultural fields [1]. The concentration of CO_2 in the atmosphere has substantially exceeded, by about 40%, that of pre-industrial levels [1–5]. Forests, agricultural crops, soil and water bodies contribute towards the reduction in CO_2 and act as major sinks [6–9]. Quantification of exchange of CO_2 between agricultural soils and the atmosphere is important in order to assess the global carbon budgets as CO_2 accounted for 76% of total anthropogenic GHG emissions in 2010 [1–5,10]. Rice (*Oriyza sativa*) and wheat (*Triticum aestivum*) are the two important crops that are essential for global food security. The rice–wheat cropping system is the largest agricultural production system in the world, occupying around 13.5 million hectares (M ha) in South Asia [11] and around 13 M ha in China [12]. In China, the rice harvest is followed by winter wheat (7–8 month duration), whereas in South Asia, where the winters are milder, the wheat grown during the winter is called spring wheat and is of shorter duration (4 months). In South Asia, 85% of the rice-wheat area lies in the Indo-Gangetic Plains (IGP) which produce about 50% of the total food grains from this region [13].

In the rice–wheat system of the study region, rice is grown during the summer months from July to October when the climate is warm and sub-humid, whereas the successive wheat is grown from December to April in a relatively dry winter [14]. There is a high level of variation in the temperatures and the amount of rainfall received between the rice and wheat crop growing seasons. Rice and wheat are grown not only in contrasting climatic conditions but the agricultural practices of tillage, manure, fertilizer, seeding, transplanting and water management also differ [15]. Rice is transplanted and grown in submerged anaerobic soil, whereas wheat is grown in upland aerobic soil [16]. Rice is a tropical crop, while wheat is a winter crop, with temperature playing a key role in their growth due to its influence on photosynthetic efficiency [17]. The contrasting soil environments influence the carbon and nitrogen dynamics, affecting the degradation of soil organic matter and, thus, the net ecosystem exchange (NEE) of CO_2.

Net ecosystem exchange (NEE) is a measure of the net exchange of C between an ecosystem and per unit ground area of the atmosphere and is a primary gauge of ecosystem C sink strength [18]. GPP represents the gross CO_2 uptake through photosynthesis by plants, whereas ecosystem respiration (RE) is the gross CO_2 release from the ecosystem through both autotrophic and heterotrophic respiration [19]. Rice and wheat ecosystems may be significant sinks/sources of CO_2, depending on the soil type, crop duration, management and the climate of the region. In Asian regions, measurements of net ecosystem CO_2 exchange have been conducted in rice and wheat [12,20,21]. The crop duration of both rice and spring wheat in the South Asian region is about 100 to 120 days, unlike the winter wheat which has a much longer duration. Winter wheat is mostly grown as a rain-fed crop, whereas spring wheat is grown under irrigated conditions in the IGP [22].

Chen et al. [12] quantified the NEE of CO_2 in a rice-winter wheat cropping system in the North China Plain, however, no measurements have been collected of NEE in South Asian rice-spring wheat systems, where two contrasting crops are grown in one annual cycle. The factors controlling CO_2 gas exchange between the rice canopy and atmosphere are different from those in upland wheat fields, as rice is flooded during most of its cultivation period, thus, leading to a great deal of uncertainty in the net CO_2 budgets [23]. Thus, to reduce the uncertainties that exist in the global CO_2 budgets of rice-spring wheat systems, that occupy nearly 13.5 M ha of land area in South Asia, the present investigation was conducted with the objective to understand (1) the differences in NEE, RE and GPP between rice and spring wheat grown in rice-wheat rotation systems, (2) compare the characteristics of diurnal and daily variations in the CO_2 flux, (3) analyze the environmental factors that affect these variations in NEE, RE and GPP in the north-western Indo-Gangetic Plains.

2. Material and Methods

2.1. Site Description

This study was conducted in a Typic Ustochrept soil at the experimental farm of the Indian Agricultural Research Institute (IARI) (latitude 28°38′37.7″ N, longitude 77°09′09.8″ E, 207 MSL) Delhi, India, where an eddy covariance (EC) flux tower was installed in a rice–wheat crop rotation system. The area provided a sufficient upwind fetch of homogenous vegetation required for measuring fluxes using the EC technique. The climatic variables during the rice and wheat season are shown in Figure 1.

2.2. Climatic Description

The climate of the region is subtropical and semi-arid, receiving about 750 mm of rainfall annually, 80% of which occurs during June to October. The mean maximum and minimum temperatures are 33 and 23 °C during the rice and 25 and 11 °C during the wheat growth periods.

2.3. Soil Properties

The alluvial soil of the study site had a bulk density of 1.42 g cm^{-3}, pH (1:2 soil:water) of 8.0, electrical conductivity of 0.47 dS m^{-1}, and was loamy in texture. It had total N, organic carbon, Olsen P, and ammonium acetate extractable K contents of 0.31, 4.6, 0.008, and 0.14 g kg^{-1}, respectively.

2.4. Crop Management

A conventionally puddled rice field was uniformly transplanted on 11 July 2013, with a spacing of 20 cm (row to row) × 15 cm (hill to hill) distance. The field was irrigated whenever the moisture content dropped below the saturation level during the growing season, except during the last two weeks before harvesting. Wheat was sown on 16 December 2013, maintaining a plant spacing of 20 cm (row to row) × 5 cm (plant to plant) distance. Urea, at a rate of 120 kg N ha^{-1}, was applied in three splits of 60, 30, and 30 kg N ha^{-1} at −1, 47, and 66 days after transplanting (DAT) in rice and −1, 25, and 67 days after sowing (DAS) in wheat. The wheat crop was irrigated at the crown root (22 DAS), tillering (44 DAS), late jointing (62 DAS), flowering (85 DAS) and milking (97 DAS). Phosphorus (26.2 kg ha^{-1}) and K (50 kg ha^{-1}) were incorporated into the soil at the time of sowing using single super phosphate (SSP) and muriate of potash (KCl), respectively, in both rice and wheat. No pesticides were applied during crop cultivation. The crop leaf area index was measured using a plant canopy analyzer (LAI-2000) at the key growth stages. The yield related parameters were measured at the harvest of the crops. Total plant biomass of 1335 and 1392 gm^{-2} was recorded at harvest in rice and wheat, respectively.

2.5. Eddy Covariance Measurements

Flux densities of CO_2, momentum, as well as the friction velocities over the rice and wheat canopy were measured by the EC technique from 1 July 2013 to 30 June 2014 from pre-transplanting/sowing to harvesting in both the crops and also during the fallow periods after rice and wheat. A sonic anemometer (CSAT3, Campbell Scientific, Logan, UT, USA) measured the three-dimensional wind and the sonic, or virtual, temperature along the three non-orthogonal sonic axes. The fluctuations in CO_2 and H_2O density were measured using an LI-7200 (LI-COR, Lincoln, NE, USA) close-path infrared gas analyzer. The data from CSAT3 and LI-7200 were sampled at 10 Hz using a CR3000 (Campbell Scientific, Logan, UT, USA) data logger. The Edire software processed the data in real time and computed the CO_2 flux and frictional velocity along with the covariance, standard deviations, and means on a half-hourly basis. NEE is usually calculated as the sum of eddy CO_2 flux (Fc) and CO_2 storage change (Fs) within the airspace below the flux-measuring height [24]. In this study, the Fs term was neglected for the NEE calculation as the canopy height was relatively low at less than 1 m. In the current study, we use the term NEE, which indicates CO_2 flux only.

2.6. Auxiliary Measurements

The LI-7200 was installed at 2.05 m height with a northward, eastward and vertical sensor separation of −5.5 cm, +5.0 cm and 0 cm from the CSAT3, respectively. The north off set was 55°. A net radiometer (CNR4, Kipp&Zonan, Delft, the Netherlands), photosynthetically active radiation (PAR) sensor (PQS 1, Kipp&Zonan) were installed at 2.90 m height, whereas, the air temperature (AT) probe (Pt 100, Rotronic, Bassersdorf, Switzerland) and relative humidity (RH) probe (Hygro Clip S3, Rotronic) were installed at 2.0 m above the soil surface. Rainfall was measured with a tipping bucket rain gauge

(TE-525, Texas Instruments, Dallas, TX, USA). Data of these slow sensors were averaged over a 30-minute period. The system received its power from four 12 V, 100AH DC batteries and was also supported by an AC connection in parallel on foggy and cloudy days. Vapor pressure deficit (VPD) was estimated from the vapor pressure monitored by the LI-7200. The leaf area index (LAI) was periodically measured by a canopy analyzer LI 2200 (LICOR, Lincoln, NE, USA) at the vegetative, reproductive and maturity stages. Half-hourly measurements of soil moisture (SM), heat flux (SHF) and temperature (ST) were taken at 5 cm soil depth using a water content reflectometer (CS 616-L), soil heat flux sensor (HFT3 transducer) and soil and water temperature probe (107 B), respectively.

2.7. Quality Control of Flux Data

The flux data were subjected to quality control checks to screen data for instrument malfunctions, rainfall, etc. A diagnostic output of LI-7200: automatic gain control (agc_li) was used to filter out half-hourly values when the instrument was performing poorly and for identification of outliers. Spike removal was carried out and then corrections for density fluctuations [25] were performed for variations in temperature and water vapor. The coordinate rotation correction was carried out following the method by Kaimal and Finnigan [26]. Sonic temperature correction for humidity was applied as per Van Dijk et al. [27]. The time lag compensations due to the sensor separations between CSAT-3 and LI-7200 were applied as per Fan et al. [28]. Data quality was checked by applying the stationarity test and integral turbulence characteristics test [29] to the computed covariance. The rejection rate was 40% (10% for non-stationary, 40% for integral turbulence characteristics and 14% for both tests) in rice and 40% (15% for non-stationary, 40% for integral turbulence characteristics and 16% for both tests) in wheat. The rejection rate of data was larger when the canopy was small due to lower magnitude of CO_2 flux. The u* threshold of 0.1 m s^{-1} was estimated by plotting the nighttime CO_2 flux (Fc) and frictional velocity (u*). Gap-filling of missing or discarded data was done by simple linear interpolation (1–3 consecutive missing data) and mean diurnal variation method (up to 4–6 consecutive missing data) and by look-up table approach for more than 6 h missing data using standard approach and available data points.

2.8. Partitioning NEE into GPP and RE

Net ecosystem productivity is the net gain or loss of C from the ecosystem and was the difference of gross primary production (GPP) and ecosystem respiration (RE).

$$\text{NEP} = \text{GPP} - \text{RE} \quad (1)$$

The net ecosystem production (NEP) is equivalent to net ecosystem exchange (NEE) but has an opposite sign [30]. Negative NEE or positive NEP indicates CO_2 uptake by the ecosystem.

$$\text{NEP} = -\text{NEE} \quad (2)$$

GPP represents CO_2 assimilation by photosynthesis of the plant during daytime, while RE represents the CO_2 released through respiration from the soil, aboveground plant parts, and the roots of the plants during day and night. As plant respiration during the day is unknown, direct measurement of GPP and RE is difficult [31]. The plant respiration rate was assumed to be the same during the day and night [32]. We estimated the daytime respiration RE (D) using the nighttime average of respiration RE (N). To partition NEE into photosynthetic and respiratory components, we applied the conventional method of Falge et al. [33], in which GPP and RE (D) are expressed as empirical functions of air temperature. We estimated RE (D) using the relationship between nighttime NEE and air temperature. At nighttime, as GPP is zero, the NEE is taken to be equal to RE (N). The RE (N) increases as an exponential function of increasing temperature and is represented as:

$$\text{RE} = \text{A}.\text{e}^{\text{BT}} \quad (3)$$

where, A and B are empirical constants; B is related to the temperature coefficient Q_{10}, as B = ln(Q_{10})/10, and A indicates RE at 0 °C [24]. We divided the whole study period into different growth stages viz. early vegetative (EVS), tillering, stem elongation, booting, heading, ripening and harvesting. For each growth period, the nighttime RE (filtered using $u^* > 0.1$ ms^{-1}) was separated into 1 °C wide temperature bins and averaged, A and B constants were determined by fitting the nighttime NEE as a function of averaged air temperature using the least squares method. At each growth stage, a regression analysis was performed between GPP and PAR [34]. The data after gap filling were used for averaging the CO_2 fluxes at different growth stages. The duration of each stage was used as the weight while computing the CO_2 flux for the whole season. Linear Regression and partial correlation of the datasets were carried out using SPSS (16.0).

3. Results

3.1. Climatic Variability during Rice and Wheat Growing Season

Climatic conditions differed considerably between the two seasons of the study (Figure 1). Solar radiation and wind speed were generally higher during the wheat season than in rice. Higher rainfall (above seasonal average) was recorded during the rice season in 2013 and total rainfalls of 1198 and 168 mm were recorded during the kharif (rice) and rabi (wheat) season, respectively. In rice, most of the rain occurred during July, August and September due to the south-west monsoon, while in wheat, it rained mostly in February and March (non-monsoonal rainfall) due to western disturbances. The maximum air temperature during the growth of rice was higher, while the minimum air temperature was lower in wheat. The average air temperature from 1 July to 31 October 2013, was 28.4 °C and from 1 December 2013, to 30 April 2014, was 17.7 °C and was higher by 1.0 and 0.9 °C in rice and wheat than the seasonal averages [35], respectively. The average air temperature was the highest at the post transplanting stage (30.5 °C) in rice and then decreased as crop growth progressed. It decreased to 28.9 °C at maximum tillering. This decrease was accompanied by an increase in precipitation. The average air temperature again increased at the booting stage in rice (29.8 °C). During wheat, air temperature decreased initially till early in the vegetative stage and then subsequently increased. Total rainfall varied significantly between the rice and wheat crops. A total of 1053 mm and 155 mm rainfall was received during rice and wheat crop growth periods, respectively (Figure 1). The maximum rainfall in rice was received corresponding to the maximum tillering till panicle initiation (425 mm) stage, and in wheat corresponding to the stem elongation stage (54.8 mm). Average wind speeds of 4.25 and 4.67 km h^{-1} were recorded during the rice and wheat growth periods, respectively. Wind from east (E) and east-northeasterly (ENE) directions prevailed during most of the rice crop growth period, while during the wheat period, northwest (NW) and west-northwesterly (WNW) winds prevailed. The average sunshine hours during rice and wheat growth periods were 4.76 and 5.08, respectively. Low levels of solar radiation during the rice and wheat crop were associated with the rainy and winter foggy seasons in northern India, respectively.

3.2. Diurnal Variation in NEE

Throughout the crop growth period, NEE was negative during the daytime (uptake) and positive during the nighttime (release) (Figure 2). A higher negative NEE was observed during the day in wheat as compared to rice. The average diurnal NEE varied between +0.21 to −0.70 in rice and between +0.20 to −0.97 $mgm^{-2}s^{-1}$ in the wheat crop. The daytime NEE decreased with the progress of the crop season in both rice and wheat, reaching a peak average NEE of −0.7 $mgm^{-2}s^{-1}$ in rice (booting-flowering) and −0.97 $mgm^{-2}s^{-1}$ in wheat (heading–ripening). The nighttime average NEE increased marginally during the booting to ripening stages in rice, however, in wheat it showed a steady increasing trend from the early vegetative stage, reaching a peak at the heading-ripening stage. A distinct diurnal pattern in NEE was observed from the early vegetative to ripening stage in rice and wheat, which followed a reverse pattern to that of the PAR during the daytime (Figure 3A,B).

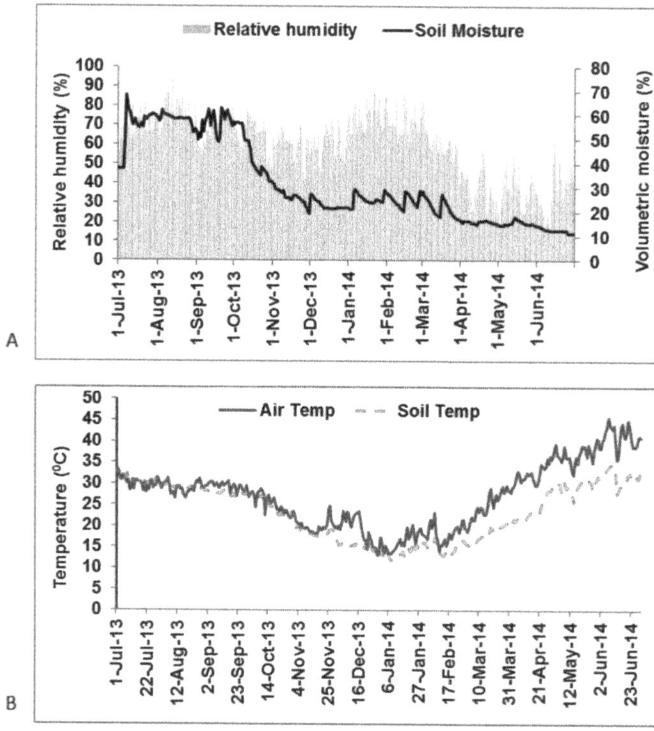

Figure 1. Variations in (**A**) relative humidity and soil moisture (**B**) air temperature and soil moisture during the study period.

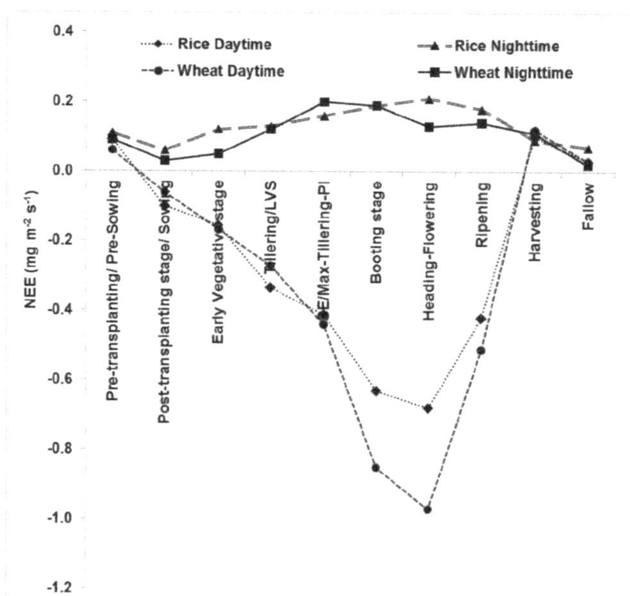

Figure 2. Daytime and nighttime net ecosystem exchange (NEE) in rice and wheat over the crop season.

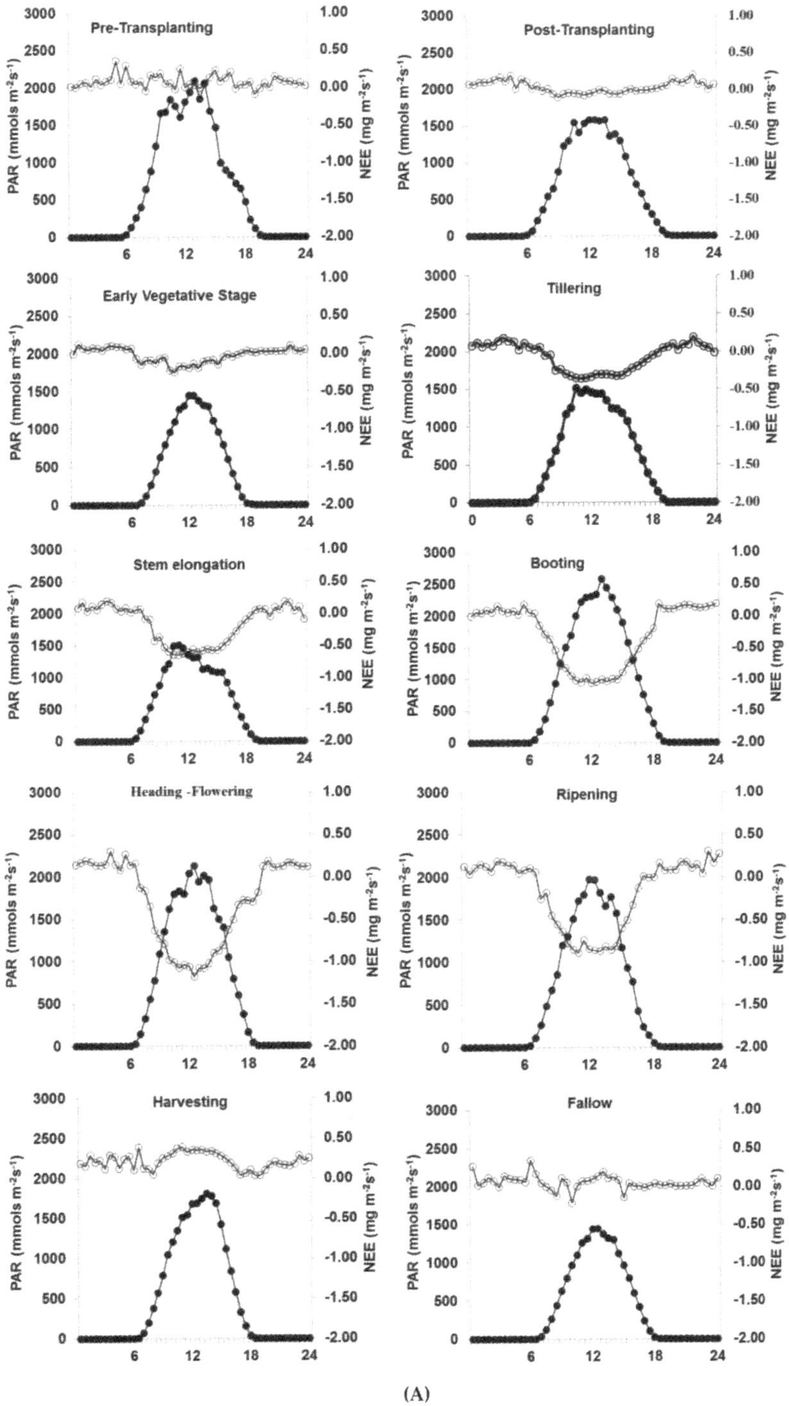

Figure 3. *Cont.*

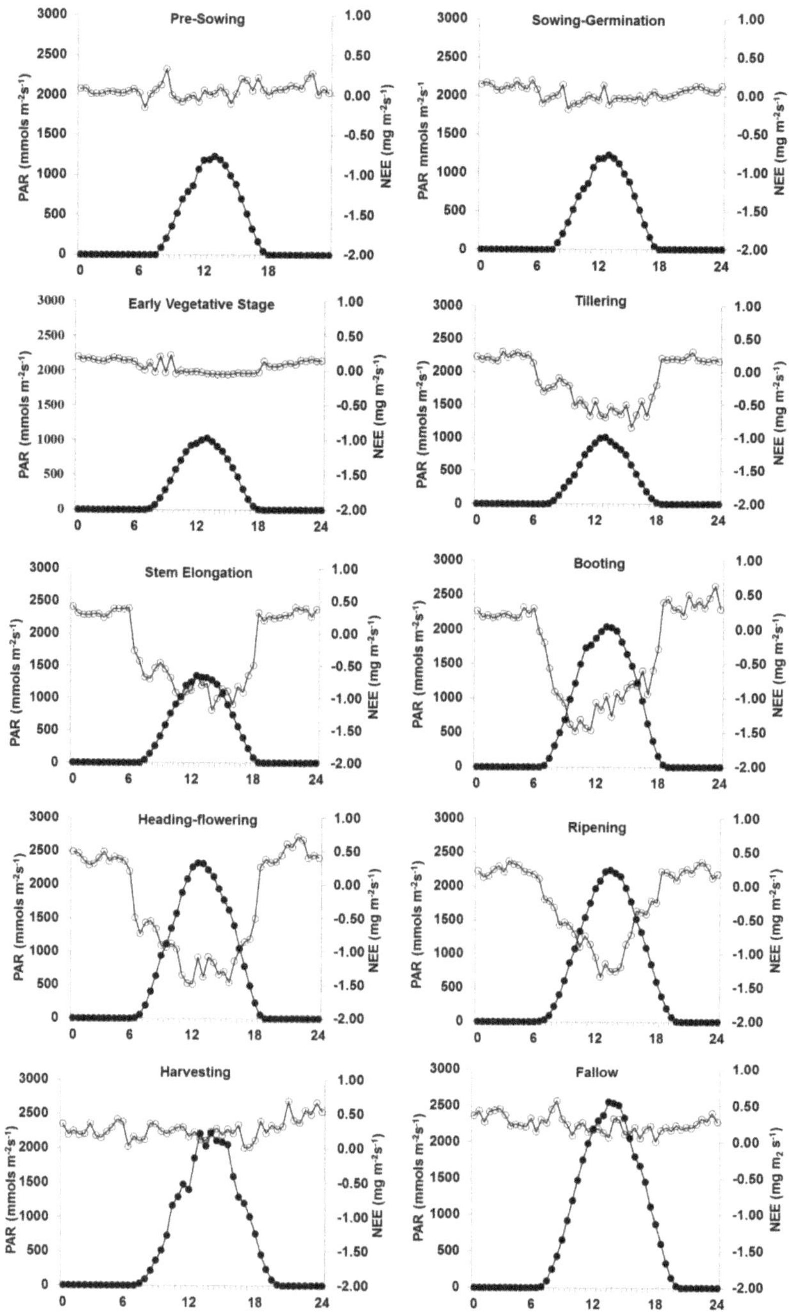

(B)

Figure 3. (**A**). Diurnal variation in NEE (empty circle) and incident photosynthetically active radiation (PAR) (filled circle) at half hourly interval, averaged for different rice growth stages. (**B**). Diurnal variation in NEE of CO_2 (empty circle) and incident PAR (filled circle) at half hourly interval, averaged for different wheat growth stages.

3.3. Seasonal Variation in Daily NEE

The daily average NEE over the crop season varied from+0.67 to −8.83 g C m^{-2} d^{-1} in rice (Figure 4A) and from +01.07 to −11.96 g C m^{-2} d^{-1} in wheat (Figure 4B). The daily average NEE increased with the rice growth stage and peaked during the reproductive stage. During the reproductive stage of heading–flowering, the NEE was −8.83 g C m^{-2} d^{-1} in rice and −11.96 g C m^{-2} d^{-1} in wheat. The average daily NEE for the crop season was 25% lower in rice at −3.74 g C m^{-2} d^{-1}, as compared to −5.01 g C m^{-2} d^{-1} in wheat (Table 1). Wheat had a cumulative NEE of −576 g C m^{-2}, as compared to rice at −368 g C m^{-2} (Table 1). The fallow period duration, after rice and wheat, was 61 and 91 days long, respectively. The NEE at the harvest of rice and wheat were 0.35 g C m^{-2} d^{-1} and 0.46 g C m^{-2} d^{-1}, respectively (Table 1). The cumulative NEE was higher in the fallow period after wheat at 32.76 g C m^{-2}, as compared to after rice at 17.08 g C m^{-2}.

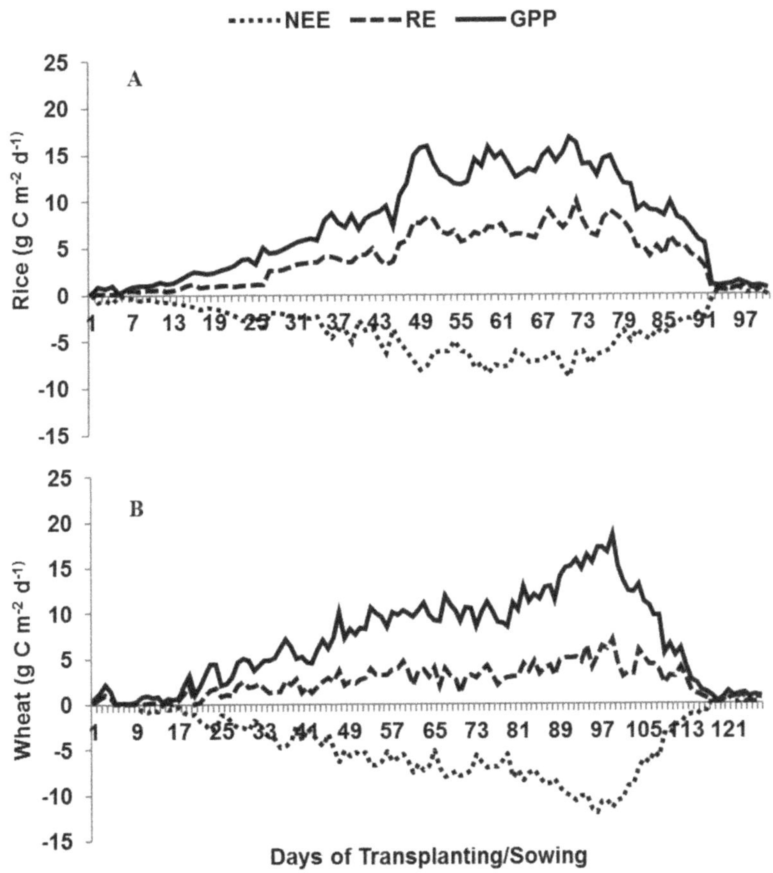

Figure 4. Daily mean net ecosystem exchange (NEE), gross primary productivity (GPP) and ecosystem respiration (RE) at different growth stages of rice (**A**) and wheat (**B**) in northwestern Indo-Gangetic Plains (IGP).

Table 1. NEE, RE and GPP at different rice and wheat growing stages.

Crop/Stage	Crop/Stage Duration (Days)		Rate ($g\ C\ m^{-2}\ d^{-1}$)						Cumulative ($g\ C\ m^{-2}$)					
	Wheat	Rice	Wheat	Rice	Wheat	Rice	Wheat	Rice	Wheat	Rice	Wheat	Rice	Wheat	Rice
			NEE		RE		GPP		NEE		RE		GPP	
Pre-sowing/Pre-transplanting	4	4	0.69	0.88	0.7	0.89	0.01	0.01	2.76	3.52	2.8	3.56	0.04	0.04
Sowing-germination/Post-Transplanting	9	6	−0.27	−0.21	0.12	0.11	0.39	0.32	−2.43	−1.26	1.08	0.66	3.51	1.92
Early Vegetative Stage	11	14	−1.28	−0.31	0.79	0.42	2.07	0.73	−14.08	−4.34	8.69	5.88	22.77	10.22
Tillering	27	11	−3.6	−2.25	2.01	0.99	5.61	3.24	−97.2	−24.75	54.27	10.89	151.47	35.64
Stem Elongation	28	19	−6.63	−3.41	3.21	3.53	9.84	6.94	−185.64	−64.79	89.88	67.07	275.52	131.86
Booting	10	13	−8.1	−6.54	3.77	6.71	11.87	13.25	−64.8	−85.02	30.16	87.23	94.96	172.25
Heading	15	18	−10.43	−7.12	5.23	7.3	15.66	14.42	−125.16	−128.16	62.76	131.44	187.92	259.56
Ripening	15	17	−4.34	−3.4	2.85	4.92	7.18	8.32	−86.8	−57.8	57.05	83.64	143.6	141.44
Harvest	2	2	0.46	0.35	1.04	0.86	0.58	0.51	0.92	0.7	2.08	1.72	1.16	1.02
Average/Total (Crop period)	115	98	−5.01	−3.74	2.64	3.22	7.65	7.68	−576	−368	304	387	880	753
Fallow (after)	91	61	0.4	0.28	0.87	1.02	0.51	0.74	32.76	17.08	79.17	62.22	46.41	45.14

3.4. RE and GPP at Different Crop Growth Stages

The RE varied from 0.07 to 9.93 $g\ C\ m^{-2}\ d^{-1}$ in rice and 0.04 to 7.08 $g\ C\ m^{-2}\ d^{-1}$ in wheat depending upon the crop stage (Figure 4A,B). The rate of RE during the pre-transplanting and pre-sowing period in rice and wheat was observed to be 0.89 and 0.7 $C\ g\ m^{-2}\ d^{-1}$, respectively (Table 1).

The average rate of GPP over the different growth stages varied from 0.32 to 14.42 $g\ C\ m^{-2}\ d^{-1}$ in rice and 0.39 to 15.66 $g\ C\ m^{-2}\ d^{-1}$ in wheat (Table 1). The highest GPP rate was observed during heading (14.42 and 15.66 $g\ C\ m^{-2}\ d^{-1}$) in both of the crops. The cumulative GPP was much higher in wheat at −888 $g\ C\ m^{-2}$ as compared to −753 $g\ C\ m^{-2}$ in rice at our study site. Total plant biomasses of 1335 and 1392 gm^{-2} were removed at harvest in rice and wheat, respectively. The GPP rate increased with increasing LAI (vegetative stage: LAI 0.9, GPP 2.42 $g\ C\ m^{-2}\ d^{-1}$; reproductive: LAI 4.3, GPP 13.83 $g\ C\ m^{-2}\ d^{-1}$; maturity: LAI 1.9, GPP 7.18 $g\ C\ m^{-2}\ d^{-1}$) in rice and in wheat (vegetative: LAI 1.3, GPP 4.51 $g\ C\ m^{-2}\ d^{-1}$; reproductive: LAI 4.6, GPP 13.76 $g\ C\ m^{-2}\ d^{-1}$; maturity: LAI 2.4, GPP 8.32 $g\ C\ m^{-2}\ d^{-1}$).

3.5. Environmental Variables in Rice and Wheat

Differences in soil temperature at 5 cm depth were observed between rice and wheat. The diurnal average soil temperature varied between 27.74 to 28.92 °C in rice and from 15.38 to 17.53 °C in wheat (Figure 5), depending on the intensity of incident solar radiation. The diurnal mean air temperature ranged from 31.8 to 38.2 °C in rice and from 16.3 to 27.6 °C in wheat, respectively (Figure 5) and the diurnal RH varied between 60–85% in rice and 40–80% in wheat. The relative humidity (RH) and air temperature (AT) showed an opposite diurnal pattern with respect to each other. Soil heat flux (SHF) at 5 cm soil depth was positive from about 12:00 hr to 20:30 hr and was negative from 21:00 hr to 11:30 hr in both rice and wheat. The SHF ranged from −12.2 Wm^{-2} (7:00 hr) to +15.9 Wm^{-2} (15:30 hr) in rice and from −15.6 Wm^{-2} (8:00 hr) to +16.2 Wm^{-2} (15:30 hr) in wheat. The diurnal values of incident PAR and Net radiation (NR) were higher during the rice than in the wheat season (Figure 5). The peak NR value in rice was 370.3 Wm^{-2} and 316.7 Wm^{-2} in wheat, with corresponding incident PAR values of 1742 $\mu\ mol\ m^{-2}s^{-1}$ in rice and 1519 $\mu\ mol\ m^{-2}s^{-1}$ in wheat. The average VPD was 1.75 kPa in rice and 1.48 kPa in wheat.

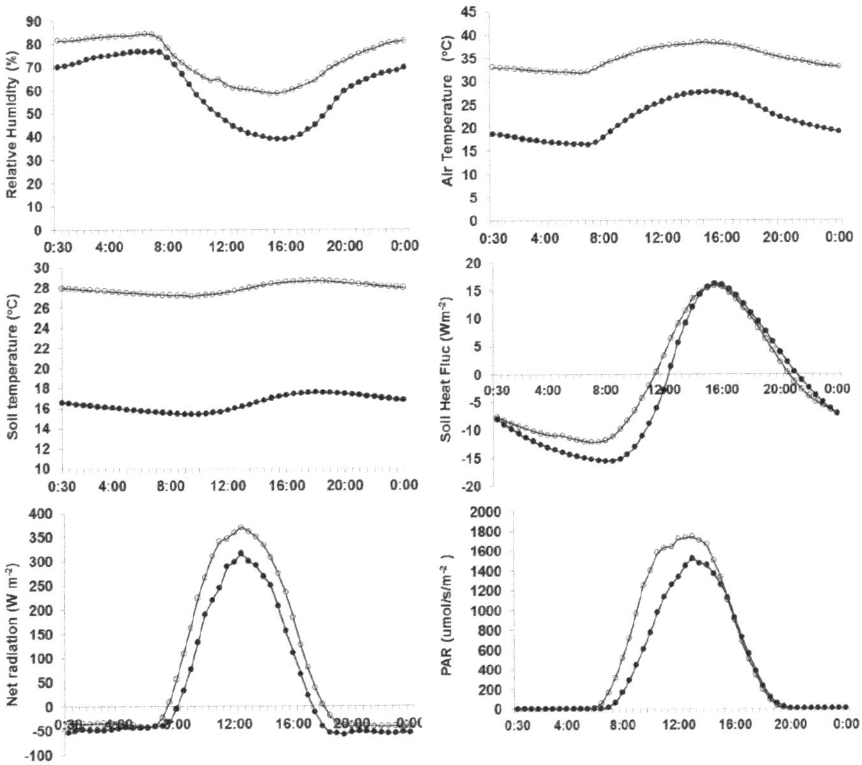

Figure 5. Average diurnal pattern of relative humidity (RH), air temperature (AT), soil temperature (ST), soil heat flux (SHF), net radiation (NR)and PAR in rice (empty circle) and wheat (filled circle).

4. Discussion

4.1. Diurnal and Seasonal Variation in NEE

The average nighttime NEE was, on average, 0.11 and 0.12 $mgm^{-2}s^{-1}$ during the rice and wheat growth periods, respectively. These values are close to those measured by Miyata et al. [36] in Japan (<0.2 $mgm^{-2}s^{-1}$) and Gao et al. [37] in China (0.12 $mgm^{-2}s^{-1}$) in intermittently flooded rice paddies. The mean diurnal variation of NEE showed that both rice and wheat crops acted as net CO_2 sinks, as CO_2 uptake during the day was higher than emission after daylight hours. The highest values of half-hourly NEE were observed during the reproductive stage and the variations in NEE followed those in incident PAR. Zhang et al. [38] also observed relatively large fluctuations in diurnal NEE in maize cropland, orchard, wetland and vegetable field ecosystems with highest CO_2 uptake levels observed in maize crops due to the differences in the light saturation point (α) and carbon absorption capacities of the crops.

The highest leaf area index in rice (4.3) and wheat (4.6) was observed during the reproductive stages of heading to flowering. As PAR values increased beyond 2000 μ mol m^{-2} s^{-1}, there was an increase in carbon assimilation in rice and wheat. As the wheat crop advanced to the heading stage, the peak NEE increased to -1.48 mg CO_2 m^{-2} s^{-1} with a PAR of 2342 μ mol m^{-2} s^{-1}. Higher positive values of NEE at harvesting could be attributed to increased CO_2 emission from soil and plant respiration than uptake level by the plants.

The magnitude of NEE in rice was lower during the vegetative stages and reached its maximum during reproductive stages of heading-flowering at -8.83 g C m^{-2} d^{-1} in rice and -11.96 g C m^{-2} d^{-1} in wheat. Similar values of daily NEE between vegetative and reproductive stages were reported for rice [34,39] and wheat [40]. With the onset of senescence during ripening and maturity, the daily NEE became less negative in rice and turned positive at harvesting.

The daily average NEE during wheat and rice in our study was −5.01 and −3.74 g C m^{-2}, respectively, showing that spring wheat was a stronger CO_2 sink than rice. Chen et. al. [12], in a rice–winter wheat rotation system in China, observed a daily NEE of −2.35 and −3.96 g C m^{-2} in wheat and rice, respectively, even though (Table 4) the winter wheat field took up more CO_2 on a seasonal basis as compared to the rice paddy fields because of the longer growing season. Much higher levels of CO_2 uptake have been reported in maize as compared to soybean and wheat in other eddy covariance studies. Baker and Griffis [41] observed a maximum CO_2 uptake of 51.3 g CO_2 m^{-2} d^{-1} in a maize crop in maize/soybean ecosystem in the USA which was similar to the maximum CO_2 uptake of 45.8 g CO_2 m^{-2} d^{-1} in a maize ecosystem in the North China Plain [42]. The CO_2 uptake levels are different for crops such as corn, as compared to soybean and wheat, due to biochemical and physiological differences between C3 (soybean, wheat) and C4 crops (maize).

Water management, duration and weather conditions are factors which differ in crops and lead to differences in the NEE and uptake of CO_2. In our study, the soil and air temperatures were significantly different between the rice and wheat crop seasons impacting the NEE. The rice duration was 98 days and was irrigated every 2–3 days, whereas wheat irrigated for a duration of 115 days was irrigated five times. The cumulative NEE at our site in the rice–wheat system was higher in wheat (−576 g C m^{-2}) as compared to rice (−368 g C m^{-2}). A much higher NEE in rice of −448 g C m^{-2} was observed by Bhattacharya et al. [19] in eastern India in a rice-rice system. Alberto et al. [34] and Saito et al. [24] reported an NEE of −258 and −398 g C m^{-2} in rice. Higher NEE values have been reported in winter wheat in Germany by Schmidt et al. [43] at −627 and −537 g C m^{-2} in (Table 4) because of the longer crop growth duration compared to spring wheat.

The NEE at the harvest of rice was positive at 0.35 g C m^{-2} d^{-1} due to the respiration losses by the root residues (Table 1). The NEE was more positive at 0.46 g C m^{-2} d^{-1} after the wheat harvest due to higher soil temperatures resulting in higher respiration losses (1.04 g C m^{-2} d^{-1}) in wheat as compared to rice. More weed growth in the fallow period after rice resulted in a higher average GPP at 0.74 g C m^{-2} d^{-1} as compared to 0.54 g C m^{-2} d^{-1} in the fallow period after wheat growth (Table 1). Higher soil temperatures and low soil moisture in the fallow period after wheat resulted in lower GPP values as compared to the fallow period after rice. Bhattacharya et al. [19] also reported a resultant NEE of 22 g C m^{-2} in the summer fallow period as the RE was balanced by photosynthetic CO_2 assimilation. Moureaux et al. [44] reported a contribution of 20 g C m^{-2} due to residues, which represents 5% of the seasonal carbon budget. The RE and GPP during the fallow periods contributed to 17% and 5% of the annual cumulative RE and GPP.

4.2. *Effect of Crop Growth, LAI and Air Temperature on GPP*

The rate of GPP was significantly higher from booting to heading due to increased daytime CO_2 uptake, resulting in higher NEE values of −6.54 and −8.10 at booting and −7.12 and −10.43 g C m^{-2} d^{-1} at heading in rice and wheat, respectively, due to increases in LAI which increased the canopy light interception efficiency. The GPP increased with LAI up to heading, but the GPP per unit leaf area later decreased because of increased shading of the lower leaves in the canopy.

The GPP was significantly correlated to AT in both rice (0.129, $p < 0.05$) and wheat (0.443, $p < 0.01$). Maximum CO_2 uptake was observed at 26–31 °C and 10–26 °C in rice and wheat, respectively. The AT started to decrease after the reproductive stage in rice, thereby reducing the plant maintenance respiration, which produced a greater abundance of assimilates available for growth and yield [45]. The rate of GPP decreased after ripening in wheat. In wheat, higher temperatures after the reproductive stage stimulated photorespiration and the wheat plants reduced their photosynthetic activity, thereby lowering the GPP after ripening. Similar to our results, higher GPP levels from the vegetative to reproductive periods were reported in rice by Campbell et al. [39] and higher net assimilation fluxes ranging between −9 and −13 g C m^{-2} d^{-1} were reported in winter wheat by Béziat et al. [46]. Plants may reduce their photosynthetic activity, thereby lowering CO_2 uptake at higher temperature due to stomatal closure in response to increased transpiration losses.

4.3. Effect of NR, PAR, SM, RH and VPD on GPP

After sunrise, the NR gradually increased, reaching its peak value at around noon and before starting to decline gradually. The same trend was also observed in PAR. During the daytime, the GPP increased as there was greater CO_2 uptake with increasing PAR. A significant positive correlation was observed between PAR and GPP in both rice and wheat (0.225, 0.348; $p < 0.01$), respectively (Table 2). The GPP is mainly dependent on the amount of PAR absorbed (APAR) by green leaf area of the crop canopy [47].

Table 2. Partial correlation coefficient between ecosystem respiration (RE), gross primary productivity (GPP) and related environmental variables.

Environmental Variables	Ecosystem Respiration (RE)		Environmental Variables	Gross Primary Productivity (GPP)	
	Rice (N = 4712)	Wheat (N = 5472)		Rice (N = 4712)	Wheat (N = 5472)
AT	0.060 ^{ns}	0.294 **	AT	0.129 *	0.443 **
ST	0.543 **	0.010 ^{ns}	PAR	0.225 **	0.348 **
SM	−0.088 ^{ns}	−0.322 **	VPD	−0.315 **	−0.147 ^{ns}
SHF	0.084 ^{ns}	0.111 ^{ns}	NR	0.248 *	0.099 ^{ns}
RH	0.140 ^{ns}	−0.121 ^{ns}	SM	0.172 ^{ns}	0.471 **
GPP	0.917 **	0.698 **			

For each pair of correlation all the other variables in the column are the covariates; * Significant at ($p < 0.05$); ** Significant at ($p < 0.01$); ^{ns} = non-significant; N is number of observations.

When the PAR value was greater than 2500 μ mol m^{-2} s^{-1}, the maximum carbon assimilation rate was observed at the booting stage in rice, whereas a PAR of more than 2000 μ mol m^{-2} s^{-1} in wheat led to a greater CO_2 uptake in wheat. These differences in the rates of CO_2 uptake were due to the different light response parameters of rice and wheat. Apparent quantum yield (α) and Pmax (maximum photosynthetic capacity at light saturation) varied at different growth stages according to the crop phenology (Table 3). The highest P max was observed at the booting stage in rice, whereas it was highest at the heading stage in wheat. The Pmax values ranged from 0.347 to 2.412 mg CO_2 $m^{-2}s^{-1}$ for rice and from 0.461 to 2.841 mg CO_2 $m^{-2}s^{-1}$ in wheat. The initial value of α was in the range of 0.0007 to 0.0008 mg CO_2 μ mole $photon^{-1}$ and started to increase at the vegetative stage as the plants started to grow rapidly. It was the highest at the heading stage in both rice and wheat. The average annual α and Pmax values in our study were a little lower than those reported by Bao et al. [48] in a maize–winter wheat rotation system and by Chen et al. [12] in a rice–winter wheat system. The ecosystem α and Pmax values, could be affected by the differences in air temperature, vapor pressure deficit (VPD), and other biotic factors, such as the green leaf area.

Table 3. Apparent ecosystem quantum yield (α), the hypothetical maximum of GPP (Pmax), at different growth stages in rice and wheat.

Growth Stage	Rice			Wheat		
	α (mg CO2 μmolephoton^{-1})	Pmax (mg CO_2 $m^{-2}s^{-1}$)	R^2	α (mg CO2 μmole photon^{-1})	Pmax (mg CO_2 $m^{-2}s^{-1}$)	R^2
EVS	0.0007	0.347	0.67	0.0008	0.461	0.61
LVS/Tillering	0.0012	0.642	0.84	0.0014	0.707	0.69
SE/MT-PI	0.0016	1.334	0.70	0.0017	1.469	0.73
Booting	0.0022	2.412	0.82	0.0022	2.203	0.89
Heading	0.0023	1.909	0.91	0.0026	2.841	0.70
Ripening	0.0018	0.938	0.76	0.0020	1.318	0.88

In the current investigation, the VPD was between 1.5 to 2.5 kPa and influenced the GPP in both the crops. The VPD increases with air temperature, increasing the plant respiration rate and leading to a reduction in CO_2 uptake [34]. However, it was difficult to evaluate the effects of high VPD on overall photosynthesis and GPP because high VPD conditions mostly persist for a short duration around midday and the stress impact may not persist throughout the rest of the day [49]. The GPP was significantly negatively correlated with VPD in rice (−0.315, $p < 0.01$) (Table 2). An increase in VPD may cause partial closure of the stomata causing a reduction in photosynthesis and thereby reducing the GPP.

Soil moisture is another key parameter affecting plant productivity. It was positively correlated with GPP in wheat (0.471, $p < 0.01$), however, no significant correlation of SM with GPP was obtained in irrigated rice in our study as there was no moisture stress (Table 2). Soil moisture generally has a positive relation with LAI and phenology, thereby controlling the seasonal variation in GPP [30]. Limiting soil moisture constrains stomatal conductance, thereby lowering carbon uptake and plant water use. In wheat, as the water supply was limited and the RH was low (62.15%), water use was reduced as compared to rice, where the RH was higher (71.85%) and water was abundant.

4.4. Effect of Crop Stage, Environmental Variables on RE

The primary cause of seasonal change in RE was changes in the above-ground biomass with crop growth. The rate of RE during the pre-transplanting and pre-sowing periods in rice and wheat was mainly due to soil and weed respiration (Table 1). The RE increased in both rice and wheat until heading and then declined at the ripening stage following the senescence of the green leaf area and old roots [30]. Baldocchi [50] observed higher RE values during the ripening stage in wheat due to the differences in canopy's radiative temperature after the heading stage subsequent to normalizing the LAI, which led to higher levels of canopy respiration. The respiration rate at the different growth stages in rice was relatively higher than in wheat, despite the higher biomass in wheat, due to higher air and soil temperature (ST) during rice. The higher value of RE during the pre-transplanting period in rice was because of the higher ST (31 °C) which enhanced the soil respiration and also due to the growth of weeds supported by rainfall and an increased number of sunshine hours. Our rice field was frequently irrigated and due to availability of dissolved oxygen in the irrigation water, there was microbial decomposition of soil organic matter leading to higher soil respiration levels in the frequently irrigated rice [34]. Mielnik and Dugas [51] observed that soil respiration increased as the soil water content decreased, but under very dry conditions the CO_2 efflux decreased due to low root and microbial activity. In our study, there was no water stress at any stage in both the crops.

The RE was significantly positively correlated with ST in rice (0.543, $p < 0.01$) (Table 2). The diurnal range of ST was smaller in rice on account of standing water which acted as a barrier to ST changes [52]. The ST during most of the rice crop was nearly constant, except at the maturity stage, while it showed comparatively larger variation in the case of wheat and was influenced by changes in net solar radiation.

The RE was negatively correlated with soil moisture (SM) in both rice and wheat, however, it was significant only in wheat (−0.322, $p < 0.01$). The RE was positively correlated with RH in rice and negatively correlated in wheat; however, it was not significant in both the crops. The RE was positively correlated with AT in both rice and wheat, however, the correlation was significant only in wheat (0.294, $p < 0.01$) (Table 2). Increased AT stimulated respiration and photorespiration in plants and a decline in photosynthetic activity occurred as leaf temperatures increased [53]. The SHF influenced the soil temperature and transferred the energy to or from the soil for maintaining the optimum conditions for soil microbial respiration [54], however, the RE was not significantly correlated with SHF in both rice and wheat (Table 2).

4.5. Relation between GPP and RE

The RE was positively correlated with GPP in both rice (0.917, $p < 0.01$) and wheat (0.698, $p < 0.01$) (Table 2). The regression relation between daily GPP and RE in rice and wheat over the crop season was positive, i.e., the higher the GPP (uptake of CO_2) the higher the RE (CO_2 release) (Figure 6). The strength of this relationship in both the crops was highly significant ($p < 0.01$), though a higher R^2 of 0.94 was obtained in rice than the R^2 of 0.74 in wheat. Thus, plants respire more with increased photosynthesis in order to achieve higher growth and maintenance, thereby increasing the GPP. The maximum GPP and RE rates of 14.42 and 5.23 g CO_2 m^{-2} d^{-1} in rice and 15.66 g C m^{-2} d^{-1} and 7.3 g CO_2 m^{-2} d^{-1}, respectively, in wheat were observed at the heading stage in our study. These were much less as compared to the maximum GPP rates of 70.8, 59.8, 41.7, 41.8 g CO_2 m^{-2} d^{-1} and the maximum RE rates of 27.5, 32.3, 19.4, 28.7 g CO_2 m^{-2} d^{-1} observed in the maize cropland, orchard, wetland and vegetable field ecosystems, respectively, by Zhang et al. [38]. The higher GPP in maize was due to the greater carbon assimilation capacities at the different growth stages of the maize. With a greater leaf area, maize had a greater ability to absorb PAR and this increased its CO_2 uptake capacity. The differences in RE were mainly due to the variations in air temperature, soil moisture and canopy radiative temperature.

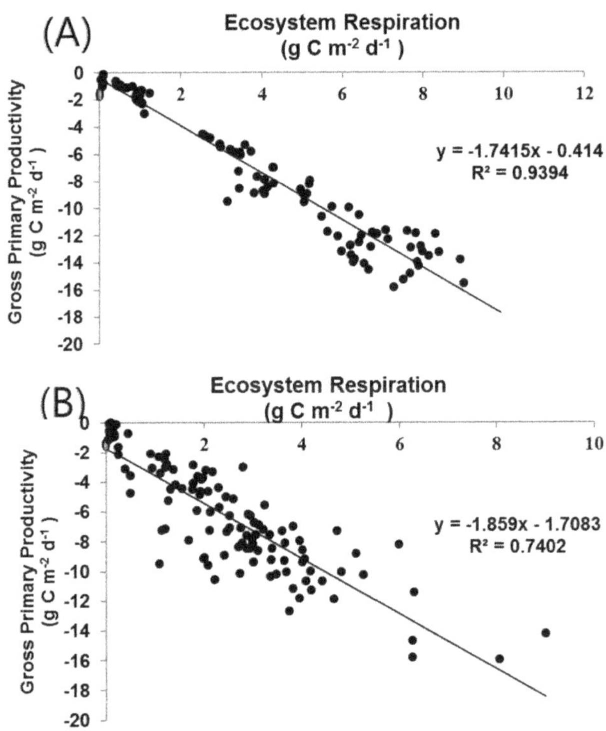

Figure 6. Scatter plot of daily ecosystem respiration (RE) and gross primary productivity (GPP) in (**A**) rice and (**B**) wheat over the crop season.

The cumulative RE at our site was 386.81 g C m^{-2} in rice, whereas it was much lower at 303.89 g C m^{-2} in wheat (Table 4), even though wheat has a longer duration of growth. Higher cumulative RE in rice of 521 and 743 was reported in the Philippines [34] and Japan [24] (Table 4). The cumulative RE in wheat was reported to be 676 and 529 g C m^{-2} in winter wheat in Germany for two consecutive years by Schmidt et al. [43] (Table 4). The much higher RE at these locations was probably due to higher soil organic carbon and longer rice and winter wheat crop durations compared to at our study site.

Table 4. Net ecosystem exchange (NEE), ecosystem respiration (RE) and gross primary productivity (GPP) at different rice and wheat sites.

	Location	Climate	NEE	GPP	RE	RE/GPP	Reference
Rice	International Rice Research Institute, Philippines	Tropical	−258	−778	521	0.67	[34]
	Mase paddy site, Japan	Tropical	−396	−1140	743	0.65	[24]
	Central Rice Research Institute, Cuttak, India	Sub-tropical monsoon	−448	−811	363	0.44	[19]
	North China Plain, China	Subtropical semi-humid monsoon	−583 −512 −451	−1220 −1135 −859	637 623 459	0.52 0.54 0.53	[12]
	Indian Agricultural Research Institute, New Delhi, India	Sub-tropical, semi-arid	−368	−753	387	0.51	Present study
Wheat	Selhausen test site, Germany (First Year)	Temperate	−627	−1304	676	0.51	[43]
	Selhausen test site, Germany (Second Year)	Temperate	−537	−1067	529	0.49	[43]
	North China Plain, China	Subtropical semi-humid monsoon climate	−438 −431	−987 −966	552 538	0.56 0.56	[12]
	Indian Agricultural Research Institute, New Delhi, India	Subtropical semi-arid	−576	−888	304	0.34	Present study

The annual GPP and RE in our study in a subtropical semi-arid rice–wheat crop rotation system was −1725 and 832 g C m^{-2}, whereas Bhattacharya et al. [19] recorded an annual GPP and RE of −1340 and +883 g C m^{-2}, respectively, in a tropical humid rice–rice system. A lower annual GPP and RE of −1174, and810 g C m^2, respectively, was reported in wheat and −1008, and 872 g C m^2 in maize in the North China Plain by Zhang et al. [55].

The RE/GPP ratio at our study site was 0.51 in rice and 0.34 in spring wheat (Table 4). The ratio obtained for rice, in the range of 0.51–0.67, is comparable in magnitude to those reported by other researchers (Table 4). The GPP/RE ratio was higher than one, which implied that the most important factor affecting the GPP was PAR and the seasonal changes in PAR explained 69% and 81% of the variability in NEE in rice and wheat, respectively. The ratio obtained for spring wheat in this study was much lower than those reported by other researchers studying winter wheat. Gifford [56] reported an RE/GPP ratio of 0.35 for spring wheat grown in Australia, which is closer to the value obtained by us.

Carbon budgets have become a powerful tool in guiding climate policy and for identifying the mitigation measures required to keep the global temperatures below 2 °C, the global average temperature increase limits set out in the United Nations Paris Agreement [57]. The Indian IGP—occupying around 10.5 Mha of area under a rice–wheat system [11]—is a carbon sink, playing an important role in the global carbon budget estimation. The global carbon budget estimation in the northern IGP has—until now—been mostly extrapolated from closed-chamber measurements in this major rice–wheat growing region. The results of our study show that when estimating the carbon sink potential in this intensively cultivated northern IGP, it must be taken into consideration that spring wheat may be a moderately stronger sink of CO_2 as compared to rice in the rice–wheat crop rotation system.

5. Conclusions

The cumulative uptake of CO_2 from the atmosphere by the transplanted rice was 25% lower than wheat as the RE/GPP in wheat was much lower than that in rice. Increased respiratory losses were noted as compared to wheat due to the warmer and wetter conditions during the rice growing season. The RE in rice was significantly affected by the soil temperature, whereas the air temperature significantly affected the respiratory losses in wheat. Fluctuations in α and Pmax were the most important drivers of CO_2 uptake at the different growth stages in both rice and wheat. The highest CO_2 uptake levels were observed at the heading stage in both the crops. Seasonal changes in PAR explained 69% and 81% of the variability in the daytime NEE in rice and wheat, respectively. The seasonal NEE fluxes were strongly determined by the development of the crop canopy and CO_2 uptake became dominant during the late vegetative to reproductive stages, turning the wheat and rice field into a carbon sink up until the ripening stage. The fields acted as a CO_2 source around the late maturity and harvesting stage of rice and wheat, when the daily

NEE turned positive due to declining photosynthetic activity and enhanced ecosystem respiration from the soil. Intercrop variability in NEE and GPP values were mainly controlled by PAR and AT. The GPP was observed to be positively correlated with RE in both rice and wheat. Rice contributed 39%, 55.9% and 46.2% to the annual totals of NEE, RE and GPP, respectively, while spring wheat contributed 61%, 44.1% and 53.8%. The integrated cumulative NEE, RE and GPP for the annual rice–spring wheat rotation including the fallow period were −894.16, 832.39 and 1724.55 g C m^{-2}, respectively. The annual GPP/RE ratio for the rice-spring wheat sequence was observed to be 0.48. The RE/GPP ratio of spring wheat was lower as compared to earlier studies in winter wheat due to its shorter crop duration and irrigated conditions, which ensured the availability of soil moisture. The present results are most valuable in the context of the global carbon budget estimation, which, so far, have mostly been extrapolated from closed-chamber measurements in this major rice–wheat growing region.

Author Contributions: The authors have made the following declaration about their contributions. Conceptualization: A.B., H.P. and V.K.S.; methodology and structure: A.K., A.B. and V.K.S.; formal analysis of data: A.B., A.K. and V.K.S.; data curation and table preparation: A.B. and A.K.; contribution in their respective specialization in different sections: A.K., A.B. and V.K.S.; funding acquisition: H.P.; project administration: H.P. and A.B.; software: N.J.; the original draft preparation: A.K., A.B. and V.K.S.; edited and finalized the manuscript: A.B., A.K., V.K.S., H.P. and R.T.; All authors have read and agreed to the published version of the manuscript.

Funding: This research was funded by the Project No. IARI-12-115 titled "National Innovations in Climate Resilient Agriculture" financed by Indian Council of Agricultural Research, Ministry of Agriculture and Farmer's Welfare, Government of India.

Acknowledgments: Authors thank the Director Indian Agricultural Research Institute, New Delhi and National Innovation on Climate Resilient Agriculture (NICRA) project, Indian Council of Agricultural research, Ministry for Environment, Forest and Climate Change, MOEFCC, Government of India for providing support to carry out this research.

Conflicts of Interest: The authors declare no conflict of interest.

References

1. Bhatia, A.; Kumar, A.; Kumar, V.; Jain, N. Low carbon option for sustainable agriculture. *Ind. Farm.* **2013**, *63*, 18–22.
2. IPCC, Climate Change. *The Physical Science Basis. Contribution of Working Group I to the Fifth Assessment Report of the Intergovernmental Panel on Climate Change*; Stocker, T.F., Qin, D., Eds.; Cambridge University Press: Cambridge, UK; New York, NY, USA, 2013; p. 1535.
3. Kumar, A.; Bhatia, A.; Fagodiya, R.K.; Malyan, S.K.; Meena, B.L. Eddy Covariance Flux Tower: A Promising Technique for Greenhouse Gases Measurement. *Adv. Plants Agric. Res.* **2017**, *7*, 00263.
4. Bhattacharyya, R.; Bhatia, A.; Das, T.K.; Lata, S.; Kumar, A.; Tomer, R.; Singh, G. Aggregate-associated N and global warming potential of conservation agriculture-based cropping of maize-wheat system in the north-western Indo-Gangetic Plains. *Soil Till. Res.* **2018**, *182*, 66–77. [CrossRef]
5. Kumar, A.; Tomer, R.; Bhatia, A.; Jain, N.; Pathak, H. *Greenhouse Gas Mitigation in Indian Agriculture in Agro-Technologies for Adaptation to Climate Change*; Pathak, H., Chakrabarti, B., Eds.; CESCRA, IARI: New Delhi, India, 2015.
6. Kumar, A.; Sharma, M.P.; Taxak, A.K. Effect of Vegetation Communities and Altitudes on the Soil Organic Carbon Stock in Kotli Bhel-1A Catchment, India. *CLEAN Soil Air Water* **2017**, *45*, 1600650. [CrossRef]
7. Kumar, A.; Sharma, M.P. Carbon stock estimation in the catchment of Kotli Bhel 1A hydroelectric reservoir, Uttarakhand, India. *Ecotoxicol. Environ. Saf.* **2016**, *134*, 365–369. [CrossRef]
8. Kumar, A.; Yang, T.; Sharma, M.P. Greenhouse gas measurement from Chinese freshwater bodies: A review. *J. Clean. Prod.* **2019**, *233*, 368–378. [CrossRef]
9. Kumar, A.; Sharma, M.P.; Yang, T. Estimation of carbon stock for greenhouse gas emissions from hydropower reservoirs. *Stoch. Environ. Res. Risk Assess.* **2018**. [CrossRef]
10. Sándor, R.; Ehrhardt, F.; Grace, P.; Recous, S.; Smith, P.; Snow, V.; Soussana, J.-F.; Basso, B.; Bhatia, A.; Brilli, L.; et al. Ensemble modelling of carbon fluxes in grasslands and croplands. *Field Crop Res.* **2020**, *252*, 10779. [CrossRef]
11. Ladha, J.K.; Pathak, H.; Tirol-Padre, A.; Dawe, D. Productivity trends in intensive rice–wheat cropping systems in Asia. In *Productivity and Sustainability of Rice–Wheat Systems: Issues and Impacts*; ASA Spec. Publ. 65; ASA, CSSA, and SSSA: Madison, WI, USA, 2003; pp. 45–76.

12. Chen, C.; Dan, L.I.; Zhiqiu, G.; Tang, J.; Xiaofeng, G.; Linlin, W.; Bingcheng, W. Seasonal and interannual variations of carbon exchange over a rice–wheat rotation system on the north china plain. *Advan. Atmos. Sci.* **2015**, *32*, 1365–1380. [CrossRef]
13. Gupta, D.K.; Bhatia, A.; Kumar, A.; Chakrabarti, B.; Jain, N.; Pathak, H. Global warming potential of rice–wheat cropping system of the IGP. *Ind. J. Agric Sci.* **2015**, *85*, 807–816.
14. Bhatia, A.; Pathak, H.; Aggarwal, P.K.; Jain, N. Trade-off between productivity enhancement and global warming potential of rice and wheat in India. *Nutr. Cycl. Agroecosys.* **2010**, *86*, 413–424. [CrossRef]
15. Gupta, D.K.; Bhatia, A.; Das, T.K.; Singh, P.; Kumar, A.; Jain, N.; Pathak, H. Economic analysis of different greenhouse gas mitigation technologies in rice–wheat cropping system of the IGP. *Curr. Sci.* **2016**, *110*, 867–874.
16. Gupta, D.K.; Bhatia, A.; Kumar, A.; Das, T.K.; Jain, N.; Tomar, R.; Malyan, S.K.; Fagodiya, R.K.; Dubey, R.; Pathak, H. Mitigation of greenhouse gas emission from rice–wheat system of the IGP: Through tillage, irrigation and fertilizer management. *Agric. Ecosyst. Environ.* **2016**, *230*, 1–9. [CrossRef]
17. Sage, R.F.; Kubien, D.S. The temperature response of C_3 and C_4 photosynthesis. *Plant Cell Environ.* **2007**, *30*, 1086–1106. [CrossRef]
18. Kramer, K.; Leinonen, I.; Bartelink, H.H.; Berbigier, P.; Borghetti, M.; Bernhofer, C.; Cienciala, E.; Dolman, A.J.; Froer, O.; Gracia, C.A.; et al. Evaluation of six process-based forest growth models using eddy-covariance measurements of CO_2 and H_2O fluxes at six forest sites in Europe. *Global Chang. Biol.* **2002**, *8*, 213–230. [CrossRef]
19. Bhattacharya, P.; Neogi, S.; Roy, K.S.; Dash, P.K.; Tripathi, R.; Rao, K.S. Net ecosystem CO_2 exchange and carbon cycling in tropical lowland flooded-rice ecosystem. *Nutr. Cycl. Agroecosys.* **2013**, *95*, 133–144. [CrossRef]
20. Alberto, M.C.R.; Hirano, T.; Miyata, A.; Wassmann, R.; Kumar, A.; Padre, A.; Amante, M. Influence of climate variability on seasonal and inter-annual variations of ecosystem CO_2 exchange in flooded and non-flooded rice fields in the Philippines. *Field Crops Res.* **2012**, *134*, 80–94. [CrossRef]
21. Tseng, K.H.; Tsai, J.L.; Alagesan, A.; Tsuang, B.J.; Yao, M.H.; Kuo, P.H. Determination of methane and CO_2 fluxes during the rice maturity period in Taiwan by combining profile and eddy covariance measurements. *Agric. For. Meteor.* **2010**, *150*, 852–859. [CrossRef]
22. Hobbs, P.R.; Gupta, R.K. *Resource-Conserving Technologies for Wheat in Rice Wheat Systems*; NRG Web Publication; CIMMYT: México-Veracruz, Mexico, 2001.
23. Bhatia, A.; Agarwal, P.K.; Jain, N.; Pathak, H. Greenhouse gas emission from rice and wheat-growing areas in India: Spatial analysis and upscaling. *Greenhouse Gas Sci. Technol.* **2012**, *2*, 115–125. [CrossRef]
24. Saito, M.; Miyata, A.; Nagai, H.; Yamada, T. Seasonal variation of CO_2 exchange in rice paddy field in Japan. *Agric. For. Meteor.* **2005**, *135*, 93–109. [CrossRef]
25. Webb, E.K.; Pearman, G.I.; Leuning, R. Correction of flux measurement for density effects due to heat and water vapour transfer. *J. Roy Meteor. Soc.* **1980**, *106*, 85–100. [CrossRef]
26. Kaimal, J.C.; Finnigan, J.J. *Atmospheric Boundary Layer Flows: Their Structure and Measurement*; Oxford University Press: New York, NY, USA, 1994; Volume 289.
27. Van Dijk, A.; Kohsiek, W.; de Bruin, H.A.R. *The Principles of Surface Flux Physics: Theory, Practice and Description of the ECPACK Library*; Meteorology and Air Quality Group, Wageningen University: Wageningen, The Netherlands, 2004.
28. Fan, S.M.; Wofsy, S.C.; Bakwin, P.S. Atmosphere biosphere exchange of CO_2 and O_3 in the central Amazon forest. *J. Geophys. Res.* **1990**, *95*, 851–864.
29. Foken, T.; Wichura, B. Tools for quality assessment of surface-based flux measurements. *Agric. For. Meteor.* **1996**, *78*, 83–105. [CrossRef]
30. Zhang, X.; Fan, C.; Ma, Y. Two approaches for net ecosystem carbon budgets and soil carbon sequestration in a rice-wheat rotation system in China. *Nutr. Cycl. Agroecosys.* **2014**, *100*, 301–313. [CrossRef]
31. Hirata, R.; Hirano, T.; Saigusa, N.; Fujinuma, Y.; Inukai, K.; Kitamori, Y.; Takahashi, Y.; Yamamoto, S. Seasonal and interannual variations in CO_2 exchange of a temperate larch forest. *Agric. For. Meteor.* **2007**, *147*, 110–124. [CrossRef]
32. Taiz, L.; Zeiger, E. *Plant Physiology*; The Benjamin/Cummings Publishing Co., Inc.: Redwood City, CA, USA, 1991; pp. 434–436.
33. Falge, E.; Baldocchi, D.; Olson, R.; Anthoni, P.; Aubinet, M.; Bernhofer, C.; Burba, G.; Ceulemans, R.; Clement, R.; Dolman, H. Gap-filling strategies for defensible annual sums of net ecosystem exchange. *Agric. For. Meteor.* **2001**, *107*, 43–69. [CrossRef]
34. Alberto, M.C.R.; Wassmann, R.; Hirano, T.; Miyata, A.; Kumar, A.; Padre, A.; Amante, M. CO_2/heat fluxes in rice fields: Comparative assessment of flooded and non-flooded fields in the Philippines. *Agric. For. Meteor.* **2009**, *149*, 1737–1750. [CrossRef]
35. Das, D.K.; Mukherjee, J.; Vashisth, A.; Sehgal, V.K.; Singh, J. *Agro-climatic Manual of Indian Agricultural Research Institute New Delhi. Division of Agricultural Physics*; ICAR: New Delhi, India, 2015.
36. Miyata, A.; Leuning, R.; Denmead, O.W.; Kim, J.; Harazano, Y. CO_2 and methane fluxes from an intermittently-flooded paddy field. *Agr. For. Meteor.* **2000**, *102*, 287–303. [CrossRef]
37. Gao, Z.; Bian, L.; Zhou, X. Measurements of turbulent transfer in the near-surface layer over a rice paddy in China. *J. Geophys. Res.* **2003**, *108*, 4387. [CrossRef]
38. Zhang, L.; Sun, R.; Xu, Z.; Qiao, C.; Jiang, G. Diurnal and Seasonal Variations in Carbon Dioxide Exchange in Ecosystems in the Zhangye Oasis Area, Northwest China. *PLoS ONE* **2015**, *10*, e0120660. [CrossRef]
39. Campbell, C.S.; Heilman, J.L.; McInnes, K.J.; Wilson, L.T.; Medley, J.C.; Wu, G.W.; Cobos, D.R. Diel and seasonal variation in CO_2 flux of irrigated-rice. *Agric. For. Meteor.* **2001**, *108*, 15–27. [CrossRef]

40. Patel, N.R.; Dadhwal, V.K.; Saha, S.K. Measurement and Scaling of Carbon Dioxide Exchanges in Wheat Using Flux-Tower and Remote Sensing. *J. Ind. Soc. Remote Sens.* **2011**, *39*, 383–391. [CrossRef]
41. Baker, J.M.; Griffis, T.J. Examining strategies to improve the carbon balance of corn/soybean agriculture using eddy covariance and mass balance techniques. *Agric. Forest Meteorol.* **2005**, *128*, 163–177. [CrossRef]
42. Li, J.; Yu, Q.; Sun, X.M.; Tong, X.J.; Ren, C.Y.; Wang, J.; Liu, E.; Zhu, Z.; Yu, G. Carbon dioxide exchange and the mechanism of environmental control in a farmland ecosystem in North China plain. *Sci. China Ser. D* **2006**, *49*, 226–240. [CrossRef]
43. Schmidt, M.; Reichenaua, T.G.; Fienerc, P.; Schneider, K. The carbon budget of a winter wheat field: An eddy covariance analysis of seasonal and inter-annual variability. *Agric. For. Meteor.* **2012**, *165*, 114–126. [CrossRef]
44. Moureaux, C.; Debacq, A.; Bodson, B.; Heinesch, B.; Aubinet, M. Annual net ecosystem carbon exchange by a sugar beet crop. *Agric. For. Meteorol.* **2006**, *139*, 25–39. [CrossRef]
45. Amthor, J.S. The McCree—de Wit—Penning de Vries—Thornley respiration paradigms: 30 years later. *Ann. Bot.* **2000**, *86*, 1–20. [CrossRef]
46. Béziat, P.; Ceschia, E.; Dedieu, G. Carbon balance of a three crop succession over two cropland sites in South West France. *Agric. For. Meteor.* **2009**, *149*, 1628–1645. [CrossRef]
47. Gitelsona, A.A.; Penga, Y.; Arkebauer, T.J.; Suyker, A.E. Productivity absorbed photosynthetically active radiation, and light use efficiency in crops: Implications for remote sensing of crop primary production. *J. Plant. Physiol.* **2015**, *177*, 100–109. [CrossRef]
48. Bao, X.; Li, Z.; Xie, F. Environmental influences on light response parameters of net carbon exchange in two rotation croplands on the North China Plain. *Sci. Rep.* **2019**, *9*, 18702. [CrossRef]
49. Shirke, P.A.; Pathre, U.V. Influence of leaf-to-air vapour pressure deficit (VPD) on the biochemistry and physiology of photosynthesis in *Prosopis juliflora*. *J. Exp. Bot.* **2004**, *55*, 2111–2120. [CrossRef]
50. Baldocchi, D. A comparative-study of mass and energy-exchange rates over a closed C_3 and an open C_4 crop. *Agric. For. Meteor.* **1994**, *67*, 291–321. [CrossRef]
51. Mielnik, P.C.; Dugas, W.A. Soil CO_2 flux in a tall grass prairie. *Soil Biol. Biochem.* **2000**, *32*, 221–228. [CrossRef]
52. Mowjood, M.I.M.; Ishiguro, K.; Kasubuchi, T. Effect of convection in ponded water on the thermal regime of a paddy field. *Soil Sci.* **1997**, *162*, 583–587. [CrossRef]
53. Long, S.P. Modification of the response of photosynthetic productivity to rising temperature by atmospheric CO_2 concentration: Has its importance been underestimated? *Plant Cell Environ.* **1991**, *14*, 729–739. [CrossRef]
54. Atkin, O.K.; Edwards, E.J.; Loveys, B.R. Response of root-respiration to changes in temperature and its relevance to global warming. *New. Phytol.* **2008**, *147*, 141–154. [CrossRef]
55. Zhang, Q.; Lei, H.; Yang, D.; Xiong, L.; Liu, P.; Fang, B. Decadal variation in CO_2 fluxes and its budget in a wheat and maize rotation cropland over the North China Plain. *Biogeosciences* **2020**, *17*, 2245–2262. [CrossRef]
56. Gifford, R.M. Whole plant respiration and photosynthesis of wheat under increased CO_2 concentration and temperature: Long-term vs. short-term distinctions for modeling. *Glob. Chang. Biol.* **1995**, *1*, 385–396. [CrossRef]
57. Rogelj, J.; Forster, P.M.; Kriegler, E.; Smith, C.J.; Séférian, R. Estimating and tracking the remaining carbon budget for stringent climate targets. *Nature* **2019**, *571*, 335–342. [CrossRef]

land

MDPI

Article

Influence of Aspect and Elevational Gradient on Vegetation Pattern, Tree Characteristics and Ecosystem Carbon Density in Northwestern Himalayas

D. R. Bhardwaj [1], Habibullah Tahiry [1,2], Prashant Sharma [1], Nazir A. Pala [3,†], Dhirender Kumar [1], Amit Kumar [4,*,†] and Bharti [5]

1 Department of Silviculture and Agroforestry, Dr. YSP University of Horticulture and Forestry, Solan 173230, India; bhardwajdruhf@yspuniversity.ac.in (D.R.B.); habibulaht@yahoo.com (H.T.); prashantsharma92749@gmail.com (P.S.); dhirenderthakur00@gmail.com (D.K.)
2 Climate Change Division, National Environmental Protection Agency, Government of Islamic Republic of Afghanistan, Kabul 209, Afghanistan
3 Department of Silviculture and Agroforestry, Faculty of Forestry, Sher-e-Kashmir University of Agricultural Sciences & Technology of Jammu (SKUAST), Jammu 191201, India; drnazirpala@skuastkashmir.ac.in
4 School of Hydrology and Water Resources, Nanjing University of Information Science and Technology, Nanjing 210044, China
5 Division of Sample Survey, ICAR-Indian Agricultural Statistics Research Institute, New Delhi 110012, India; bharti.bharti@icar.gov.in
* Correspondence: amitkdah@nuist.edu.cn; Tel.: +86-18362071057
† Contributed equally.

Citation: Bhardwaj, D.R.; Tahiry, H.; Sharma, P.; Pala, N.A.; Kumar, D.; Kumar, A.; Bharti. Influence of Aspect and Elevational Gradient on Vegetation Pattern, Tree Characteristics and Ecosystem Carbon Density in Northwestern Himalayas. *Land* **2021**, *10*, 1109. https://doi.org/10.3390/land10111109

Academic Editor: Gabriele Broll

Received: 29 July 2021
Accepted: 18 October 2021
Published: 20 October 2021

Abstract: Himalayan forest has been threatened by rapid anthropogenic activities, resulting in the loss of forest diversity and climate change. The present study was carried out on four aspects (northern, southern, western and eastern), at three different altitudinal ranges, namely, 1000–1300 m above sea level (m a.s.l.), 1300–1600 m a.s.l. and 1600–1900 m a.s.l., and at three diverse mountain ranges (Kalaghat, Barog and Nangali) of sub-temperate forest ecosystems of the mid Himalayan ranges, to elucidate their influence on vegetation, tree characteristics and ecosystem carbon density. The results revealed that *Pinus roxburghii* is the most dominant forest community of the mid Himalaya's forest, irrespective of altitudinal gradient and slope. The south-facing slopes are occupied by the xerophytic tree species frequently found in the lower Shiwalik *P. roxburghii* forest, whereas the north-facing ones are dominated by mesophyllic species, such as *Cedrus deodara* and *Quercus leucotrichophora*, which commonly grows in the northwestern Himalayan temperate forest ecosystem. The maximum stem density (211.00 Nha^{-1}) was found at 1000–1300 m a.s.l., and on the northern aspect (211.00 Nha^{-1}). The maximum stem volume (236.50 m^3 ha^{-1}) was observed on the northern aspect at 1000–1300 m a.s.l., whereas the minimum (32.167 m^3 ha^{-1}) in the southern aspect at 1300–1600 m a.s.l. The maximum carbon density (149.90 Mg ha^{-1}) was found on the northern aspect and declined with increasing elevation from 123.20 to 74.78 Mg ha^{-1}. Overall, the study establishes that the southern and western aspects are very low in carbon density, whereas the northern aspect represents higher biodiversity as well as carbon and nutrient stocks. Therefore, aspect and altitude should be given due importance for efficient managing of biodiversity and mitigating climate change.

Keywords: tree parameters; biomass density; altitude; ecosystem; vegetation community

1. Introduction

In recent decades, Himalayan forest has been threatened by rapid anthropogenic activities, resulting in the loss of forest diversity and climate change. Mountain forests cover about 23% of the total forest land area [1] and host ~12% of the human population worldwide [2]. Compared to other mountain ecosystems, the Indian Himalayan region is dynamically young [3] and harbors a wide variety of biodiversity apart from sustaining life for a significant part of the Indian subcontinent population. Vegetation cover plays a crucial role in local, regional and global climate apart from decreasing erosion in mountain areas [4]. By adopting sustainable environmental management practices, the protection of mountainous habitats (particularly, biological diversity) will help in safeguarding livelihoods and further improvement of local communities. They also shield individuals effectively from natural calamities, e.g., debris flows, floods and landslides [5]. In forest conservation and management, tree diversity plays an important role because it provides resources and shelter to forest species [6]. Besides its effect on species diversity and ecosystem functioning, it also affects tolerance and resistance to climate changes in the future [7,8]. Several edaphic factors and topographical factors (e.g., altitude, aspect and slope) control vegetation growth [9,10] and play a vital role in plant species distribution [7,11]. Altitude plays a key role in deciding the temperature system and diversification of forest species [12,13]. Moreover, co-factors such as soil type and slope propensity help in guiding forest composition at one altitude [14,15]. The determinants of species richness are altitude, physiography, productivity [16] and biotic variables. A comprehension of stand characteristics and forest structure is a prerequisite for identifying different ecological processes and modeling the functioning and dynamics of forests [17]. Several biotic (e.g., intensity of photosynthesis, leaf area index, forest types and plant architecture) and abiotic characteristics (e.g., solar radiation, temperature, soil moisture and length of growing season) affect carbon cycling and further affect the regional and/or global carbon budget [18–21].

In addition to being a crucial natural constraint to climate change, forest habitats occupy approximately 30% of land areas and play a dominant role in the exchange of carbon dioxide (CO_2) between the atmosphere and biosphere [22,23]. Major shares (81%) of the terrestrial carbon (C) biomass are found in forest ecosystem [24], where 2/3 of the C is fixed annually in terrestrial ecosystems [25]. Vegetation biomass is a central ecological element to understand the climate system's evolution and possible future changes. Therefore, biomass acts as an important indicator of climate change prediction models, and further helps to make mitigation and adaptation strategies [26]. Forests, particularly primary ones, are actively engaged in the cycling of carbon (i.e., C stock in vegetation and its underlain soil through the process of photosynthesis and respiration) [27], and the extent depends on age, tree species, location of forests and management practices [28]. Due to land-use change and tropical deforestation, nearly 1.7 billion tons of carbon is emitted annually, which impacts climate badly [29,30]. The tree biomass pool is a vital origin of uncertainty in C balance in tropical regions [31] and plays a crucial role in the global C cycle. The management of forests for maintaining or enhancing carbon stocks is receiving increasing interest from forest land owners and land management agencies [32,33] due to the impacts from a change in climate intensity [34].

The structure and functions of Himalayan forests are believed to be changing over time, resulting in loss of forest diversity [35]. In most of the studies, the parameters pertaining to forest productiveness are overestimated because of the unaccountability of slope and aspects [36]. The ecological studies and biomass estimation along altitudinal gradients in forests of western Himalaya have been well documented over the years [37–39]. However, the literature reveals that very few studies report the effect of aspect in the northwestern Himalayas on forest ecosystems. For the effective management of forest ecosystems in the mid-hill of the Himalayas, there is an urgent need to know the key indicators that impact the tree characteristics, plant communities, forest biomass and its underlying soil with respect to elevation gradients, slope and aspect [7,36,40]. The objectives of our study were to test the hypothesis that different aspects (northern, eastern, western and southern) and altitudinal ranges influence the tree characteristics, vegetation biomass, total carbon stocks and physico-chemical characteristics of the soil. By taking into consideration these objectives, we shall be able to answer these above issues for the management and conservation prospective of forest ecosystems in the mid hills of the Himalayas. This study would be helpful in designing afforestation and reforestation programs for different altitudes and aspects in the mid Himalayas, for enhanced carbon sequestration and mitigating climate change.

2. Materials and Methods

2.1. Study Area

The present study was carried out at three altitudinal ranges: AR_1 (1000–1300 m above mean sea level (m a.s.l.)), AR_2 (1300–1600 m a.s.l.) and AR_3 (1600–1900 m a.s.l.); at four aspects: northern (N), southern (S), eastern (E) and western (W); and at three separate mountainous ranges (Kalaghat, Barog and Nangali) located in the Solan district of Himachal Pradesh (H.P), India, situated in the middle of the NW Himalayas. The area lies between 30°51′16″ to 30°53′40″ N latitudes and 77°9′27″ to 77°3′51″ E longitude (Survey of India Toposheet No. 53F/1) and lies in a transitional zone between the sub-tropical to moist temperate Himalayas and falls under the Champion and Seth forest type of lower Shiwalik *P. roxburghii* forest ($9/C_{1a}$), Upper Himalayan *P. roxburghii* forest ($9/C_{1b}$) and Lower western Himalayan temperate forest ($12/C_1$). The seasons are well marked, and there are distinct pre-monsoon, monsoon, post-monsoon and winter seasons. On average, the annual rainfall ranges from 1000 to 1400 mm, with little pre-monsoon precipitation, and July–September contribute the most to precipitation. Winter precipitation of low quantum is common. The mean minimum and maximum temperatures differ between 1 °C in the winter and 35 °C in the summer with a mean annual temperature (MAT) of 20 °C. The parent material is composed of shale, schist, quartzite and slate, with a sandy loam to loam and clayey texture in discrete pockets. Brown and podzol soil forms commonly occur in the upper regions. The soils are rich in nutrients and organic matter (OM) except sites that are too exposed and eroded.

2.2. Vegetation Sampling and Data Analysis

During the winter season of 2017, a reconnaissance survey of the region was carried out to identify the mountain ranges for the present investigation (Figure 1). In this study area, the three altitudinal ranges, AR_1, AR_2 and AR_3, of the three separate mountainous (Kalaghat, Barog and Nagali) regions were selected. At every altitude, the four aspects, i.e., the N-, S-, W- and E-facing slopes, each had three sample plots of 0.1 ha (31.62 m × 31.62 m),which were randomly laid out on the surveyed area to analyze the tree composition (12 elevation stands × 3 sample plot each with a total of 36 sample plots). For the shrub- and herb-related characteristics, four subplots of size 5 m × 5 m and eight plots of 1 m × 1 m were laid out within each tree sample plot, respectively. Further, the density of the trees was determined by counting trees in each sample plot. In each sample plot, trees and shrubs were marked and counted to determine the community parameters viz. the frequency, stand density and Importance Value Index (IVI) [41]. The forest associations and forest types were named based on tree composition and their stand density values. The identification of plants was done with the help of the Herbarium of Dr. YSP University of Horticulture and Forestry, Nauni, Solan, and the current taxonomic literature. Tree height was measured using a Ravi multimeter (BlumeLeiss Hypsometer) from the uphill side in the slopy areas [42], whereas the diameter of trees was measured at breast height (DBH, about 1.37 m). The diameter of the grasses, herbs, shrubs and regeneration were derived from the girth measured at the collar portion. Crown height (m) was taken from the ground level to the point halfway between the lowest green branch and the green branches forming a green crown all around. Similarly, crown length (m) was the vertical length from the top to the point halfway between the lowest green branch and the green branches forming a green crown all around. The diameter of the crown was determined in two directions (N–S and E–W) and the mean diameter was estimated as per Assmann [43] and Chaturvedi and Khanna [42].

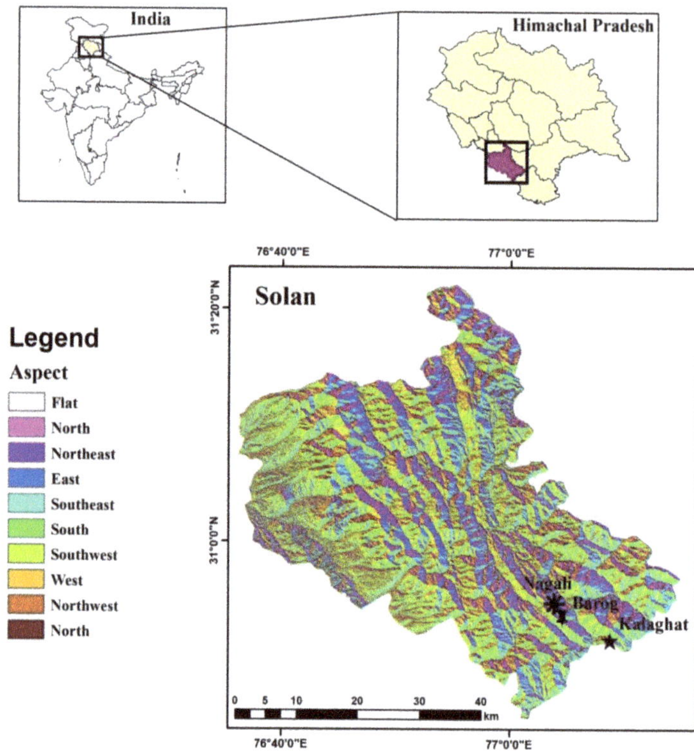

Figure 1. Map of the study area. The images were procured from Google Earth Pro 7.3.4.

2.3. Biomass and Carbon

The allometric equations proposed by the Forest Survey of India (FSI) and Forest Research Institute (FRI) for respective forest tree species were used to estimate the growing stock volume density (GSVD; m^3 ha^{-1}) [44]. After that, the estimated GSVD of the individual tree species was converted into total tree biomass as the Above Ground Biomass Density (AGBD; Mg ha^{-1}) by multiplying the GSVD of the tree species by the Biomass Expansion Factor (BEF) expressed below [45]. Specific gravity (SG) was calculated by the method adopted by Rajput et al. [46].

$$\text{Stem biomass (SB)} = \text{Average SG wood} \times \text{Volume} \quad (1)$$

The BEF for hardwood is expressed by BEF = exp {1.91–0.34 × ln (GSVD)} (for GSVD ≤ 200 m^3 ha^{-1}), but in the case of pine species, it is used as a fixed value based on the GSVD (i.e., BEF for GSVD < 10, 10–100 and >100 m^3 ha^{-1} are used as 1.68, 0.95 and 0.81 respectively) [47].

$$\text{AGBD} = \text{GSVD} \times \text{BEF} \quad (2)$$

The tree volume could be transformed into biomass by multiplying it with specific gravity using the maximum moisture method [48]. Belowground biomass density (BGBD) of trees was calculated by following IPCC [49] guidelines [22,50]. Further, total biomass of the tree was finally computed by adding AGB and BGB. Moreover, the biomass of the herbs and shrubs were calculated by the oven-dried weight. Thus, each shrub and herb sample collected were brought to the laboratory followed by segregation into leaves, branches and a stem portion after washing and oven-dried at 70 °C for 72 h till a constant dry weight was achieved. Each sample was weighed to estimate AGB (including the stem, branch and leaves) of each species. BGB is estimated by extracting the roots of the sample plant (shrubs). The extract of the root sample was packaged in the paper and brought to the laboratory to know the biomass. First of all, root hairs of individual species were washed thoroughly and weighed to know their fresh weight (Wi) and after that dried at 70 ± 5 °C for 72 h to estimate their dry weight. Similarly, total biomass of a shrub species was calculated by adding its AGB and BGB components.

2.4. Total Carbon Density

Total carbon density (or stocks) of forest vegetation was estimated by vegetation biomass (AGB+BGB) multiplied by a default value of 0.50 [49].

2.5. Vegetation Carbon Density

Vegetation carbon density (VCD) was calculated by adding all the biomass (tree, shrubs and herbs).

$$\text{VCD} = \text{Tree biomass} + \text{Herb biomass} + \text{Shrub Biomass} \tag{3}$$

2.6. Determination of Soil Characteristics

The soil samples were taken from the soil profile at 0–20 cm and 20–40 cm depths in three replications from each 0.1 ha sample plot. These samples were then sun-dried and sieved to a 2-mm size. The soil pH was determined by a combined glass–calomel electrode of aqueous suspensions (1:2.5 soil:water ratio). Electrical conductivity (dS m^{-1}) was estimated by a conductivity bridge [51] of aqueous suspensions (1:2 soil:water ratio). Soil organic carbon (SOC) was estimated by a well-known method of wet digestion [22,52]. Potassium (K) was measured by a flame photometer [53] and available nitrogen (N) and phosphorus (P) by the alkaline permanganate method [54].

2.7. Ecosystem Carbon Density

The total ecosystem carbon density (ECD) is referred to as the sum of all carbon pools, including the soil carbon density or stocks, and was measured as

$$\text{ECD} = \text{SOC} + \text{Detritus carbon density} + \text{Tree carbon density} \tag{4}$$

Soil carbon density was calculated as follows: Soil bulk density (g cm^{-3}) × Depth of soil (cm) × Carbon (%) [55]. Solar radiation was estimated using a Lux meter.

2.8. Statistical Analysis

In this study, the data obtained on variations in vegetation, tree characteristics and ecosystem carbon density were subject to statistical analysis by Factorial Randomized Block Design using R Studio data analysis software, as follows:

Package: doebioresource
Function: frbd2fact
Syntax: frbd2fact (data, replicationvector, fact.A, fact.B, Multiple.comparison.test)
Arguments:

> *Data: dependent variables; Replication vector: vector containing replications; fact.A: vector containing levels of first factor; fact.B: vector containing levels of second factor; Multiple.comparison.test: 0 for no test, 1 for lsd test, 2 for Duncan test and 3 for HSD test*

The significance of the effect of aspect and altitudinal ranges on different parameters was judged using an LSD post-hoc test.

3. Results

3.1. Floristic Composition of Vegetation Species

The distribution of the floristic composition of the tree species growing at the study sites varied with altitudinal range (Table S1). Among the tree species, *Pinus roxburghii* was dominant followed by *Quercus leucotrichophora* and *Erythrina indica*. At the 1000–1300 m a.s.l. altitudinal range (AR_1), the maximum number of tree species (4) was recorded on the northern aspect followed by the eastern, southern and western slopes, respectively. At the 1300–1600 m a.s.l. (AR_2) altitudinal range, the maximum number of species (6) was recorded on the southern (S) aspect followed by the eastern (E) and northern (N) ones, respectively. However, at the 1600–1900 m a.s.l. altitudinal range (AR_3), the maximum number of species (3) were recorded on the northern aspect followed by the western aspect. Among the three altitudinal ranges, the AR_2 altitude range displayed the highest number of tree species, i.e., 14, while the lowest number was at AR_3 (7). Similarly, 21 shrub species represent the mid-hills of the northwestern Himalayas, out of which the prominent species are *Berberis aristata*, *Carissa carandas* and *Lantana camara*, etc., whereas *Cassia floribunda*, *Hypericum oblongifolium* and *Sarcocoas aligna* are the least prominent ones (Table S2). At the AR_1 altitudinal range, the maximum numbers of shrub species were found on the southern aspect followed by the western, northern and eastern aspects, respectively. At the AR_2 altitudinal range, the northern and southern aspects were represented by 7 species of shrubs, whereas the eastern and western aspects were represented by six species each. At the AR_3 altitudinal ranges, the number of species on the western and southern aspects were slightly more than the eastern and northern aspects. In general, the number of shrub species declined with increasing altitudinal ranges.

The overall status of plant diversity composition is presented in Figure S1. It reveals that the mid-hills of the northwestern Himalayas are richer in shrub diversity than the trees as measured by family, genus and species presence. Maximum diversity in trees and shrubs was measured on the S-facing slope at 1300–1600 m a.s.l. and 1000–1300 m a.s.l., respectively. Maximum biodiversity was found on the S-facing slope, followed by north, east and west, respectively.

3.2. Distribution of Vegetation Communities and Forest Types

The floristic composition concerning vegetation communities and forest type identified revealed that *P. roxburghii* is an important component of all the communities at all the combinations of altitude and aspect (Table S3). Lower (1000–1300 m a.s.l.) and upper (1600–1900 m a.s.l.) altitudinal ranges of the E- and W-facing slopes of the study area were represented by the *P. roxburghii* community, whereas the mid-altitude (1300–1600 m a.s.l.) by the *P. roxburghii*–*Quercus leucotrichophora* and *P. roxburghii*–*Myrica esculenta* community, respectively. The N-facing slopes are represented by the *P. roxburghii*–*Q. leucotrichophora* community at all the altitudinal gradients. However, the S-facing slopes are represented by the *P. roxburghii*–*A. catechu* community at a low altitude; *P. roxburghii*, *Acacia catechu* and a diverse mixture of broad-leaved vegetation at mid altitude; and a pure *P. roxburghii* community at a higher altitude. The vegetation of the E- and W-facing slopes is classified into Himalayan subtropical pine forest ($9/C_1$), subtype Upper Himalayan *P. roxburghii* forest ($9/C_{1b}$). S-facing slopes belong to the same forest type but a different subtype, i.e., Lower Shiwalik *P. roxburghii* forest ($9/C_{1a}$). However, the N-facing slopes belong to entirely different forest types, i.e., Lower western Himalayan temperate Ban *Q. leucotrichophora* forest (12 C_1).

3.3. Variation in Tree Characteristics

The various tree characteristics, namely, DBH, height, crown length, stem density, stem volume and stem biomass, of the studied forest ecosystems are depicted in Table 1.

Table 1. Tree parameters and biomass density on the different aspects and altitudinal ranges.

Altitudinal Ranges (AR) (m a.s.l.)	Aspect (A)				
	E	W	N	S	Mean
	DBH (cm)				
AR_1 (1000–1300)	25.00	19.00	36.30	23.40	26.00 [a]
AR_2 (1000–1300)	22.00	18.60	22.50	22.80	22.00 [b]
AR_3 (1000–1300)	26.20	20.00	26.00	19.40	20.00 [b]
Mean	28.00 [a]	21.80 [b]	22.00 [b]	21.16 [b]	
LSD	AR = 3.70		A = 4.30	AR × A = 3.90	
	Height (m)				
AR_1 (1000–1300)	14.50	10.22	22.40	16.48	16.26 [a]
AR_2 (1000–1300)	13.00	9.30	11.60	10.00	12.18 [b]
AR_3 (1000–1300)	12.00	12.30	13.00	10.00	11.38 [b]
Mean	13.90 [ab]	12.40 [bc]	16.30 [a]	10.50 [c]	
LSD	AR = 2.70		A = 3.20	AR × A = NS	
	Crown length (m)				
AR_1 (1000–1300)	4.70	5.50	7.90	4.40	6.00 [a]
AR_2 (1000–1300)	6.00	4.60	5.30	3.50	5.00 [a]
AR_3 (1000–1300)	5.00	3.00	3.15	2.20	3.00 [b]
Mean	5.16	3.70	5.20	4.40	
LSD	AR = 1.90		A = NS	AR × A = 4.50	
	Stem density (N ha^{-1})				
AR_1 (1000–1300)	258.00	198.00	217.00	170.00	211.00 [a]
AR_2 (1000–1300)	228.00	175.00	283.00	165.00	188.00 [a]
AR_3 (1000–1300)	116.00	188.00	133.00	143.00	146.00 [b]
Mean	200.00 [ab]	154.60 [c]	211.00 [a]	166.00 [bc]	
LSD	AR = 37.00		A = 43.00	AR × A = NS	
	Stem volume (m^3 ha^{-1})				
AR_1 (1000–1300)	118.83	40.90	236.50	74.40	117.68 [a]
AR_2 (1000–1300)	132.16	39.50	105.40	32.17	77.33 [b]
AR_3 (1000–1300)	101.58	35.30	40.90	53.00	57.89 [b]
Mean	127.60 [a]	53.00 [b]	117.50 [a]	38.60 [b]	
LSD	AR = 38.80		A = 44.80	AR × A = 53.23	
	Stem biomass (Mg ha^{-1})				
AR_1 (1000–1300)	120.50	36.58	72.75	21.58	82.85
AR_2 (1000–1300)	58.92	16.25	132.16	18.66	56.50
AR_3 (1000–1300)	20.08	35.50	63.91	21.91	35.35
Mean	66.50 [a]	29.44 [b]	89.61 [a]	20.72 [b]	
LSD	AR = NS		A = 30.73	AR × A = 53.23	

E—Eastern, W—Western, N—Northern, S—Southern, NS—non-significant. Mean values in the same column followed by a different letter differ significantly at $p \leq 0.05$ according to lsd post-hoc test.

3.3.1. Diameter at Breast Height (DBH, cm)

The maximum average DBH of 26.00 cm was recorded at AR_1, which was found to be significantly ($p < 0.05$) higher than both AR_2 and AR_3 (Table 1). Irrespective of altitudinal ranges, maximum DBH was recorded on the eastern aspect and was significantly higher ($p < 0.05$) than the other aspects. The order of DBH among different aspect followed the following trend: eastern (28.00 cm) > northern (22.00 cm) > southern (21.16 cm) > western (21.00 cm). In the interaction, the maximum DBH (36.30 cm) was recorded on the northern aspect at AR_1, whereas the minimum (18.60 cm) was found on the western aspect at AR_2.

3.3.2. Height (m)

Height (Table 1) showed a declining trend with an increase in altitude. AR_1 displayed a significant difference ($p < 0.05$) over both AR_2 and AR_3. Among the studied aspects, the N-facing slopes demonstrated superiority over the other aspects. In the interaction, an inconsistent trend was observed at the aspect level with an increase in altitudinal range. The maximum (22.40 m) height growth was noticed on the northern aspect at AR_1 and the minimum (9.30 m) was on the western aspect (AR_2).

3.3.3. Crown Length (m)

It is evident from Table 1 that among the altitudinal ranges, the maximum crown length (6.00 m) was recorded at AR_1 and was significantly higher ($p < 0.05$) than AR_3, only. The maximum crown length (5.20 m) was recorded on the northern aspect and minimum (3.70 m) on the western aspect. The maximum crown length (7.90 m) in the interaction was seen on the northern aspect at AR_1, whereas the minimum (2.20 m) was on the northern aspect at AR_3.

3.3.4. Stem Density (N ha^{-1})

Stem density (N ha^{-1}) was found to be significantly influenced by altitudinal variation and aspect (Table 1). It declined with increasing altitude, and both AR_1 and AR_2 evinced markedly higher values over AR_3. Stem density was found to be maximum (211.00 Nha^{-1}) on the northern aspect, which was statistically at par with eastern aspect. However, interaction effect between the altitude and aspect could not exercise a significant influence on the stem density.

3.3.5. Stem Volume (m^3 ha^{-1})

Stem volume (m^3 ha^{-1}) was also significantly influenced by altitude, aspect (Table 1), and it declined with the increase in the altitude. A significant influence was observed between AR_1 over AR_2 and AR_3. Irrespective of altitudinal gradients, maximum stem volume (127.60 m^3 ha^{-1}) was recorded on eastern aspect but was significantly at par with the northern aspect. Minimum stem volume (38.60 m^3 ha^{-1}) was recorded on the southern aspect, however it remains at par with western one. In the interaction, maximum stem volume (236.50 m^3 ha^{-1}) was recorded on northern aspect at AR_1, whereas minimum stem volume (32.17 m^3 ha^{-1}) was recorded on southern aspect at AR_2.

3.4. *Vegetation Biomass and Carbon Density*

3.4.1. Above Ground-Below Ground Biomass (Mg ha^{-1})

Table 2 evinced that significantly maximum AGB (116.25 Mg ha^{-1}) was recorded at AR_1. Among the aspects, the maximum AGB (166.22 Mg ha^{-1}) was recorded on the northern aspect and the minimum (39.64 Mg ha^{-1}) on southern ones. The trend observed for AGB on different aspects followed the order: N > E > W > S. In the interaction between the aspect and altitudinal ranges, northern aspect displayed maximum biomass accumulation (252.31 Mg ha^{-1}) at 1000–1600 m a.s.l. range, and was found to be considerably higher than others. The results in respect of BGB and total biomass for altitudinal gradient, aspects, and their interaction followed more or less similar trend as that of AGB.

Table 2. Vegetation biomass and carbon density at different aspects and altitudinal ranges.

Altitudinal Ranges (AR) (m a.s.l.)	Aspect (A)				
	E	W	N	S	Mean
	Above ground biomass (t ha^{-1})				
AR_1 (1000–1300)	222.05	69.88	130.61	42.49	116.25 ^{a}
AR_2 (1000–1300)	112.44	31.14	252.31	43.36	109.81 ^{a}
AR_3 (1000–1300)	38.38	67.87	115.75	33.07	63.76 ^{b}
Mean	124.28 ^{a}	56.29 ^{b}	166.22 ^{a}	39.64 ^{b}	
lsd	AR = 40.00		A = 55.25	AR × A = 95.693	
	Below ground biomass (t ha^{-1})				
AR_1 (1000–1300)	47.88	14.67	41.65	9.39	28.40
AR_2 (1000–1300)	29.77	6.54	90.31	7.6	33.56
AR_3 (1000–1300)	8.06	22.96	32.24	6.94	17.5
Mean	28.57 ^{b}	14.72 ^{bc}	54.73 ^{a}	7.98 ^{c}	
lsd	AR = NS		A = 18.74	AR × A = 32.46	
	Total tree biomass (t ha^{-1})				
AR_1 (1000–1300)	315.65	84.56	172.26	51.88	156.08 ^{a}
AR_2 (1000–1300)	142.21	37.68	342.62	53.09	143.90 ^{a}
AR_3 (1000–1300)	46.44	90.83	147.99	40.01	81.32 ^{b}
Mean	168.09 ^{a}	71.02 ^{b}	220.95 ^{a}	48.32 ^{b}	
lsd	AR = 58.69		A = 67.77	AR × A = 117.388	
	Shrub biomass (Mg ha^{-1})				
AR_1 (1000–1300)	1.84	1.75	1.85	2.37	1.95 ^{a}
AR_2 (1000–1300)	1.60	1.76	1.56	1.79	1.67 ^{ab}
AR_3 (1000–1300)	1.40	1.33	1.38	1.72	1.45 ^{b}
Mean	1.61	1.62	1.59	1.96	
lsd	AR = 0.303		A = NS	AR × A = NS	
	Herb biomass (Mg ha^{-1})				
AR_1 (1000–1300)	2.93	3.27	3.10	3.92	3.30
AR_2 (1000–1300)	3.24	2.77	2.80	3.94	3.19
AR_3 (1000–1300)	2.55	2.46	2.22	3.49	2.68
Mean	2.90 ^{b}	2.83 ^{b}	2.71 ^{b}	3.78 ^{a}	
lsd	AR = NS		A = 0.63	AR × A = NS	
	Vegetation biomass density (Mg ha^{-1})				
AR_1 (1000–1300)	320.41	89.57	177.22	57.97	161.29 ^{a}
AR_2 (1000–1300)	146.82	42.11	347.33	58.66	148.73 ^{a}
AR_3 (1000–1300)	50.39	94.63	151.86	45.16	85.50 ^{b}
Mean	172.62 ^{a}	75.46 ^{b}	225.27 ^{a}	53.93 ^{b}	
lsd	AR = 58.78		A = 67.85	AR × A = 117.52	

E—eastern, W—western, N—northern, S—southern, NS—non-significant. Mean values in the same column followed by a different letter differ significantly at $p \leq 0.05$ according to the LSD post-hoc test.

3.4.2. Shrub and Herb Biomass (Mg ha^{-1})

Table 2 shows that shrub biomass was significantly influenced by altitudinal ranges, whereas herb biomass significantly differed between aspects. This shrub biomass decreased with an increase in altitude (highest at AR_1: 1.95 Mg ha^{-1} and lowest at AR_3: 1.45 Mg ha^{-1}). Among the different aspects, the maximum shrub biomass (1.96 Mg ha^{-1}) was estimated on southern aspect and the minimum (1.59 Mg ha^{-1}) on the northern ones. Similarly, the maximum biomass of herbs (2.90 Mg ha^{-1}) was recorded on the eastern aspect, and the minimum (2.71 Mg ha^{-1}) on northern.

3.4.3. Vegetation Biomass Density (Mg ha^{-1})

Vegetation biomass density declined with an increase in elevation; however, significant differences ($p < 0.05$) were seen between AR_1 and AR_3 (Table 2). Vegetation biomass (225.27 Mg ha^{-1}) was maximum on the northern aspect followed by eastern, western and southern, respectively. The interaction between altitudinal gradient and aspect also exercised a significant influence on the vegetation biomass density. On the eastern, western and southern aspects, the maximum values for this parameter were at AR_1; however, on the northern aspects, the maximum vegetation biomass density (347.32 Mg ha^{-1}) was at the altitudinal range of 1600–1900 m a.s.l.

Vegetation, soil and total ecosystem carbon density declined with the increase in altitudinal range. However, significant differences were seen between AR_1 and AR_3 (Table 3). Carbon stored in the form of humus was maximum at AR_2, while the carbon stored in the leaf litter increased significantly with rising altitude. Among the four aspects, VCD and ECD were significantly ($p < 0.05$) at their maximum on the northern aspects.

In contrast, carbon stored in the form of humus was significantly higher ($p < 0.05$) on the eastern aspect and followed the following trend: E > S > N > W. Interaction between altitudinal gradient and aspects also exercised a significant influence on the carbon density of different pools, except the humus. On the eastern aspects, the maximum values for the vegetation carbon pool were at AR_1; however, on the northern aspects, the maximum vegetation biomass was at the AR_2 altitudinal range. Soil carbon density also varied significantly ($p < 0.05$) at different altitudes with aspects. In general, the values on the different aspects declined with an increase in altitude. Carbon density stored in the leaf litter on the eastern and western aspects declined rapidly with an increase in altitude from AR_1 to AR_3. On the northern and southern aspects, the maximum value was observed at AR_2. The maximum ECD (214.13 Mg ha^{-1}) was recorded on the northern aspect, which remained statistically identical to AR_1 of the eastern aspect. The ECD did not vary significantly ($p < 0.05$) at the western and southern aspects with the variation in altitudinal range.

3.5. *Soil Physico-Chemical Characteristics*

The soil physico-chemical characteristics of the sites under investigation is depicted in Table 4. pH value (6.31) was found significantly ($p < 0.05$) higher at AR_2 (1300–1600 m a.s.l.) than both AR_1 and AR_3. Among the aspects, the maximum pH (6.34) was recorded on the southern aspect and followed the following trend: southern > northern > western > eastern. The organic carbon (OC) and available nutrients (particularly, N, P and K), declined significantly ($p < 0.05$) with the increasing altitudinal ranges.

OC, available N and P were found to be highest on the northern aspect and lowest on the southern aspect, whereas available K was the highest on the eastern aspect. Bulk density showed an increasing trend with elevational range. Bulk density was maximum (0.96 g cm^3) on the southern aspect, followed by western, eastern and northern, respectively. The values of the pH and bulk density increased significantly ($p < 0.05$) from L_1 (0–20 cm) to L_2 (20–40 cm) layer. However, OC, available N, P and K declined significantly from the L_1 to L_2 soil layer.

Table 3. Carbon density in different pools, ecosystem carbon density and solar radiation at different aspects and altitudinal ranges.

Altitudinal Ranges (AR) (m a.s.l.)	Aspect (A)				
	E	W	N	S	Mean
	Vegetation carbon density (Mg ha^{-1})				
AR_1 (1000–1300)	160.21	44.79	88.61	28.98	80.64 ^{a}
AR_2 (1000–1300)	73.41	21.05	173.67	29.33	74.36 ^{ab}
AR_3 (1000–1300)	25.20	47.31	75.93	22.58	42.75 ^{b}
Mean	86.27 ^{a}	37.71 ^{b}	112.73 ^{a}	26.96 ^{b}	
LSD	AR = 33.92		A = 29.38	AR × A = 58.76	
	Soil carbon density (Mg ha^{-1})				
AR_1 (1000–1300)	43.23	40.83	33.83	39.77	39.42 ^{a}
AR_2 (1000–1300)	40.93	36.90	36.67	37.50	38.00 ^{a}
AR_3 (1000–1300)	25.53	28.20	31.43	29.40	28.64 ^{b}
Mean	36.56	35.31	33.97	35.55	
LSD	AR = 36.56		A = NS	AR × A = 5.94	
	Soil humus carbon density (Mg ha^{-1})				
AR_1 (1000–1300)	1.57	0.73	1.23	1.60	1.28 ^{b}
AR_2 (1000–1300)	3.10	1.33	1.73	2.57	2.18 ^{a}
AR_3 (1000–1300)	2.17	1.27	1.17	1.80	1.60 ^{b}
Mean	2.28 ^{a}	1.11 ^{c}	1.38 ^{bc}	1.99 ^{ab}	
LSD	AR = 0.55		A = 0.64	AR × A = NS	
	Leaf litter carbon (Mg ha^{-1})				
AR_1 (1000–1300)	2.22	2.02	1.25	1.92	1.85 ^{b}
AR_2 (1000–1300)	2.17	1.94	2.06	2.11	2.07 ^{a}
AR_3 (1000–1300)	1.98	1.78	1.71	1.95	1.86 ^{b}
Mean	2.12 ^{a}	1.91 ^{c}	1.67 ^{d}	1.99 ^{a}	
LSD	AR = 0.02		A = 0.051	AR × A = 0.11	
	Ecosystem carbon density (Mg ha^{-1})				
AR_1 (1000–1300)	207.23	88.37	125.73	71.58	123.20 ^{a}
AR_2 (1000–1300)	119.61	61.22	214.13	70.66	116.40 ^{a}
AR_3 (1000–1300)	54.88	78.56	110.52	55.20	74.78 ^{b}
Mean	127.20 ^{a}	76.11 ^{b}	149.90 ^{a}	65.80 ^{b}	
LSD	AR = 29.60		A = 34.18	AR × A = 59.21	

E—eastern, W—western, N—northern, S—southern, NS—non-significant. Mean values in the same column followed by a different letter differ significantly at $p \leq 0.05$ according to the LSD post-hoc test.

Table 4. Soil physico-chemical characteristics in different aspects and altitudinal ranges.

Treatment	Parameters						
	pH	Bulk Density (g cm^{-3})	Organic Carbon (g kg^{-1})	Available N (kg ha^{-1})	Available P (kg ha^{-1})	Available K (kg ha^{-1})	EC (dS m^{-2})
			Altitudinal range (AR) (m a.s.l.)				
AR_1 (1000–1300 m a.s.l.)	6.09 ^{b}	0.86 ^{c}	22.8 ^{a}	413.22 ^{a}	19.78 ^{a}	242.37 ^{a}	0.31 ^{a}
AR_2 (1300–1600 m a.s.l.)	6.31 ^{a}	0.90 ^{bc}	21.2 ^{b}	355.68 ^{b}	16.70 ^{b}	223.38 ^{b}	0.30 ^{b}
AR_3 (1600–1900 m a.s.l.)	5.98 ^{c}	0.94 ^{a}	15.3 ^{c}	327.11 ^{c}	13.42 ^{c}	209.20 ^{c}	0.29 ^{c}
LSD (E)	0.09	0.051	0. 89	21.23	1.33	6.87	0.009
			Aspect (A)				
E (Eastern)	5.98 ^{c}	0.87 ^{bc}	19.9 ^{b}	347.28 ^{b}	15.87 ^{b}	229.63 ^{a}	0.29 ^{b}
W (Western)	5.99 ^{c}	0.92 ^{ab}	19.3 ^{bc}	330.86 ^{c}	18.43 ^{a}	213.54 ^{b}	0.29 ^{b}
N (Northern)	6.18 ^{b}	0.85 ^{c}	21.2 ^{a}	480.80 ^{a}	18.62 ^{a}	195.79 ^{c}	0.31 ^{a}
S (Southern)	6.34 ^{a}	0.96 ^{a}	18.6 ^{c}	304.39 ^{d}	13.60 ^{c}	200.98 ^{c}	0.30 ^{ab}
LSD (A)	0.11	0.059	1.03	24.51	1.54	7.93	0.01
			Soil layers (cm)				
L_1 (0–20 cm)	5.95 ^{b}	0.79 ^{a}	21.0 ^{a}	380.19 ^{a}	18.25 ^{a}	235.89 ^{a}	0.31
L_2 (20–40 cm)	6.30 ^{a}	1.01 ^{b}	18.5 ^{b}	350.48 ^{b}	15.01 ^{b}	214.08 ^{b}	0.30
LSD (L)	0.07	0.04	0. 73	17.33	1.09	5.61	NS

* EC = electric conductivity. Mean values in the same column followed by a different letter differ significantly at $p \leq 0.05$ according to the LSD post-hoc test.

4. Discussion

4.1. Floristic Diversity and Distribution of Vegetation Community

The present study revealed a marked contrast among the four aspects for the composition of the species (trees as well shrub species), forest communities and forest types of the mid-hill ecosystem of the northwestern Himalayas. S-facing slopes bear the tree species, namely, *Acacia catechu*, *Bauhinia variegata*, *Bombax ceiba*, *Butea monosperma*, *Erythrina indica* and *Pinus roxburghii*, which are xerophytic and are characteristic of the low hill forest ecosystem of the Shiwalik region. The forest community and forest types were identified based on the composition and importance value index, which are the *P. roxburghii–A. catechu* community, pure *P. roxburghii* and *A. catechu*. The other broad-leaved tree species are characteristics forest communities of the Low level Southern tropical sub-tropical pine forest type [56]. N-facing slopes apart from *P. roxbughii* also supports the mesophilic tree species, such as *Acer oblongum*, *Cedrus deodara* and *Quercus leucotrichophra*, etc. The forest community identified on the N-facing slope is the *P. roxburghii–Q. leucotrichophora* community. However, on the E-facing slope, pure *P. roxburghii* occurs on the low and upper altitudinal ranges and *P. roxburghii–Q. leucotrichophora* in the middle ones. These forest communities have their presence across the sub-temperate region of the northwestern Himalayas. The variation in the vegetation is mainly due to differential exposure to the sun and the water-holding capacities of the underlying soil. The S-facing slopes are drier than other slopes as they are exposed to sun rays during the day time and directly intercept the moisture-laden southwest monsoon, leading to excessive soil erosion on these slopes. In turn, the N-facing slopes support the moisture-loving tree species as they hold more water because of indirect exposure to the sun and low-intensity north–east rains during the winter season. The species richness of both trees as well as shrubs is found to be maximum on the S-facing slopes followed by N-facing. The higher species richness of the S-facing slopes may be mainly due to the dryness of the site on which no particular species have a competitive advantage over others.

The diversity was recorded in terms of the distribution of the families, genera and species at different altitudinal gradients. The species richness on the eastern, western and southern aspects followed a hump shape, whereas, on the S-facing slopes, the species richness declined with the increase in altitudinal gradient. Similarly, a higher species richness of vascular plants also was observed at the middle elevations in the western Himalayas [57]. The main reason for the low species richness and diversity in the lower altitudinal ranges could be justified as this range was a more accessible area for livestock grazing and human-made disturbance. It was also found that the E-, S- and W-facing slope of the lower and upper altitudinal gradients are occupied by the xerophytic vegetation communities and middle altitudes by mesophyllic ones; this indicates the prevalence of optimum growing conditions for temperature, rainfall and nutrients, etc. In turn, the N-facing slopes are occupied by the *P. roxburghii*–*Q. leucotrichophora* community at all the altitudinal gradients. Compared to moderate slopes, the reasons for a hump-shaped distribution pattern may be due to changes in inclinations, exposures and geological substrates in the mountains. However, the available area for plant establishment decreases disproportionately at higher altitudes [58] and thus the ability for organisms to survive also decrease. Such hump-shaped patterns may result from the perspective of population ecology due to the dispersal of species from lower and higher elevations, resulting in the highest overlap at mid-elevations of such dispersing populations. Contrary to this, only dispersing species are obtained in the extremes of the gradient [59]. Anthropogenic activity, primarily at lower altitudes (altitude < 1000 m a.s.l.), also plays a crucial role in decreased species diversity [57,60]. According to Sang [61], water supply can be important at lower altitudes in a continental climate, but a lower temperature is relevant at higher altitudes. In general, environmental parameters [62,63] and management practices [64] are closely associated with the diversity of forest species.

4.2. Variations in Tree Characteristics

Regional factors also play a vital role in understanding the ecological and evolutionary responses of biota at different eco-regions [65,66]. In the present study, it was observed that the tree parameters, such as diameter, top height, crown length and stem density, were in the order $AR_1 > AR_2 > AR_3$. This trend in the growth and developmental parameters could be attributed to the nutrient and water conservation capacity of these altitudes. Lower altitudes have a better level of OC, available N, P and K because of high temperature and moisture conditions. Besides, higher altitudinal ranges can also suffer from both nutrient leaching and water shortages [17,67]. In this study, the diameter of the tree species followed the following order: eastern (28.00 cm) > northern (22.00 cm) > southern (21.16 cm) > western (21.00 cm). Better growth on the eastern aspect can be due to the occurrence of the dominative *P. roxburghii* community, which is best adapted to the forest ecosystems of the mid-hills of the northwestern Himalayas. Tree parameters such as top height, crown length, stem density and stem volume were found to be higher on the northern aspect followed by the eastern, southern and western aspects. N-facing slopes of the northern hemisphere receive indirect sunlight, which leads to taller trees and a longer crown length [68]. Significant ($p < 0.05$) variation in stem density was observed on the four studied aspects, the maximum (211.00 Nha^{-1}) being on the northern and minimum (154.00 Nha^{-1}) on the western ones; this could be due to the better ability of this aspect to support a greater number of individuals because of its better water- and nutrient-supplying capacity. Mong and Vetaas [69] also recorded more tree species and higher tree densities on N-facing forests slopes than S-facing ones and owed it to the pronounced aridity of the area, where higher solar radiation results in faster drying of S-facing slopes, hence reducing tree growth.

The tree diameter and height reported in the present study are more or less similar to those from Banday et al. [70], also from the northwestern Himalayas. Missanjo et al. [71] also reported significant differences ($p < 0.05$) in mean height (DBH) and biomass of pine species at different altitudes. The stem density in the present investigation ranged from 211 individuals per hectare at 1000–1300 m altitude to 146 individuals per hectare at 1600–1900 m a.s.l. Similarly, the SD varied from 283 to 116 in northern and southern aspects, respectively. The SD in the present study is on the lower side than the values reported by Bhardwaj et al. [37], but more or less comparable with the findings of Banday et al. [70] under similar forest types of the Indian Himalayas. Pala et al. [72] have also reported higher tree densities for the central Himalayan region, which may be due to variation in species composition and other environmental factors. The reported values of the SD under study are also on the lower side than the other reported values globally [70,73,74], which can be due to difference in the slope, aspect, or even in the same type of vegetation [75]. *P. roxburghii* is the dominant plant community and requires more space because of its light-demanding nature and hence may be the reason for their lower densities. Banday et al. [70] have also reported on *P. roxburghii* having the lowest density amongst the five forest types of the northwestern Himalayas.

4.3. Vegetation Biomass and Carbon Density

It was observed that biomass stored in different components, viz., above, below and total tree biomass, declined with increasing elevation. Similarly, a decline in the aboveground biomass with increasing elevation has been reported by earlier worker [76–78]. The decline in the aboveground traits with increasing elevation could be owed to the progressive fall in the level of OC, available N, P and K with an increasing elevation gradient, thereby limiting biomass accumulation [76,79]. The elevational gradient we investigated follow the classical pattern of nutrient availability. However, Alves et al. [80] have also reported an opposite pattern of aboveground biomass accumulation for a short elevation gradient of tropical Atlantic moist forest in Brazil. In addition to better storage of biomass in different components of the vegetation, such as trees, shrubs, and herbs at the lower altitudinal range, are also the optimum conditions of moisture, temperature, light, better nutrient storage and less soil erosion. Biomass in different components, as well as biomass carbon density, was maximum on the northern aspect followed by the eastern, western and southern aspects, respectively. It again could be owing to the ability of this aspect to support a higher tree density and bigger-sized trees because of better moisture and nutrient storage ability. Contrary to this, shrub and herb biomass density was maximum (1.96 Mg ha^{-1}) on the southern aspect and minimum (1.59 Mg ha^{-1}) on the northern aspect. Furthermore, Måren et al. [68], in the semiarid ecosystem of inner Himalaya, reported that bush cover was highest on S-facing slopes, and litter was highest on N-facing slopes ($p < 0.01$). Stem volume and biomass decreased with an increase in altitude from 117.68 to 57.89 m^3 ha^{-1} and from 89.61 to 20.72 Mg ha^{-1}, respectively. Stem biomass was also reported lowest (SB: 20.72 Mg ha^{-1}) on the southern aspect and highest (SB: 89.61 Mg ha^{-1}) on the northern ones. The estimated value reported in this study was also supported by the reported value of Leuschner and Moser's [81], and a decrease in total biomass (AGB + BGB) from the lower altitude to higher from 317 to 174 Mg dry mass ha^{-1} between 1050 and 3060 m a.s.l.. The argument is supported by the reason that the net primary productivity (NPP) of the forest could also decrease with an increase in elevation because of an increase in respiration loss [81]. The present study reported a significantly ($p < 0.05$) maximum ECD (149.90 Mg ha^{-1}) on the northern aspect, which declined with descending altitudinal ranges. Thus, the maximum ECD (123.20 Mg ha^{-1}) was recorded at lower altitudes and the minimum (74.78 Mg ha^{-1}) at the highest altitudinal range. Sharma et al. [36] had also reported the occurrence of moister and other favorable environmental factors on the northern aspects for higher growth and biomass. In the northern aspect, vegetation carbon density was high because of optimum growing conditions, such as moisture and temperature. However, the lower biomass and carbon density on the S- and W-facing forest slopes may partly be due to biotic disturbance as these slopes are the perfect habitat for human habitation and livestock grazing owing to their warmer temperature. The lower altitudinal range had the highest ecosystem carbon density, which may be because of a higher optimum growing condition, such as moisture, organic matter, soil condition and temperature, etc., for trees and low soil erosion.

4.4. Soil Physico-Chemical Properties and Correlation

Higher levels of OC, available N, P and K at the lower altitudinal range may have favored more biomass production and better nutrient cycling because of higher temperatures, reduced soil erosion, and nutrient deposition. OC, available N and P displayed their highest values on the northern aspect and lowest values on the southern aspect, whereas available K was maximum on the eastern aspect. These are some of the reasons along with indirect sunlight for the higher biomass and carbon density of the N-facing slopes compared to the S-facing slopes. However, Måren et al. [68] did not found any notable contrast between slope and aspect, except for soil available K, which was reported to be higher in S-facing slopes.

5. Conclusions

The results of the present investigation suggest that topographic factors, mainly the aspect of the slope and altitude, play a substantial role in vegetation composition and stand characteristics, carbon density and nutrient status of the soil. There is marked variation in the vegetation composition, tree characteristics, carbon density and soil physicochemical characteristics among the four aspects and altitudinal gradients. Therefore, they should be given due importance for the efficient management of these forest ecosystems from a biodiversity conservation and carbon mitigation point of view. Biodiversity of the tree and shrub species, particularly light-demanding xerophytic species, such as *A. catechu* and *E. indica*, are prevalent on the S-facing slope, which are otherwise very poor on account of the C sequestration potential due to less accumulation of stem biomass and growth rate. Aspect and slope can be the basis for fixing the boundary of the compartment/sub-compartment (the fundamental unit of forest management) and their allotment under a particular working circle/silvicultural system. S-and W-facing slopes are very poor on account of their carbon density; therefore, they should be the focus of the attention of plantation programs, such as joint forest management and REDD (reducing emissions from deforestation in developing countries)/REDD+ (reduce emissions from deforestation and forest degradation in developing countries). In turn, E-facing slopes are extensively occupied by *P. roxburghii* communities, which should be managed intensively for the production of timber and resin, etc. N-facing slopes harbor mesophyllic tree communities, such as *Q. leucotrichophora–P. roxburghii*, and are a store house of diversity, carbon and nutrients, and therefore should be conserved and enriched with valuable tree species such as *C deodara*. The present study provides sufficient information related to the distribution pattern and density of different species at different aspects and altitudinal ranges of mid-Himalayan forest ecosystems. Based on the species distribution and density, we can go for enrichment of these sites through various interventions, which will be quite useful in biodiversity conservation and mitigating climate change.

Supplementary Materials: The following are available online at https://www.mdpi.com/article/10.3390/land10111109/s1, Table S1: Floristic composition of tree species at different altitude and aspects; Table S2: Floristic compositions of shrub species at different altitudes and aspects; Table S3: Distribution of vegetation communities and forest types with aspects and altitudinal ranges, Figure S1: The total richness of Family, Genus, and species of trees and shrubs to aspects and altitudinal ranges on mid-hill of Himachal Pradesh.

Author Contributions: Conceptualization, methodology, validation, formal analysis, investigation, D.R.B. and H.T.; data curation, writing—original draft, D.R.B. and H.T.; writing—review and editing, software visualization, P.S., N.A.P., D.K., A.K. and B. All authors have read and agreed to the published version of the manuscript.

Funding: This research is partially supported by high-level Nanjing University of Information Science and Technology (NUIST)-talent startup Fee-2020r037, Project number: 1521052101002.

Institutional Review Board Statement: Not applicable.

Informed Consent Statement: Not applicable.

Data Availability Statement: Data could be provided on reasonable request from the first author.

Acknowledgments: The authors are grateful to the Head of the Department of Silviculture and Agroforestry, Y.S. Parmar, University of Horticulture and Forestry, Solan (HP), India, for providing the necessary facilities during the study.

Conflicts of Interest: The authors declare no conflict of interest.

References

1. Price, M.F.; Georg, G.; Lalsa, A.D.; Thomas, K.; Daniel, M.; Rosalaura, R. *Mountain Forests in a Changing World: Realizing Values, Addressing Challenges*; Food and Agriculture Organisations of the United Nations (FAO) with the Support of the Swiss Agency for Development and Cooperation (SDC): Rome, Italy, 2011.
2. FAO. *The state of Food Insecurity in the World*; Food and agriculture Organisation of the United Nations (FAO): Rome, Italy, 2002.
3. Singh, J.S. Sustainable development of the Indian Himalayan region: Linking ecological and economic concern. *Curr. Sci.* **2006**, *90*, 784–788.
4. Kumar, A.; Sharma, M.P.; Taxak, A.K. Effect of vegetation communities and altitudes on the soil organic carbon stock in Kotli Bhel-1A Catchment, India. *CLEAN—Soil Air Water* **2017**, *45*, 1600650. [CrossRef]
5. Brang, P.; Schönenberger, W.; Ott, E.; Gardner, B. Forests as protection from natural hazards. In *The Forests Handbook*; Evans, J., Ed.; Blackwell Science: Oxford, UK, 2001; Volume 2, pp. 53–81.
6. Huang, W.; Pohjonen, V.; Johansson, S.; Nashanda, M.; Katigula, M.; Luukkanen, O. Species diversity, forest structure and species composition in Tanzanian tropical forests. *For. Ecol. Manag.* **2003**, *173*, 11–24. [CrossRef]
7. Bhat, J.A.; Kumar, M.; Negi, A.; Todaria, N.; Malik, Z.A.; Pala, N.A.; Kumar, A.; Shukla, G. Altitudinal gradient of Species diversity and community of woody vegetation along altitudinal gradient of the Western Himalayas. *Glob. Ecol. Conserv.* **2020**, *24*, e01302. [CrossRef]
8. Chapin, F.S., III; Zavaleta, E.S.; Eviner, V.; Naylor, R.L.; Vitousek, P.M.; Reynolds, H.L.; Hooper, D.U.; Lavorel, S.; Sala, O.E.; Hobbie, S.; et al. Consequences of changing biodiversity. *Nature* **2000**, *405*, 234–242. [CrossRef]
9. Wood, E.; Sivapalan, M.; Beven, K.; Band, L. Effects of spatial variability and scale with implications to hydrologic modeling. *J. Hydrol.* **1988**, *102*, 29–47. [CrossRef]
10. Dawes, W.R.; Short, D. The significance of topology for modeling the surface hydrology of fluvial landscapes. *Water Resour. Res.* **1994**, *30*, 1045–1055. [CrossRef]
11. O'Loughlin, E. Saturation regions in catchments and their relations to soil and topographic properties. *J. Hydrol.* **1981**, *53*, 229–246. [CrossRef]
12. Enright, N.; Miller, B.; Akhter, R. Desert vegetation and vegetation-environment relationships in Kirthar National Park, Sindh, Pakistan. *J. Arid. Environ.* **2005**, *61*, 397–418. [CrossRef]
13. Brown, J.H. Mammals on mountainsides: Elevational patterns of diversity. *Glob. Ecol. Biogeogr.* **2001**, *10*, 101–109. [CrossRef]
14. Kumar, A.; Pinto, M.C.; Candeias, C.; Dinis, P.A. Baseline maps of potentially toxic elements in the soils of Garhwal Himalayas, India: Assessment of their eco-environmental and human health risks. *Land Degrad. Dev.* **2021**, *32*, 3856–3869. [CrossRef]
15. Holland, P.G.; Steyn, D.G. Vegetational Responses to Latitudinal Variations in Slope Angle and Aspect. *J. Biogeogr.* **1975**, *2*, 179–183. [CrossRef]
16. Sharma, C.M.; Suyal, S.; Gairola, S.; Ghildiyal, S.K. Species richness and diversity along an altitudinal gradient in moist temperate forest of Garhwal Himalaya. *J. Am. Sci.* **2009**, *5*, 119–128.
17. Singh, S.; Malik, Z.A.; Sharma, C.M. Tree species richness, diversity, and regeneration status in different oak (*Quercus* spp.) dominated forests of Garhwal Himalaya, India. *J. Asia-Pac. Biodivers.* **2016**, *9*, 293–300. [CrossRef]
18. Chen, W.J.; Black, T.A.; Yang, P.C.; Barr, A.; Neumann, H.H.; Nesic, Z.; Blanken, P.D.; Novak, M.D.; Eley, J.; Ketler, R.J.; et al. Effects of climatic variability on the annual carbon sequestration by a boreal aspen forest. *Glob. Chang. Biol.* **1999**, *5*, 41–53. [CrossRef]
19. Wang, H.; Saigusa, N.; Yamamoto, S.; Kondo, H.; Hirano, T.; Toriyama, A.; Fujinuma, Y. Net ecosystem CO2 exchange over a larch forest in Hokkaido, Japan. *Atmos. Environ.* **2004**, *38*, 7021–7032. [CrossRef]
20. Li, S.G.; Asanuma, J.; Kotani, A.; Eugster, W.; Davaa, G.; Oyunbaatar, D.; Sugita, M. Year-round measurements of net eco-system CO2 flux over a montane larch forest in Mongolia. *J. Geophys. Res. Atmos.* **2005**, *110*. [CrossRef]
21. Kumar, A.; Sharma, M.P.; Yang, T. Estimation of carbon stock for greenhouse gas emissions from hydropower reservoirs. *Stoch. Environ. Res. Risk Assess.* **2018**, *32*, 3183–3193. [CrossRef]
22. Brown, J.H.; Stevens, G.C.; Kaufman, D.M. the geographic range: Size, shape, boundaries, and internal structure. *Annu. Rev. Ecol. Syst.* **1996**, *27*, 597–623. [CrossRef]
23. Kumar, A.; Kumar, M. Estimation of Biomass and Soil Carbon Stock in the Hydroelectric Catchment of India and its Implementation to Climate Change. *J. Sustain. For.* **2020**, *36*, 1–16. [CrossRef]
24. Sandrine, L.; Claude, N.Y.S.; Christian, W.; Françoise, F.; Sandrine, H.; Paula, R.; Stéphane, F. Estimation of carbon stocks in a beech forest (Fougères Forest—W. France): Extrapolation from the plots to the whole forest. *Ann. For. Sci.* **2006**, *63*, 139–148. [CrossRef]
25. Zhang, N.; Yu, Z.; Yu, G.; Wu, J. Scaling up ecosystem productivity from patch to landscape: A case study of Changbai Mountain Nature Reserve, China. *Landsc. Ecol.* **2007**, *22*, 303–315. [CrossRef]
26. FAO. *Climate-Smart Agriculture: Policies, Practices and Financing for Food Security, Adaptation and Mitigation*; Food and Agriculture Organisation of the United Nations (FAO). Rome, Italy, 2010.
27. McMahon, S.M.; Parker, G.; Miller, D.R. Evidence for a recent increase in forest growth. *Proc. Natl. Acad. Sci. USA* **2010**, *107*, 3611–3615. [CrossRef]
28. Mwakisunga, B.; Majule, A.E. The influence of altitude and management on carbon stock quantities in rung we forest, southern highland of Tanzania. *Open J. Ecol.* **2012**, *02*, 214–221. [CrossRef]

29. MacDicken, K.G. *A Guide to Monitoring Carbon Emissions from Deforestation and Degradation in Developing Countries: An Examination of Issues Facing the Incorporation of REDD into Market-Based Climate Policies*; Resource for Future: Washington, DC, USA, 1997.
30. IPCC. *Climate Change: A Report of the Intergovernmental Panel on Climate Change*; Intergovernmental Panel on Climate Change: Geneva, Switzerland, 1995.
31. Chauhan, M.; Kumar, M.; Kumar, A. Impact of carbon stocks of Anogeissus latifolia on climate change and socio-economic development: A case study of Garhwal Himalaya, India. *Water Air Soil Pollut.* **2020**, *231*, 436. [CrossRef]
32. McNulty, S.; Treasure, E.; Jennings, L.; Meriwether, D.; Harris, D.; Arndt, P. Translating national level forest service goals to local level land management: Carbon sequestration. *Clim. Chang.* **2018**, *146*, 133–144. [CrossRef]
33. St-Laurent, G.P.; Hagerman, S.; Kozak, R.; Hoberg, G. Public perceptions about climate change mitigation in British Columbia's forest sector. *PLoS ONE* **2018**, *13*, e0195999. [CrossRef]
34. A Ontl, T.; Janowiak, M.K.; Swanston, C.W.; Daley, J.; Handler, S.; Cornett, M.; Hagenbuch, S.; Handrick, C.; McCarthy, L.; Patch, N. Forest Management for Carbon Sequestration and Climate Adaptation. *J. For.* **2020**, *118*, 86–101. [CrossRef]
35. Sharma, C.M.; Mishra, A.K.; Tiwari, O.P.; Krishan, R.; Rana, Y.S. Effect of altitudinal gradients on forest structure and composition on ridge tops in Garhwal Himalaya. *Energy Ecol. Environ.* **2017**, *2*, 404–417. [CrossRef]
36. Sharma, C.M.; Gairola, S.; Baduni, N.P.; Ghildiyal, S.K.; Suyal, S. Variation in carbon stocks on different slope aspects in seven major forest types of temperate region of Garhwal Himalaya, India. *J. Biosci.* **2011**, *36*, 701–708. [CrossRef]
37. Bhardwaj, D.R.; Banday, M.; Pala, N.A.; Rajput, B.S. Variation of biomass and carbon pool with NDVI and altitude in sub-tropical forests of northwestern Himalaya. *Environ. Monit. Assess.* **2016**, *188*, 635. [CrossRef]
38. Banday, M.; Bhardwaj, D.R.; Pala, N.A.; Rajput, B.S. Quantitative analysis of woody vegetation in subtropical forests of Himachal Pradesh, India. *Indian For.* **2017**, *143*, 617–629.
39. Shah, S.; Sharma, D.P.; Pala, N.A.; Tripathi, P.; Kumar, M. Temporal variations in carbon stock of Pinus roxburghii Sargent forests of Himachal Pradesh, India. *J. Mt. Sci.* **2014**, *11*, 959–966. [CrossRef]
40. Kumar, A.; Kumar, M.; Pandey, R.; Yu, Z.; Cabral-Pinto, M. Forest soil nutrient stocks along with an altitudinal range of Uttarakhand Himalayas: An aid to Nature Based Climate Solutions. *CATENA* **2021**, *207*, 105667. [CrossRef]
41. Cottam, G.; Curtis, J.T. The use of distance measures in phytosociological sampling. *Ecology* **1956**, *35*, 451–460. [CrossRef]
42. Chaturvedi, A.N.; Khanna, L.S. *Forest Mensuration*; International Book Distributors: Dehradun, India, 1982.
43. Assmann, E. *The Principles of Forest Yield Study*, 2nd ed.; Oxford Pregmon Press Ltd.: Oxford, UK, 1970.
44. FSI. *Volume Equations for Forests of India, Nepal, and Bhutan*; Forest Survey of India, Ministry of Environment and Forests, Government of India: Dehradun, India, 1996.
45. Brown, S.L.; Schroeder, P.; Kern, J.S. Spatial distribution of biomass in forests of the eastern USA. *For. Ecol. Manag.* **1999**, *123*, 81–90. [CrossRef]
46. Rajput, S.S.; Shukla, N.K.; Gupta, V.K. Specific gravity of Indian timbers. *J. Timber Dev. Assoc. India* **1985**, *31*, 12–41.
47. IPCC. *Guidelines for National Greenhouse Gas Inventories*; Intergovernmental Panel on Climate Change: Geneva, Switzerland, 2006; Volume 4.
48. Smith, D.M. *Maximum Moisture Content Method for Determining Specific Gravity of Small Wood Samples*; US Department of Agriculture, Forest Service, Forest Products Laboratory: Madison, WI, USA, 1954; pp. 110–120.
49. IPCC. Impacts, adaptations and mitigation of climate: Scientific-technical analyses. In *Contribution of the Second Assessment Report of the Intergovernmental Panel on Climate Change*; Cambridge University Press: Cambridge, UK, 1996.
50. Cairns, M.A.; Brown, S.; Helmer, E.H.; Baumgardner, G.A. Root biomass allocation in the world's upland forests. *Oecologia* **1997**, *111*, 1–11. [CrossRef]
51. Richards, L.A. *Diagnosis and Improvement of Saline and Alkali Soils*; USDA Handbook 60, United States Deptartment of Agriculture: Washington, DC, USA, 1954.
52. Walkley, A.; Black, J.A. Estimation of soil organic carbon by chromic acid filtration method. *Soil Sci.* **1934**, *37*, 38–39.
53. Rich, C.I. Elemental analysis by flame photometry. In *Methods of Soil Analysis. Part 2: Chemical and Microbiological Properties*; Black, C.A., Ed.; American Society of Agronomy: Madison, WI, USA, 1965; pp. 849–864. [CrossRef]
54. Subbiah, B.V.; Asija, G.L. A rapid procedure for the determination of available nitrogen in soil. *Curr. Sci.* **1956**, *25*, 259–260.
55. Nelson, D.W.; Sommers, L.E. Total carbon, organic carbon and organic matter. In *Method of Soil Analysis. Part 3. Chemical Methods*; Sparks, D.L., Page, A.L., Helmke, P.A., Loeppert, R.H., Soltanpour, P.N., Tabatabai, M.A., Johnston, C.T., Summer, M.E., Eds.; Soil Science Society of America: Madison, WI, USA, 1996; pp. 961–1010.
56. Champion, H.G.; Seth, S.K. *A Revised Survey of Forest Types of India*; Government of India Press: New Delhi, India, 1968.
57. Chawla, A.; Rajkumar, S.; Singh, K.N.; Lal, B.; Singh, R.D.; Thukral, A.K. Plant species diversity along an altitudinal gradient of Bhabha Valley in western Himalaya. *J. Mt. Sci.* **2008**, *5*, 157–177. [CrossRef]
58. Körner, C. Why are there global gradients in species richness? Mountains might hold the answer. *Trends Ecol. Evol.* **2000**, *15*, 513–514. [CrossRef]
59. Krashevs'ka, V. Diversity and Community Structure of Testate Amoebae (Protista) in Tropical Montane Rain Forests of Southern Ecuador: Altitudinal Gradient, Aboveground Habitats and Nutrient Limitation. Doctoral Thesis, Vom Fachbereich Biologie der Technischen Universität, Darmstadt, Germany, 2008.
60. Noguesbravo, D.; Araújo, M.B.; Romdal, T.S.; Rahbek, C. Scale effects and human impact on the elevational species richness gradients. *Nature* **2008**, *453*, 216–219. [CrossRef] [PubMed]

61. Sang, W. Plant diversity patterns and their relationships with soil and climatic factors along an altitudinal gradient in the middle Tianshan Mountain area, Xinjiang, China. *Ecol. Res.* **2009**, *24*, 303–314. [CrossRef]
62. McCain, C.M. Elevational gradients in diversity of small mammals. *Ecology* **2005**, *86*, 366–372. [CrossRef]
63. Bässler, C.; Müller, J.; Hothorn, T.; Kneib, T.; Badeck, F.-W.; Dziock, F. Estimation of the extinction risk for high-montane species as a consequence of global warming and assessment of their suitability as cross-taxon indicators. *Ecol. Indic.* **2010**, *10*, 341–352. [CrossRef]
64. Monarrez-Gonzalez, J.C.; Gonzalez-Elizondo, M.S.; Marquez-Linares, M.A.; Gutierrez-Yurrita, P.J.; Perez-Verdin, G. Effect of forest management on tree diversity in temperate ecosystem forests in northern Mexico. *PLoS ONE* **2020**, *15*, e0233292. [CrossRef] [PubMed]
65. Cui, H.T.; Liu, H.Y.; Dai, J.U. *Research on Mountain Ecology and Alpine Treeline*; Science Press: Beijing, China, 2005.
66. Körner, C. The use of 'altitude' in ecological research. *Trends Ecol. Evol.* **2007**, *22*, 569–574. [CrossRef] [PubMed]
67. Fang, J.; Shen, Z.; Cui, H. Ecological characteristics of mountains and research issues of mountain ecology. *Biodivers. Sci.* **2004**, *12*, 10–19.
68. Måren, I.; Karki, S.; Prajapati, C.; Yadav, R.K.; Shrestha, B.B. Facing north or south: Does slope aspect impact forest stand characteristics and soil properties in a semiarid trans-Himalayan valley? *J. Arid. Environ.* **2015**, *121*, 112–123. [CrossRef]
69. Mong, C.E.; Vetaas, O.R. Establishment of Pinuswallichiana on a Himalayan glacier foreland: Stochastic distribution or safe sites? *Arct. Antarct. Alp. Res.* **2006**, *38*, 584–592. [CrossRef]
70. Banday, M.; Bhardwaj, D.R.; Pala, N.A. Variation of stem density and vegetation carbon pool in subtropical forests of Northwestern Himalaya. *J. Sustain. For.* **2018**, *37*, 389–402. [CrossRef]
71. Missanjo, E.; Kamanga-Thole, G.; Manda, V. Estimation of Genetic and Phenotypic Parameters for Growth Traits in a Clonal Seed Orchard of Pinus kesiya in Malawi. *ISRN For.* **2013**, *2013*, 346982. [CrossRef]
72. Pala, N.A.; Negi, A.K.; Gokhale, Y.; Aziem, S.; Vikrant, K.K.; Todaria, N.P. Carbon stock estimation for tree species of Sem Mukhem sacred forest in Garhwal Himalaya, India. *J. For. Res.* **2013**, *24*, 457–460. [CrossRef]
73. Iqbal, K.; Pala, N.A.; Bhat, J.A.; Negi, A.K. Regeneration status of trees around Khoh River in Garhwal Himalaya. *Indian J. For.* **2012**, *35*, 471–476.
74. Iqbal, K.; Bhat, J.A.; Pala, N.A.; Hussain, A.; Negi, A.K. Carbon and biomass density of trees in Duggada area of Garhwal Himalaya, India. *Indian For.* **2014**, *140*, 18–22.
75. Srivastava, R.K.; Khanduri, V.P.; Sharma, C.M.P. Structure, diversity and regeneration potential of Q. leucotrichophora domi-nant conifer mixed forest along an altitudinal gradient of Garhwal Himalaya. *Indian For.* **2005**, *131*, 1537–1553.
76. Kitayama, K.; Aiba, S.-I. Ecosystem structure and productivity of tropical rain forests along altitudinal gradients with contrasting soil phosphorus pools on Mount Kinabalu, Borneo. *J. Ecol.* **2002**, *90*, 37–51. [CrossRef]
77. Moser, G.; Hertel, D.; Leuschner, C. Altitudinal Change in LAI and Stand Leaf Biomass in Tropical Montane Forests: A Transect Study in Ecuador and a Pan-Tropical Meta-Analysis. *Ecosystems* **2007**, *10*, 924–935. [CrossRef]
78. Leuschner, C.; Moser, G.; Bertsch, C.; Röderstein, M.; Hertel, D. Large altitudinal increase in tree root/shoot ratio in tropical mountain forests of Ecuador. *Basic Appl. Ecol.* **2007**, *8*, 219–230. [CrossRef]
79. Raich, J.W.; Russell, A.E.; Kitayama, K.; Parton, W.J.; Vitousek, P.M. Temperature influences carbon accumulation in moist tropical forests. *Ecology* **2006**, *87*, 76–87. [CrossRef] [PubMed]
80. Alves, L.F.; Vieira, S.A.; Scaranello, M.A.; de Camargo, P.B.; Santos, F.; Joly, C.A.; Martinelli, L. Forest structure and live aboveground biomass variation along an elevational gradient of tropical Atlantic moist forest (Brazil). *For. Ecol. Manag.* **2010**, *260*, 679–691. [CrossRef]
81. Moser, G.; Leuschner, C.; Hertel, D.; Graefe, S.; Soethe, N.; Iost, S. Elevation effects on the carbon budget of tropical mountain forests (S Ecuador): The role of the belowground compartment. *Glob. Chang. Biol.* **2010**, *17*, 2211–2226. [CrossRef]

land

Article

Stand Structure, Biomass and Carbon Storage in *Gmelina arborea* Plantation at Agricultural Landscape in Foothills of Eastern Himalayas

Mendup Tamang [1], Roman Chettri [1], Vineeta [1], Gopal Shukla [1], Jahangeer A. Bhat [2], Amit Kumar [3,*], Munesh Kumar [4], Arpit Suryawanshi [5], Marina Cabral-Pinto [6] and Sumit Chakravarty [1]

1 Department of Forestry, Uttar Banga Krishi Viswavidyalaya, Pundibari, Cooch Behar, West Bengal 736165, India; dupmentamang99@gmail.com (M.T.); romanchettri18@gmail.com (R.C.); vineeta@ubkv.ac.in (V.); gopal@ubkv.ac.in (G.S.); sumit@ubkv.ac.in (S.C.)
2 Department of Forest Products and Utilization, College of Horticulture and Forestry, Rani Lakshmi Bai Central Agricultural University, Jhansi 284003, India; jahan191@gmail.com
3 School of Hydrology and Water Resources, Nanjing University of Information Science and Technology, Nanjing 210044, China
4 Department of Forestry and Natural Resources, H.N.B. Garhwal University (A Central University), Srinagar Garhwal, Uttarakhand 249161, India; muneshmzu@yahoo.com
5 Department of Soil Science and Agricultural Chemistry, College of Agriculture, Rani Lakshmi Bai Central Agricultural University, Jhansi 284003, India; arpitsurya226@gmail.com
6 Department of Geosciences, University of Aveiro, Campus de Santiago, 3810-193 Aveiro, Portugal; marinacp@ua.pt
* Correspondence: amitkdah@nuist.edu.cn

Citation: Tamang, M.; Chettri, R.; Vineeta; Shukla, G.; Bhat, J.A.; Kumar, A.; Kumar, M.; Suryawanshi, A.; Cabral-Pinto, M.; Chakravarty, S. Stand Structure, Biomass and Carbon Storage in *Gmelina arborea* Plantation at Agricultural Landscape in Foothills of Eastern Himalayas. *Land* **2021**, *10*, 387. https://doi.org/10.3390/land10040387

Academic Editor: Bruno Marino

Received: 23 February 2021
Accepted: 31 March 2021
Published: 7 April 2021

Abstract: In the modern era, *Gmelina arborea* plantations are a hotspot of future research because of their high carbon sequestration potential. The present work was conducted during 2018 to 2020 on a young unmanaged *Gmelina* farm to understand the ecosystem's carbon and its dynamics. The study area was categorized into three age classes: ≤5, 5–10, and 10–15 years. In a plantation, *Gmelina* trees (10%) were randomly selected while other trees (90%) were also taken into the consideration for ecosystem carbon. A stratified random nested quadrate sampling method was adopted for analyzing other vegetation forms under study. Overall, 51 individual species in the studied *Gmelina* farm were found which includes 23 tree species, 7 shrub species, 16 herbs, 2 climbers, and 3 species of ferns. The estimated quantitative vegetation parameters and diversity indices indicate that the plant assemblages were heterogeneous with native diverse species evenly distributed with fairly higher densities, frequencies, and abundance. Herbs were the most important species followed by shrubs and trees. Consequently, with the increasing age of plantation, the richness of plant species increased. Soil properties were significantly influenced by the age of the plantation but exhibited no discreet trend. Total biomass density and total carbon density increased with increasing plantation age while no drastic variation was found in available soil organic carbon (SOC) because of insignificant variability in litter production. Total carbon, available SOC (up to 60 cm depth) and ecosystem carbon in the three age class plantations fell in the ranges of 54.51–59.91, 48.18–55.73, and 104.81–110.77 Mg ha^{-1}, respectively. The carbon sequestration potential of *Gmelina arborea* is higher compared to other reported species and highly supportive of converting unutilized agricultural landscapes to reduce the atmospheric carbon dioxide in future.

Keywords: plantation; climate change; land use management; carbon sequestration; soil

1. Introduction

Climate change is a global concern and forests play a vital role in regulation as they are a viable option for offsetting terrestrial carbon dioxide emissions [1,2]. Unfortunately, forests alone are not enough to offset all the terrestrial emissions [3] and there is a need is to find an alternate viable option to bridge this gap [4]. The accepted viable alternative is trees outside forest (TOFs) in both agricultural and human-dominated landscapes, which will not only meet timber, industrial, and livelihood demands but also effectively and viably facilitate offsetting carbon dioxide (CO_2) emissions with the forest ecosystems [5]. Tree-based land use systems including those in the nonforested landscapes such as agricultural land play an important role in global carbon (C) cycling since these are one of the largest C pools which act as a potential C sink and also as one of the major sources of CO_2 [4]. The productivity of the plantations is higher (3.2 Mg ha^{-1} yr^{-1}) than the productivity of natural forests (1.1 Mg ha^{-1} yr^{-1}) [6]. This is because plantation forestry has an added advantage over the natural forest in terms of better silvicultural practices [7]. Thus, the past decades have witnessed increased interest in tree plantations in both agricultural- and non-agricultural-dominated landscapes in the country, especially Teak (*Tectona grandis*), *Gmelina* (*Gmelina arborea*), Deccan Neem (*Melia azadirach*), Champa (*Michelia champaca*), and Sal (*Shorea robusta*) among the farmers and entrepreneurs, particularly in the sub-Himalayan, i.e., Terai, region of West Bengal [8]. *Gmelina arborea*, which is native to India and a prime fast growing species in farm forests in India, has the potential to replace and act as a substitute for exotic timbers in the country [9]. The species has the potential to store C and is also remunerative due to its multiple uses [8,10]. In addition to timber, wood of the tree is used for fuel wood, paper and pulp making, and is used in other forest-based industries [8].

Gmelina plantations have been established and encouraged in small woodlots, home gardens, and agroforestry settings in the tropics and subtropics [8], including the Terai region of West Bengal. The potential of these trees to offset C emission needs to be assessed and monitored properly for which local, regional or national inventories are required [4]. Understanding the diverse and complex tree-based land use systems for C sequestration and nutrient cycling has become a global research interest [11]. The United Nations Framework Convention on Climate Change (UNFCCC) has recognized the importance of plantation forests as a greenhouse gas (GHG) mitigation option, as well as the need to monitor, preserve and enhance terrestrial C stocks [12]. Studies on associated plant species biodiversity, quantification of biomass, and C are available for *Gmelina* farms in India but few attempts have been made regarding chronosequencing of C sequestration potential of these plantations along with associated plant biodiversity [8], while none have been carried out for the Terai region of West Bengal. The study was thus attempted with the hypothesis that there will be chronosequence variations in terms of diversity, biomass, carbon storage, and soil properties of the unmanaged *Gmelina* farm in Terai region of West Bengal with the following objectives: (i) to assess physio-chemical characteristics of soil and (ii) to assess phyto-diversity, biomass, and carbon storage of the *Gmelina* farm. The present study was the first attempt in the region to assess the potential of *Gmelina* farms, which will be helpful in the conversion of the unutilized land for C farming to create additional C sinks and may further assist in trapping the available carbon dioxide in the atmosphere.

2. Materials and Methods

2.1. Site Description

The present study was conducted in the Terai region of West Bengal at Cooch Behar district from September 2018 to February 2020. The study site is a sub-Himalayan region located between 26°30′–26°56′ N latitude and 88°7′–89°53′ E longitude. The area around (within 10 km) the University (Uttar Banga Krishi Viswavidyalaya, Pundibari—UBKV) campus was surveyed for sampling plantations of *Gmelina* in the agricultural landscape which was a predominantly rice-based cropping system with potato and/or fallow. Moreover, the cropping system is not intensive in the area and the land of most of the farmers was marginal; therefore, they have poor resources. *Gmelina* or any other tree plantations were not normally planted as a block in crop land by the farmers in the Terai region except some scattered plantations mostly in home gardens or as boundary or roadside plantations. The plantations were generally developed on land normally unsuitable for annual cropping by farmers and nonfarmers, in addition to crop land of the owners, and were kept fallow until the planting of *Gmelina* or any other tree plantations. These plantation owners were normally absentee growers. *Gmelina* was densely planted with a squared geometry of 3 × 3 m. The plantations were normally kept aside undisturbed without replanting the gaps and allowing spontaneous plant growth. The area of most of the plantations was at least one acre.

2.2. Climate and Weather

The study area is under moist tropical conditions [13], where average minimum and maximum temperatures varied from 22.8 (during winter, January) to 32.32 °C (during September). On average, the annual rainfall varied from 2000 to 3500 mm, the bulk of which occurred during the premonsoon and monsoon periods—i.e., from May to September. The relative humidity (RH) varied from 55 to 90%. The study area is warm and humid except with a sort of spell of winter from December to March (Supplementary Figure S1).

2.3. Sampling and Inventory

As there were only 21 plantations in the study area, we were able to include them all in the study. The selected plantations were marked using GPS (Model Garmin 72) and based on their availability, plantations of three age classes (seven plantations each) were categorized—i.e., ≤5 (Age class I (AC I)), 5–10 (AC II) and 10–15 years (AC III; Supplementary Table S1). Each selected plantation was of about one acre in area. At each plantation, the outer rows were excluded from measurement as a buffer. From the interior rows, all Gmelina stems with diameters at breast height (dbh) of more than 10 cm were recorded and additionally 10% of these trees were selected randomly and their heights were recorded. Species other than *Gmelina*—all trees with a dbh > 10 cm—were recorded along with their tree heights. Similarly, other tree species with dbhs ≥ 10 cm in the plantation were also selected. Shrubs and herbs were sampled following a stratified random nested method, where three (5 × 5 m) quadrats were marked diagonally across the plantation (two at the corners and one at the center) for shrubs, while for herbs five 1 × 1 m plots were marked at all corners and one at the center of the plantation. Most of the plant species were identified in the field itself, while those which could not be identified were preserved by mounting in herbarium sheets following standard procedures for identification. The mounted specimens were cross-checked with the available herbarium in the Department of Forestry UBKV Pundibari, West Bengal. A full inventory of the plant species found at the plantations was prepared including trees, shrubs, and herbs. The biomass of plantations was separately estimated for trees, shrubs, and herbs. Tree biomass was estimated separately for all the species. Litter samples were collected once during January from ten sampling locations with plots size of 1 × 1 m distributed throughout the plantation. To estimate the plantation soil properties, a total of 63 representative soil samples were collected from each plantation site—i.e., ≤ 5 (AC I), 5–10 (AC II) and 10–15 years (AC III) at three different depths (0–20, 20–40, and 40–60) using Dutch augur. Prior to estimation, collected soil samples were air-dried, grinded with a wooden pestle, passed through a 2 mm sieve, and stored in cloth bags for further analysis.

2.4. Quantitative Parameters

For each plantation, individual species were recorded for quantitative parameters following standard procedures. Importance value index (IVI) as a summation of relative frequency (RF), relative density (RD), and relative abundance (RA), as suggested by Cintron and Novelli [14], was estimated to analyze the sociability of the plant assemblages in the plantations.

2.5. Species Diversity Indices

Various diversity indices were estimated to analyze plant diversity of the plantations. Species richness was described by the available species number in a studied plantation.

Menhinick's index (D′) [15] is based on the total number of species and the total number of individuals of all species (N). $D' := S/\sqrt{N}$. This index, unlike the Shannon–Wiener index, gives more weightage to the rare species. The concentration of dominance was used to evaluate species dominance within a community [16]. This provides information on number of times a particular species was encountered during the sampling. Higher values are indicative of less diverse community and concentration of dominance is calculated using the following formula:

$$C = \sum (n_i / N)^2 \quad (1)$$

where n_i denotes the number of individuals of a species

The Shannon–Wiener diversity index (H′) [17] of species diversity was used to describe diversity, where higher values suggest more diverse nature of the plantations. The index was estimated using following formula:

$$H' = -\sum (n_i / N) \ln (n_i / N) \quad (2)$$

The method of Pielou [18] was used to estimate the species evenness index:

$$EI = H' / \ln N. \quad (3)$$

2.6. Biomass and Carbon Stock Estimation

A nondestructive method was adopted for quantifying above-ground biomass (AGB) of the trees using a recent allometric model developed for forest types in northeast India [19].

$$AGB = 0.18D^{2.16} \times 1.32 \quad (4)$$

where D is the diameter of tree at breast height.

Below-ground biomass (BGB) was estimated as 15% of the AGB [20]. Total tree biomass was the summation of AGB and BGB, which was estimated for each tree in the sampled area and then summed up. The total herbs, shrubs, and tree biomass was further converted into carbon by multiplying a factor of 0.50 [20].

2.7. Soil Properties

Soil properties such as bulk density (core sampler method), moisture (volumetric method), electrical conductivity or electrical conductivity (EC) (soil water suspension method), pH (Beckman's pH meter), soil organic carbon (Walkley and Black's rapid titration method), available nitrogen (Modified Kjeldahl method), available phosphorus (Bray's method) and available potassium (Flame Photometer method) were analyzed following standard methods [21–23]. Soil organic carbon (SOC) stock was estimated by multiplying the SOC with weight of the soil (bulk density and depth) and is expressed as mega grams per hectare (Mg ha^{-1}) [24].

2.8. Statistical Analysis

The data were analyzed using the software package Gen Stat Eleventh Edition (VSN International, Oxford, UK). One-way analysis and a Duncan multiple range test (DMRT) test were also employed.

3. Results and Discussion

3.1. Diversity Indices and Species Composition

The diversity indices of *Gmelina* farm are given in Table 1. Overall, we found 51 species including 23 tree species, 7 shrub species, 16 herbs, 2 climbers and 3 species of ferns. Four species were not identified (see Supplementary Tables S2–S5).

Table 1. Diversity indices of *Gmelina* farm forestry plantations.

Parameters	Overall	Age Classes (Year)		
		≤5	5–10	10–15
Species richness	51	35	37	43
Genera richness	46	35	35	42
Family richness	33	25	25	30
Species diversity index	1.18	0.54	0.62	0.59
Concentration of dominance	0.03	0.04	0.05	0.04
Shannon–Wiener index	1.29	1.54	1.59	2.00
Evenness index	8.69	3.35	3.92	3.65

Amongst the plant species found to be associated with *Gmelina* plantations of all age classes, herbs were prominently found followed by trees, ferns and climbers. Similar studies on associated plant species diversity in tree plantations in agricultural landscapes have been previously reported with either shrubs or herbs dominating [25]. Dominance of one stratum generally suppresses the diversity of another [26].

The different diversity indices estimated for different age class plantations are given in Table 1. Overall, the plant assemblages in the plantations were more diverse, evenly distributed, heterogeneous, and stable compared to the different age class plantations. The concentration of dominance of the different age classes estimated separately was much less but indicated a similar and higher probability of a species being encountered during sampling in these plantations. The H′ index reflects structure and heterogeneity of plant assemblages in an ecosystem—i.e., higher the index value, the more diverse and stable the community is [27]. The index values estimated for the different age class plantations were much less but increased with increasing plantation age—i.e., the plantations with increasing age became more heterogeneous and stable. Species in the different age class plantations were more or less evenly distributed.

This is a considerably higher diversity of associated plant communities developed in the *Gmelina* farm as there were no disturbances. The plantations were not managed silviculturally by the owners after planting as they were kept aside. A similar increase in plant diversity was also reported for farm forestry plantations and other nonforested landscapes when they were not managed or kept aside [28]. The diversity of the associated species in the *Gmelina* plantations increased with the increasing plantation age, reflecting the compatibility of the associated species with *Gmelina*, which is in contrast to the studies reporting on teak plantation due to its allelopathic effect [25]. Moreover, canopy gaps in the plantation allowed enough sunlight to favor understory growth for early successional species [29]. The undisturbed *Gmelina* farm forestry plantations in the Terai zone of West plantations aided the rehabilitation of fallow crop land in an agricultural landscape, allowing homeostatic capability of the system [30].

The studied *Gmelina* undisturbed farm forestry plantations with increasing age promoted succession along with resetting of many ecosystem processes such as improving microclimate, soil fertility through litter input, microbial diversity and activity, biomass production, and sequestration capacity [31]. Studies have confirmed succession in sole tree species plantation resetting the disrupted processes associated with diversity [32]. Species richness listed in the plant assemblages of the *Gmelina* plantations was comparable with an earlier study involving plantations of *Tectona grandis*, *Shorea robusta*, *Michelia champaca*, and *Lagerstroemia speciosa*, but less so than miscellaneous species stands in the Chilapatta Reserve Forest not more than 10 km from the present study sites [33].

Generally, tree plantations outside the forest were recolonized with forest species of the regional species pool [34], which promotes rehabilitation because of improved site quality factors suitable for growth of native species [31]. It was shown in all the earlier studies that in absence of disturbance, structural homogeneity of the plantation with succession gradually leads to heterogeneous multilayered secondary forests with more heliophytes and sciophytes in the understory. In the present study, 51 plant species including trees, shrubs, herbs, climbers, and ferns were documented from the *Gmelina* farm and if left to grow undisturbed until their natural rotation, a more heterogeneous secondary forest can be expected in this agricultural landscape. Rehabilitation of landscapes is a time-consuming process which needs at least as long as 100 years to reset back, akin to the earlier native forest [35] but not the same in terms of species richness and structure [25]. This study, along with previous studies, supports the policy of using tree planting to restore degraded agricultural landscapes particularly unsuitable for annual cropping with tree plantations of *Gmelina* or other timber species for biodiversity, ecosystem functions, sustainable forestry, and ecosystem services [25].

3.2. Vegetation Analysis

Frequency, abundance, density, and IVI estimated for the plantations are provided in Supplementary Tables S2–S5. In central and eastern Himalayan forests, frequency of tree layer was reported with a range of 10–100%, shrub layer 10–80%, herb layer 10–100% [27,36,37]. A similar frequency range was also observed for the species in the present study. The species documented in the study sites were native species of the region and were mostly used by the local people for food, medicine, and fodder [38,39]. Frequency or the degree of distribution of the species indicates the chance of occurrence of a species while sampling. *Acacia auriculiformis*, *Albizia lebbeck* and 15 other species had lowest representations in the sampled plots, while *Ageratum houstonianum* had the highest representation. Correspondingly, these species were also observed with the lowest and highest relative representations, respectively, while sampling. Generally, the representation of species and their relative representations during sampling initially increased—i.e., from AC I to AC II—but then decreased from AC II to AC III.

Higher chance of occurrence of the associated plant species in *Gmelina* farm forestry plantations also generally increased the numerical strength and abundance of the species, which resulted in the easier establishment of these species in the plantations. These species were initial colonizers as they adapted well as understory strata in the *Gmelina* farm forestry plantations. *Dalbergia sissoo*, *Lagerstroemia speciosa*, *Albizia lebbeck*, *Acacia auriculiformis*, and *Moringa oleifera* (all trees) were estimated as having the lowest densities, while *Clerodendron infortunatum*, a shrub with the highest density in the plantations amongst the associated species. Species density varies with forest community type, forest age class, tree species, size class, site history, site quality factors, and disturbance [40]. *Acacia auriculiformis*, *Albizia lebbeck*, *Dalbergia sissoo*, *Lagerstroemia speciosa*, and *Moringa oleifera* were the least abundant, while *Clerodendron infortunatum*, a shrub, was the most abundant associated species in the plantations.

Based on estimated IVI values, the most important of all the species associated with *Gmelina* plantations was *Clerodendron infortunatum*, a shrub, and the least were *Acacia auriculiformis*, *Albizia lebbeck*, *Dalbergia sissoo*, *Lagerstroemia speciosa*, and *Moringa oleifera*. The other important associated species in the plantations were *Clerodendron infortunatum*, *Ageratum houstonianum*, *Cynodon dactylon*, *Ageratum conyzoides*, *Lantana camara*, *Diplazium esculentum*, *Fragaria vesca*, *Lucas aspera*, *Tabernaemontana divaricata*, *Mikania micrantha*, *Colocasia esculenta*, *Pouzolzia zeylanica*, *Bombax ceiba*, *Paspalum distichum*, *Matteuccia struthiopteris*, and *Oxalis corniculata* (see Supplementary Tables S2–S5). Based on the higher IVI values of these species as compared to other recorded species, it can be concluded that these species were successful primary colonizers in the *Gmelina* farm forestry plantations, forming a definite structure with vertical understory strata comprising trees, shrubs, and herbs [30].

3.3. Soil Moisture, pH, Electrical Conductivity and Bulk Density

Soil moisture and EC (Table 2; Supplementary Tables S6 and S7) decreased gradually with increase in soil depth from 0–20 cm to 40–60 cm in plantations of all age class series but soil pH (Table 2; Supplementary Table S7) exhibited a reverse trend, while soil bulk density (Table 2) exhibited no trend with depth. None of these soil physical parameters show any consistent trend with increased plantation age, which is also evidenced by the staggered relationship observed between plantations of different age classes and their soil physical parameters (Table 3). This indicates that the soil of *Gmelina* plantations was highly inconstant and had unstable characteristics as these were planted in agricultural landscapes [41–43].

Table 2. Soil moisture, pH, electrical conductivity (EC) and bulk density in *Gmelina* plantations.

AC	Moisture (%)			pH			EC (m mhos cm^{-1})			Bulk Density (g cm^{-3})		
	D1	D2	D3	D1	D2	D3	D1	D2	D3	D1	D2	D3
I	28.23	24.56	21.22	5.56	6.19	6.59	0.31	0.29	0.17	1.73	1.66	1.56
II	26.94	23.36	21.01	5.28	5.65	6.24	0.25	0.16	0.09	1.54	1.62	1.55
III	29.59	25.84	22.93	5.39	5.81	6.27	0.28	0.20	0.17	1.53	1.58	1.63
Mean	**28.25**	**24.59**	**21.72**	**5.42**	**5.88**	**6.37**	**0.28**	**0.22**	**0.14**	**1.6**	**1.62**	**1.58**
S_{em}	**1.10**	**1.15**	**1.02**	**0.22**	**0.21**	**0.19**	**0.05**	**0.05**	**0.05**	**0.65**	**0.23**	**0.25**
CD	**3.38**	**3.54**	**3.13**	**0.66**	**0.65**	**0.59**	**0.15**	**0.17**	**0.15**	**NS**	**NS**	**NS**

AC I (Age class I—≤5 years); AC II (5–10 years); AC III (10–15 years); D1—soil depth 0–20 cm; D2—20–40 cm; D3—40–60 cm; Sem—standard error mean; CD—Critical difference.

Table 3. Pearson correlation matrix of stand, soil properties Total Biomass Density (TBD) and TCD.

	AC	pH	EC	N	P	K	SOC	TBD
AC	1							
pH	−0.266 *	1						
EC	−0.120	0.616 **	1					
N	0.060	−0.445 **	−0.170	1				
P	−0.077	−0.072	−0.142	0.002	1			
K	0.253	−0.045	−0.018	0.063	0.020	1		
SOC	−0.051	0.043	0.382 **	−0.121	−0.120	−0.043	1	
TBD	0.795 **	−0.128	−0.009	0.025	−0.135	−0.107	−0.067	1

** Significant at 0.01 level; * Significant at 0.05 level (2 tailed); AC—age class; EC—electrical conductivity; N—nitrogen; P—phosphorus; K—potassium; SOC—soil organic carbon; TBD—Total Biomass Density (total biomass having both ABG and BGB).

Prior to the establishment of plantations, lands were unused or were agricultural fallow in and around crop lands. The plantations were not at all managed except during planting with farmyard manure (FYM) application or at the most once after one year of planting. All these soil physical parameters except bulk density were significantly influenced by the age of the plantations which, however, did not exhibit any discreet trend with increasing age of the plantations. Soil water holding capacity (WHC) in tree-based land use systems is influenced by rainfall, temperature, humidity, amount of incident radiation on the soil floor, structure, and function of plant cover [44]. The Terai region of West Bengal located in the foothills of eastern Himalayas has a tropical moist climate with high rainfall and acidic soil [45]. High humidity and rainfall (Supplementary Figure S1) increased the soil water retention by reducing evaporation rates and increasing the infiltration of water [33]. Moreover, tree-based land use systems were reported with higher soil organic matter on the surface soil layer due to litter input increasing the EC, thus making the soil more acidic and these soils can also absorb and hold substantial quantities of water as compared to subsurface layers [38]. Higher acidity of surface soil is due to accumulation and subsequent slow decomposition of organic matter releasing acids [46]. Lower pH at soil surface inactivates the soil fauna resulting in slower humus decomposition with more nondecomposed matter on the soil floor [47]. The undisturbed *Gmelina* plantations increased the soil organic matter continuously without being removed, which efficiently regulated the soil physical properties by increasing leaching of bases and weathering process due to decomposition of litter [48].

Soil moisture, pH, and EC in the plantations of different age classes (AC I—≤ 5 years, AC II—5–10 years, and AC III—10–15 years) varied significantly because of site quality factors at the landscape level (Supplementary Tables S6 and S7) arising due to microlandscape differences resulting from differences in developing understory vegetation and gaps in the *Gmelina* canopy. There was mortality of *Gmelina* after its planting which resulted in its discontinuous canopy. Differences in the development of understory vegetation in the *Gmelina* plantations are indicated from the differences in its community parameters. The skewed variability in soil properties in the plantations of different age classes was due to no silvicultural operations performed and was thus strongly influenced by socio-ecological conditions [49] such as vegetation structure of the plantations as well as interculture operations performed on surrounding crop land [50].

3.4. Soil Available Organic Carbon, Nitrogen, Potassium and Phosphorus

The amount of available SOC, N, P, and K decreased with increasing soil depth, highest at surface layer and lowest in the deepest layer analyzed (Table 4; Supplementary Tables S8–S11).

Table 4. Soil organic carbon and available nitrogen, phosphorus and potassium.

AC.	SOC ($Mgha^{-1}$)			Available N ($Kg\ ha^{-1}$)			Available P ($Kg\ ha^{-1}$)			Available K ($Kg\ ha^{-1}$)		
	D1	D2	D3	D1	D2	D3	D1	D2	D3	D1	D2	D3
I	25.64	18.74	11.35	157.1	93.63	60.78	14.63	15.08	14.88	65.11	57.34	54.36
II	22.15	14.75	11.28	172.0	126.8	73.62	14.15	13.23	11.84	86.37	71.68	72.28
III	22.11	15.57	13.18	141.7	108.7	73.42	14.76	16.76	13.76	86.98	78.58	71.20
Mean	**23.3**	**16.35**	**11.94**	**156.9**	**109.7**	**69.27**	**14.51**	**15.02**	**13.49**	**79.49**	**69.2**	**65.95**
S_{em}	**1.87**	**1.46**	**0.58**	**15.45**	**17.60**	**16.39**	**0.92**	**1.93**	**0.89**	**8.30**	**10.14**	**9.79**
CD	**5.77**	**4.50**	**1.79**	**47.59**	**54.22**	**50.50**	**2.84**	**5.95**	**2.76**	**25.59**	**31.24**	**30.18**

AC I (Age class I—$\leq$ 5 years); AC II (5–10 years); AC III (10–15 years); D1—soil depth 0–20 cm; D2—20–40 cm; D3—40–60 cm.

Regular litter input along with moderate temperature and humidity in the *Gmelina*-based land use caused higher availability of organic carbon (OC) and nutrients on the surface than the subsurface layers [51]. The availability of soil primary nutrients in plantations was in the order N > K > P [36,52]. The estimated available amount of these primary nutrients indicates that soil in the *Gmelina* farm was low-medium in available nitrogen, low-high in available phosphorus, and low in available potassium [53]. Available soil nitrogen and potassium were synergistic to each other [54]. Forest and cultivated unmanaged land were estimated with highest available amount of nitrogen and organic carbon, while these are medium in well-managed cultivated soil and lowest in barren land [55]. Similar to the soil properties in the plantations, the amount of available SOC and soil primary nutrients also differed significantly but without any visible trends with increasing age of the plantations indicating local differences (Supplementary Tables S8–S11). These differences were due to variations at the microlandscape level in site quality factors associated with socio-ecological conditions such as vegetation parameters and management of surrounding crop fields [50].

The estimated amount of available SOC and primary nutrients was staggered at different soil depths with no discreet trend, indicating highly inconsistent nature and unstable characteristics of soil in the *Gmelina* farm [42]. Similar heterogeneity of soil properties with no discreet trend observed in the present study area was also reported from urban plantations [56]. Adequate soil management is crucial for nutrient availability and OC [57]. Forests were converted to agricultural lands in the Terai region of West Bengal and generally were not adequately managed, which resulted in the inconsistent and unstable nature of soil indicated by the unpredictable behavior of soil properties including less nutrient availability and OC build up than the natural forests [58]. The plantations were established in unused or fallow crop land and kept undisturbed except FYM application during planting and at the most once after one year of planting. The growing plantation used nutrients from the soil with no or very little replenishment in the soil from litter input initially. Moreover, there was also lesser understory vegetation during the initial years of plantation establishment adding less organic matter to the soil. With the continuous growth of trees and increase in understory vegetation due to no management of the plantations, there was nutrient build by organic matter (litter) input and thus the amount of soil available nutrients did not exhibit any discreet trend with increasing age of the plantations.

Factors such as topography, climate (temperature, precipitation, RH), weathering process, vegetation cover, and microbial activities influence the build-up of OC and nutrients in the soil [48]. Differences in the replenishment of nutrients back to the soil due to varied litter input, soil water content, aeration, temperature, microorganisms, and efficiency of the root system to absorb nutrients caused variation of feedback mechanism of various ecosystem processes in the plantations [59]. *Gmelina* farm forestry plantations of different age groups were thus vegetation with heterogeneous structures and compositions that caused significant variation among their SOC build up and nutrient availability [60]. The structures and compositions of vegetation and soil properties are positively correlated with available SOC and nutrients [61]. There are abundant reports on negative influence of forest conversion to crop land or plantation [62]. Tree-based land uses and management practices, however, positively change the soil properties including SOC build up [36].

Contrary to the reports of higher availability of soil primary nutrients due to higher soil organic matter in the tree-based land uses [33], the present study found no such trends with the availability of soil primary nutrients with available C build up in the plantation soil. Application of FYM at the early stage of plantation establishment might be the reason for no such trends in the plantations. Further, human activities in and around the plantations were also responsible for the unpredictable behavior of soil properties in the plantations [63]. Trends or no trends, soils under *Gmelina* farm in the Terai region of West Bengal without any management also accumulated a considerable amount of carbon and primary nutrients sometimes comparable to forest and well-managed agricultural land uses [64]. However, the process of carbon accumulation in soils of *Gmelina* farm forestry plantations was still left largely unexplained akin to other studies [64].

3.5. Biomass Production and Biomass Carbon Stock

The AGB, BGB, and total biomass with their corresponding C quantified for the *Gmelina* farm are given in Table 5. The overall contribution of AGB in the plantations was 87.37% to the total biomass. Overall in the plantations, trees contributed 96.72%, litter contributed 3.17% and understory shrubs and herbs contributed only 0.11% of the total biomass in the plantations, while overall in the plantations only *Gmelina* trees contributed 34.35% and associated trees contributed 62.37% of the total biomass. The biomass of the plantation increased gradually with its increasing age, exhibiting a strong positive correlation ($r = 0.795$; Table 3)—i.e., the biomass of the plantations increased with increasing age. The quantum of biomass increases were higher from AC II to AC III (21.19% for above ground, 21.26% for below ground, and 21.20% for total biomass) than from AC I to AC II (4.07% for above ground, 2.66% for below ground and 3.89% for total biomass). ABG, BGB, and TB of *Gmelina* were increased by 9.37, 9.28 and 9.36% from AC I to AC II, respectively; while the increases were 19.39, 19.5 and 19.4% from AC II to AC III, respectively. In age class I, mean contributions of total AGB, litter biomass and total shrubs + herbs biomass to mean total biomass of the plantations were 87.27, 2.43, and 0.1%, respectively, while contributions of *Gmelina* and associated trees AGB, BGB, and TB to mean total biomass of the plantations were 28.98, 4.35, 33.33, 55.77, 8.34, and 64.1%, respectively.

Table 5. Biomass and carbon stock (Mg ha^{-1}) in *Gmelina* farm forestry plantations.

Component	AGB	BGB	TB	AGC	BGC	TC
	Age class I (≤ 5 years)					
Gmelina arborea	31.6	4.74	36.34	15.8	2.37	18.17
Bombax ceiba	12.24	1.84	14.08	6.12	0.92	7.04
Melia azaderach	2.20	0.33	2.53	1.1	0.16	1.26
Ailanthus grandis	5.36	0.80	6.16	2.68	0.4	3.08
Chukrasia velutina	2.15	0.32	2.47	1.07	0.16	1.23
Tectona grandis	10.26	1.54	11.8	5.13	0.77	5.59
Swietenia macrophylla	13.45	2.02	15.47	6.72	1.01	7.73
Dalbergia sissoo	2.72	0.41	3.13	1.36	0.20	1.56
Albizia lebbeck	12.42	1.86	14.28	6.21	0.93	7.14
Shrub	0.045	0.007	0.052	0.0225	0.0035	0.026
Herb	0.053	0.008	0.061	0.0265	0.004	0.031
Litter	2.65	-	2.65	1.32	-	1.32
Total	**95.15**	**13.88**	**109.03**	**47.57**	**6.94**	**54.51**
	Age class II (5–10 years)					
Gmelina arborea	34.56	5.18	39.74	17.28	2.59	19.87
Melia azaderach	2.68	0.4	3.08	1.34	0.2	1.36
Chukrasia velutina	9.1	1.36	10.46	4.55	0.68	5.23
Bombax ceiba	17.22	2.58	19.8	8.61	1.29	9.9
Tectona grandis	20.64	3.1	23.74	10.32	1.55	11.87
Ailanthus grandis	7.16	1.07	8.23	3.58	0.53	4.11
Syzygium cumini	3.59	0.54	4.13	1.79	0.27	2.06
Shrub	0.054	0.008	0.062	0.027	0.004	0.031
Herb	0.062	0.009	0.071	0.031	0.004	0.035
Litter	3.95	-	3.95	1.97	-	1.97
Total	**99.02**	**14.25**	**113.27**	**49.51**	**7.12**	**56.63**
	Age class III (10–15 years)					
Gmelina arborea	41.26	6.19	47.45	20.63	3.09	23.72
Melia azaderach	7.21	1.08	8.29	3.6	0.54	4.14
Chukrasia velutina	6.32	0.95	7.27	3.16	0.47	3.63
Bombax ceiba	16.68	2.50	19.18	8.34	1.25	9.59
Tectona grandis	20.36	3.05	23.41	10.18	1.52	11.7
Ailanthus grandis	12.4	1.86	14.26	6.2	0.93	7.13
Syzygium cumini	10.84	1.63	12.47	5.42	0.81	6.23
Shrub	0.057	0.008	0.065	0.028	0.004	0.032
Herb	0.073	0.011	0.084	0.036	0.0055	0.042
Litter	4.80	–	4.80	2.40	–	2.40
Total	**120.0**	**17.28**	**137.28**	**60.0**	**8.64**	**68.64**
Mean of all age classes	**104.72**	**15.14**	**119.86**	**52.36**	**7.55**	**59.91**

AGB: above-ground biomass; BGB: below-ground biomass; TB: total biomass; AGC: above-ground carbon; BGC: below-ground carbon; TC: total carbon.

Prominent tree species associated with AC I were Bombax ceiba, Melia azaderach, Ailanthus grandis, Chukrasia velutina, Tectona grandis, Swietenia macrophylla, Dalbergia sissoo and Albizia lebbeck, contributing 12.88, 2.32, 5.65, 2.26, 10.82, 14.19, 2.87, and 13.10%, respectively, to the total plantation biomass. Similarly, in AC II the contributions of total plantation AGB, litter biomass, total shrub + herb biomass, and above-ground, below-ground and total biomass of Gmelina and associated trees to total plantation biomass were 87.42, 3.49, 0.12, 30.51, 4.57, 35.08 and 61.30%, respectively, while in AC III the contributions were 87.41, 3.5, 0.11, 30.06, 4.51, 34.56 and 61.83%, respectively. Melia azaderach, Chukrasia velutina, Bombax ceiba, Tectona grandis, Ailanthus grandis, and Syzygium cumini were associated with both AC II and AC III age group plantations and their contributions of total biomass to the total plantation biomass were 2.72 and 6.04%, 9.23 and 5.29%, 17.48 and 13.97%, 20.96 and 17.06%, 7.26 and 10.39% and 3.65 and 9.08%, respectively. The amount of biomass estimated in the Gmelina farm forestry plantations was less than that reported from plantation or stands in forest landscape from the same study area—i.e., Terai zone of West Bengal [65]. Negligible contribution of understory vegetation and significant contribution by above-ground parts to the total biomass of the tree-based land uses was also reported by many previous works [66].

With increasing age, biomass increased with no change in the contribution of both AGB and BGB towards total biomass from AC I to AC II and AC II to AC III. The contributory trend of the different components of the *Gmelina* plantation towards total mean biomass with increasing age also remained unchanged with increasing age of the plantation. Litter production though increased with increasing age of the plantation but also exhibited the same contributory behavior towards total mean biomass. *Bombax ceiba, Melia azaderach, Ailanthus grandis, Chukrasia velutina,* and *Tectona grandis* were found in all the age classes of *Gmelina* farm forestry plantations. The total biomasses of these associated tree species in ACs I, II, and III were 37.04, 65.31 and 72.41 Mg ha^{-1} which were 33.97, 57.66, and 52.74%, respectively, of the total plantation biomass. From AC I to AC II, the contribution of these five associated species increased by 23.69% but from AC II to AC III their contribution decreased by 4.92%. The total biomass contribution of *Gmelina* was 33.33%, 35.08%, and 34.56% towards total plantation biomass in AC I, AC II, and AC III, respectively. The trend in contribution of total biomass towards total plantation mean biomass to the next higher age class by the five common associated species and *Gmelina* was similar but the quantum of change was more for the five associated species considered together than the *Gmelina*.

The change and dynamics of contribution by the components of *Gmelina* farm towards total biomass can be explained by the increase in and intensity of both inter- and intraspecific competition. Carbon is considered half of biomass, so any factor (biomass and carbon) change that influences both [20] thus exhibits the same trends as exhibited by the biomass with increasing age of the plantation. Biomass and biomass carbon varies with land use, climatic conditions, edaphic conditions, topography, site quality, age, species diversity, stem density, stem size distribution, density, structure, litter production, management practices, and disturbance history along with variations in canopy height and wood density [67]. Similar quantification of biomass accumulation and carbon storage in eucalyptus plantations was also reported by Kumar et al. [68]. Quantification of biomass in tree plantations at agricultural landscape will aid in formulating sustainable management strategies for increasing carbon pool build up outside forest land use [20].

3.6. Ecosystem Carbon Stock

The overall ecosystem C values estimated in the three age classes were 110.24, 104.81 and 110.77 $Mgha^{-1}$ (Tables 4 and 5). The present study was unable to make a direct and accurate estimation of C uptake by the vegetation because of high variability in tree distribution and species causing uncertainties, as was also earlier reported [69,70].

Promoting plantations of suitable site-specific tree species in less or unproductive and degraded agricultural lands is a recognized management action for offsetting terrestrial C emission because of longer duration C storage both in biomass and soil [71]. Forests are now net emitters due to degradation and deforestation [72]. Changing the forests to net sink again from net emitter will need a supplement of additional C emission offset by the best available land management options through promoting afforestation/reforestation of available degraded and deforested lands [33]. Managing soil and biomass C in an agricultural landscape by promoting *Gmelina arborea* or any other tree species plantation will both be an avoided emission and net addition of C to terrestrial pools, thereby fulfilling the global four per mile initiative [71]. The studied *Gmelina* farm forestry plantations with only three age class series had considerable vegetation heterogeneity due to no disturbance or management which if allowed growing full normal rotation period with selective logging, development of seminatural secondary forest is expected [73].

4. Conclusions

Farm forestry is now globally recognized as a low-cost viable option to supplement forests in an effort to offset carbon (C) emissions. Younger *Gmelina* farm plantations (up to 15 years of age) left aside after planting without silvicultural management in the Terai region of West Bengal has the great potential of carbon sequestration due to their biomass and soil. The ecosystem C in the three age classes (ACs I, II, and III) of plantations was in the range of 104.81–110.77 Mg ha^{-1}, of which carbon storage by tree and soil carbon as SOC (up to 60 cm depth) were in the range of 54.51–59.91 and 48.18–55.73 Mg ha^{-1}, respectively. This C storage is mainly from young plantations and can play an important role if allowed to complete a normal rotation cycle of 60–80 years. Further, a rotation cycle without any disturbance with selective logging can develop considerable vegetation heterogeneity which might lead to the conversion of homogenous plantations into seminatural secondary forests. These microlandscapes within agricultural or human-dominated landscapes will act as an oasis for biodiversity conservation. Further, carbon sequestration potential of *Gmelina arborea* is reported to be higher compared to other species and very much supportive for atmospheric carbon reduction in future under higher temperatures by implementing a strategic plant diversity conservation plan.

5. Recommendations and Future Directions

This study recommends popular plantation programs through mission mode with these high value timber species as C farming initiatives either in the unproductive and degraded nonforested or agricultural landscapes. Popularizing such plantation programs needs policy decisions and action with suitable site-specific tree species in participatory mode with intensive growth. Plantations were generally thought to limit biodiversity and are developed by the owner for economic benefits [74]. These allegations can be cleared up by adopting different site-specific management strategies by removing disturbance factors to allow heterogeneity of the landscape so that seminatural forest vegetation within the agricultural or any other nonforested landscape is developed without compromising the timber demands while bringing social and ecological benefits [75]. This requires further studies to understand the plantations at various successional stages throughout the age classes of natural rotation of species. The effect of plant community composition on ecosystem functioning and services is yet to be understood [25]. Establishing plantations with higher diversity of indigenous tree species is required for studying this relationship. In farm tree plantations, plant life-history strategies require clear understanding to analyze patterns of biomass allocation and partitioning in various tree species for sustainable tree-based land management strategies and identifying the most productive tree species for C sequestration [76]. Driving mechanisms of terrestrial C sinks and/or sources with their regional patterns and magnitudes are unclear [77]. Therefore, there is a need to work for the success of these plantations for C reduction. Even now, uncertainties prevail over quantifying C fluxes in and out of a system due to insufficient pieces of information about land use and land cover changes [78]. Thus, information on C exchange between these plantations and atmosphere needs urgent attention for efficient C budgeting for viable policy support and strategic decisions.

Supplementary Materials: The following are available online at https://www.mdpi.com/article/10.3390/land10040387/s1.

Author Contributions: Conceptualization, G.S. and S.C.; methodology, G.S. and V.; software, resources and formal analysis, M.T. and R.C.; data curation, V., M.T. and R.C.; writing—original draft preparation, G.S., J.A.B., and S.C.; writing—review and editing, G.S., A.S., J.A.B., S.C., M.K., A.K., and; supervision, G.S. and S.C.; project administration, G.S., V., S.C., M.T. and R.C.; funding acquisition, A.K. and M.C.-P. All authors have read and agreed to the published version of the manuscript.

Funding: This research received no external funding.

Institutional Review Board Statement: Not applicable.

Informed Consent Statement: Not applicable.

Data Availability Statement: Data could be provided on reasonable request from the first author or corresponding author.

Acknowledgments: We would like to express our gratitude to all those who helped us during the writing of this article. The authors are thankful to the reviewers for their constructive comments to improve the quality of the paper and farmers of the study area are highly acknowledged.

Conflicts of Interest: The authors declare no conflict of interest.

References

1. Raj, A.; Jhariya, M.K.; Harne, S.S. Threats to biodiversity and conservation strategies. In *Forests, Climate Change and Biodiversity*; Sood, K.K., Mahajan, V., Eds.; Kalyani: New Delhi, India, 2018; pp. 304–320.
2. Kongmeesup, I.; Boonyanuphap, J. Estimation of carbon offset for teak plantation in lower northern Thailand. *Songklanakarin J. Sci. Technol.* **2019**, *41*, 580–586.
3. Chakravarty, S.; Ghosh, S.K.; Suresh, C.P.; Dey, A.N.; Shukla, G. Deforestation: Causes, effects and control strategies. In *Global Perspectives on Sustainable Forest Management*; Okia, C.A., Ed.; Intech Publishers: Croatia, Rijeka, 2012; pp. 3–28.
4. Kumar, A.; Sharma, M.P.; Taxak, A.K. Effect of vegetation communities and altitudes on the soc stock in Kotli Bhel-1a catchment, India. *CSAWAC* **2017**, *45*, 1–8.
5. Chauhan, M.; Kumar, M.; Kumar, A. Impact of carbon stocks of Anogeissus latifolia on climate change and socio-economic development: A case study of Garhwal Himalaya, India. *Water Air Soil Pollut.* **2020**, 231. [CrossRef]
6. Lal, M.; Singh, R. Carbon sequestration potential of Indian forests. *Environ. Monit. Assess.* **2000**, *60*, 315–327. [CrossRef]
7. Baishya, R.; Barik, S.K.; Upadhaya, K. Distribution pattern of aboveground biomass in natural and plantation forests of humid tropics in northeast India. *Trop. Ecol.* **2009**, *50*, 295–304.
8. Verma, P.; Bijalwan, A.; Shankhwar, A.K.; Dobriyal, M.J.R.; Jacob, V.; Rathaude, S.K. Scaling up an indigenous tree based agroforestry system India. *Int. J. Sci. Qual. Anal.* **2017**, *3*, 73–77.
9. Sarangle, S.; Rajasekaran, A.; Benbi, D.K.; Chauhan, S.K. Biomass and carbon stock, carbon sequestration potential under selected land use systems in Punjab. *For. Res. Eng. Int. J.* **2018**, *9*, 75–80. [CrossRef]
10. Gyanaranjan, S.; Majid, W.A. Scaling up an indigenous tree based agroforestry system India. *Ann. Hortic.* **2019**, *12*, 139–149.
11. Chauhan, S.K.; Sharma, R.; Singh, B.; Sharma, S.C. Biomass production, carbon sequestration and economics in on-farm poplar plantations in Punjab, India. *J. Appl. Nat. Sci.* **2015**, *7*, 452–458. [CrossRef]
12. Secretariat of the United Nations Framework Convention on Climate Change. *Report on the Workshop on Methodological Issues Relating to Reducing Emissions from Deforestation and Forest Degradation in Developing Countries*; 29th Session; UNFCC: Tokyo, Japan, 2008; Volume 1, pp. 89–96. Available online: https://digitallibrary.un.org/record/640384?ln=en (accessed on 18 November 2020).
13. Anonymous, I.V. *Working Plan for the Forests of Cooch Behar District and Jalpaiguri District (Part) Comprising Cooch Behar Forest Division and Cooch Behar, S.F. Division Volume–I, 2000-01 to 2009-10*; Divisional Forest Officer, Working Plans (North) Division: Darjeeling, India, 2001; p. 87.
14. Cintron, G.; Novelli, Y.S. Methods for studying mangrove structure. In *The Mangrove Ecosystem: Research Methods*; Snedaker, S.C., Snedaker, J.G., Eds.; United Nations Educational Scientific and Cultural Organization: Paris, France, 1984; pp. 91–113.
15. Menhinick, E.F. A comparison of some species diversity indices applied to samples of field insects. *Ecology* **1964**, *45*, 858–861. [CrossRef]
16. Simpson, E.H. Measurement of diversity. *Nature* **1949**, *163*, 688. [CrossRef]
17. Shannon, C.E.; Wiener, W. *The Mathematical Theory of Communication*; University of Illinois Press, Urbana: Chicago, IL, USA, 1963.
18. Pielou, E.C. The measurement of diversity in different types of biological collections. *J. Theor. Biol.* **1966**, *13*, 131–144. [CrossRef]
19. Nath, A.J.; Tiwari, B.K.; Sileshi, G.W.; Sahoo, U.K.; Brahma, B.; Deb, S.; Devi, N.B.; Das, A.K.; Reang, D.; Chaturvedi, S.S.; et al. Allometric models for estimation of forest biomass in north east India. *Forests* **2019**, *10*, 103. [CrossRef]
20. Shukla, G.; Pala, N.A.; Moonis, M.; Chakravarty, S. Carbon accumulation and partitioning in sub-humid forest stands of West Bengal India. *Indian Forest.* **2018**, *144*, 229–233.
21. Piper, C.S. *Soil and Plant Analysis*; International Public Inc.: New York, NY, USA, 1950.
22. Piper, C.S. *Soil and Plant Analysis*; Maver Publisher: Bombay, India, 1966.
23. Jackson, M.L. *Soil Chemistry Analysis*; Prentice Hall of India Pvt: New Delhi, India, 1967.
24. Kumar, A.; Sharma, M.P. Estimation of carbon stocks of Balganga reserved forest, Uttarakhand, India. *For. Sci. Technol.* **2015**, *11*, 177–181. [CrossRef]
25. Udayana, C.; Andreassen, H.P.; Skarpe, C. Understory diversity and composition after planting of teak and mahogany in Yogyakarta, Indonesia. *J. Sustain. For.* **2019**, *39*, 494–510. [CrossRef]
26. Whittaker, R.H. Evolution and measurement of species diversity. *Taxonomy* **1972**, *21*, 213–251. [CrossRef]
27. Shukla, G.K.; Manohar, A.; Rai, P.; Chakravarty, S. Diversity, biomass accumulation and carbon storage of shrub community in a foothill forest of Indian Eastern Himalayas. *Indian Forest.* **2019**, *146*, 309–314.

28. Manhas, R.K.; Chauhan, P.S.; Singh, M.L.; Negi, J.D.S. Structure and diversity of 80-yr-old plantations after successional colonization of the natives. *Curr. Sci.* **2011**, *100*, 714–725.
29. Whitmore, T.C. Tropical rain forest dynamics and its implications for management. In *Rain Forest Regeneration and Management*; Man and Biosphere Series 6; UNESCO: Paris, France; Parthenon Publishing Group: New Jersey, NJ, USA, 1991; pp. 67–90.
30. Odum, E.P. *Fundamentals of Ecology*, 3rd ed.; W.B. Saunders Co.: Philadelphia, PA, USA, 1971.
31. Bhat, J.A.; Kumar, M.; Negi, A.K.; Todaria, N.P.; Malik, Z.A.; Pala, N.A.; Kumar, A.; Shukla, G. Species diversity of woody vegetation along altitudinal gradient of the Western Himalayas. *Glob. Ecol. Conserv.* **2020**, *24*, e01302. [CrossRef]
32. Campos-Arceiz, A.; Blake, S. Megagardeners of the forest–The role of elephants in seed dispersal. *Acta Oecol.* **2011**, *37*, 542–553. [CrossRef]
33. Rai, P. Nutrient Status and Carbon Stock of Different Stands in a Foothill Humid Tropical Forest of Indian Eastern Himalaya. Ph. D. Thesis, Uttar Banga Krishi Viswavidyalaya, Pundibari, India, May 2017. Unpublished.
34. Flinn, K.M.; Vellend, M. Recovery of forest plant communities in post-agricultural landscapes. *Front. Ecol. Environ.* **2005**, *3*, 243–250. [CrossRef]
35. Brunet, J. Plant colonization in heterogeneous landscapes: An 80- year perspective on restoration of broadleaved forest vegetation. *J. Appl. Ecol.* **2007**, *44*, 563–572. [CrossRef]
36. Shukla, G.; Pala, N.A.; Chakravarty, S. Quantification of organic carbon and primary nutrients in litter and soil at foothill forest stands of Eastern Himalaya. *J. Forest. Res.* **2012**, *28*, 1195–1202. [CrossRef]
37. Shukla, G.; Biswas, R.; Das, A.P.; Chakravarty, S. Floristic diversity of Chilapatta Reserve Forest of West Bengal India. *Int. J. Forest Usufructs Manag.* **2016**, *17*, 46–57.
38. Shukla, G.; Chakravarty, S. *Biodiversity Cause and Pattern in the Foot Hills of Eastern Himalaya*; Lambert Academic Publishing: Saarbrücken, Germany, 2012; p. 200.
39. Kumar, M.; Rawat, S.; Nagar, B.; Kumar, A.; Pala, N.A.; Bhat, J.A.; Bussmann, R.W.; Cabral-Pinto, M.; Kunwar, R. Implementation of the use of ethnomedicinal plants for curing diseases in the Indian Himalayas and its role in sustainability of livelihoods and socioeconomic development. *Int. J. Environ. Res. Public Health* **2021**, *18*, 1509. [CrossRef]
40. Kumar, A.; Marcot, B.G.; Saxena, A. Tree species diversity and distribution patterns in tropical forests of Garo hills. *Curr. Sci.* **2006**, *91*, 1370–1381.
41. Pouyat, R.V.; Yesilonis, I.D.; Russell-Anelli, J.; Neerchal, K.N. Soil chemical and physical properties that differentiate urban land-use and cover types. *Soil Sci.* **2007**, *71*, 1010–1019. [CrossRef]
42. Pouyat, R.V.; Szlavecz., K.; Yesilonis, I.D.; Groffman, P.M.; Schwarz, K. Chemical, physical, and biological characteristics of urban soils. In *Agronomy Monograph 55. Urban. Ecosystem Ecology*; Aitkenhead-Peterson, J., Volder, A., Eds.; American Society of Agronomy: Madison, WI, USA, 2010; pp. 119–152.
43. Brady, N.C.; Weil, R.R. *The Nature and Properties of Soil*, 15th ed.; Pearson Education: London, UK, 2016.
44. Pande, P.K.; Negi, J.D.S.; Sharma, S.C. Plant species diversity and vegetation analysis in moist temperate Himalayan forest. *Indian J. For.* **2020**, *24*, 456–470.
45. Ghosh, S.; Scharenbroch, B.; Ow, L.F. Soil organic carbon distribution in roadside soils of Singapore. *Chemosphere* **2019**, *165*, 163–172. [CrossRef] [PubMed]
46. De Hann, S. Humus, its formation, its relation with the mineral part of the soil and its significance for soil productivity. In *Organic Matter Studies*; International Atomic Energy Agency: Vienna, Austria, 1977; Volume 1, pp. 21–30.
47. Thuille, A.; Schulze, E.D. Carbon dynamics in succession and afforested spruce stands in Thuringia and the Alps. *Glob. Chang. Biol.* **2006**, *12*, 325–342. [CrossRef]
48. Paudel, S.; Sah, J.P. Physiochemical characteristic of soil in Sal (*Shorea robusta*) forests in eastern Nepal. *Himalayan J. Sci.* **2003**, *1*, 107–110. [CrossRef]
49. Alberti, M. *Advances in Urban Ecology: Integrating Humans and Ecological Processes in Urban Ecosystems*; Springer: New York, NY, USA, 2008; p. 366.
50. Bardgett, R.D.; Wardle, D.A. Above ground-below ground linkages. In *Biotic Interactions, Ecosystem Processes and Global Change*; Oxford University Press: New York, NY, USA, 2010; p. 301.
51. Deb, S.; Kumar, D.; Chakraborty, S.; Weindorf, D.C.; Choudhury, A.; Banik, P.; Deb, D.; De, P.; Saha, S.; Patra, A.K.; et al. Comparative carbon stability in surface soils and sub soils under submerged rice and upland non-rice crop ecologies: A physical fractionation study. *Catena* **2019**, *7*, 400–410. [CrossRef]
52. Kumar, A.; Sharma, M.P.; Yang, T. Estimation of carbon stock for greenhouse gas emissions from hydropower reservoirs. *Stoch. Environ. Res. Risk Asses.* **2018**, *32*, 3183–3319. [CrossRef]
53. Tandon, H.L.S. *Methods of Analysis of Soils, Plants, Waters, Fertilizers and Organic Manures*; Fertilizer Development and Consultation Organization: New Delhi, India, 2005; p. 204.
54. Gairola, S.; Sharma, C.M.; Ghildiyal, S.K.; Suyal, S. Chemical properties of soils in relation to forest composition in moist temperate valley slopes of Garhwal Himalaya, India. *Environmentalist* **2012**, *32*, 512–523. [CrossRef]
55. Gupta, J.P.; Sharma, M.P.; Gupta, G.D. Characterization of Kandi belt soils of Jammu region as affected by different land use patterns. *J. Indian Soc. Soil Sci.* **2001**, *49*, 770–773.
56. Kumar, A.; Sharma, M.P. Estimation of soil organic carbon in the forest catchment of two hydroelectric reservoirs in Uttarakhand, India. *Hum. Ecol. Risk Assess. An Int. J.* **2017**, *22*, 991–1001. [CrossRef]

57. Fernández-Moya, J.; Alvarado, A.; Miguel-Ayanz, A.S.; Marchamalo-Sacristán, M. Forest nutrition and fertilization in teak (*Tectona grandis* L. f.) plantations in Central America. *N. Z. J. Forest. Sci.* **2014**, *44*, 1–8. [CrossRef]
58. Pal, D.K.; Wani, S.P.; Sahrawat, K.L. Carbon sequestration in Indian soils: Present status and the potential. *Proc. Natl. Acad. Sci. India Sect. B Biol. Sci.* **2015**, *85*, 337–358. [CrossRef]
59. Jerabkova, L.; Prescott, C.E.; Kishchuk, B.E. Nitrogen availability in soil and forest floor of contrasting types of boreal mixed wood forests. *Can. J. Forest. Res.* **2006**, *36*, 112–122. [CrossRef]
60. Scotti, R.; Bonanomi, G.; Scelza, R.; Zoina, A.; Rao, M.A. Organic amendments as sustainable tool to recovery fertility in intensive agricultural systems. *J. Soil Sci. Plant. Nutr.* **2015**, *15*, 333–352. [CrossRef]
61. Sariyildiz, T.; Kuçuk, M. Litter mass loss rates in deciduous and coniferous trees in Artvin, northeast Turkey: Relationships with litter quality, microclimate and soil characteristics. *Turk. J. Agric. Forest.* **2008**, *32*, 547–559.
62. Sidhu, G.S.; Bhattacharyya, T.; Sarkar, D.; Ray, S.K.; Chandran, P.; Mandal, D.K.; Prasad, J.; Mandal, C.; Srivastava, R.; Sen, T.K.; et al. Impact of management levels and land-use changes on soil properties in rice-wheat cropping system of the Indo-Gangetic plains. *Curr. Sci.* **2014**, *107*, 1487–1501.
63. Vrscaj, B.; Poggio, L.; Marsan, F.A. A method for soil environmental quality evaluation for management and planning in urban areas. *Landsc. Urban Plan.* **2008**, *88*, 81–94. [CrossRef]
64. Livesley, S.J.; Ossola, A.; Threlfall, C.G.; Hahs, A.K.; Williams, N.S.G. Soil carbon and carbon/nitrogen ratio change under tree canopy, tall grass, and turf grass areas of urban green space. *J. Environ. Qual.* **2016**, *45*, 215–223. [CrossRef]
65. Subba, M.; Pala, A.N.; Shukla, G.; Chakravarty, S. Study of the variability of home gardens influencing carbon stock under sub-humid tropical zone of West Bengal, India. *Indian Forest.* **2018**, *144*, 66–72.
66. Tamang, B.; Sarkar, B.C.; Pala, N.A.; Shukla, G.; Patra, P.S.; Bhat, J.A.; Dey, A.N.; Chakravarty, S. Uses and ecosystem services of trees outside forest (TOF)-A case study from Uttar Banga Krishi Viswavidyalaya, West Bengal, India. *Acta Ecol. Sinica* **2019**, *39*, 431–437. [CrossRef]
67. Tang, Y.J.; Chen, A.P.; Zhao, S.Q. Carbon storage and sequestration of urban street trees in Beijing, China. *Front. Ecol. Evol.* **2016**, *4*, 53–61.
68. Kumar, P.; Mishra, A.K.; Chaudhari, S.K.; Sharma, D.K.; Rai, A.K.; Singh, K.; Rai, P.; Singh, R. Carbon sequestration and soil carbon build-up under Eucalyptus plantation in semi-arid regions of north west India. *J. Sustain. Forestry.* **2020**. [CrossRef]
69. Velasco, E.; Roth, M.; Tan, S.H.; Quak, M.; Nabarro, S.D.; Norford, L. The role of vegetation in the CO_2 flux from a tropical urban neighbourhood. *Atmos. Chem. Phys.* **2013**, *13*, 185–202. [CrossRef]
70. Velesco, E.; Matthias, R.; Luisa, T.M. Does urban vegetation enhance carbon sequestration? *Landsc. Urban. Plan.* **2013**, *148*, 99–107. [CrossRef]
71. Minasny, B.; Malone, B.P.; McBratney, A.B.; Angers, D.A.; Arrouays, D.; Chambers, A.; Chaplot, V.; Chen, Z.-S.; Cheng, K.; Das, B.S.; et al. Soil carbon 4 per mile. *Geoderma* **2017**, *292*, 59–86. [CrossRef]
72. Kumar, A.; Kumar, M. Assessment of biomass and soil carbon stock in the hydroelectric catchment of Uttarakhand Himalayas, India. *J. Sustan. Forest.* **2020**, *36*. [CrossRef]
73. Martin, P.H.; Sherman, R.E.; Fahey, T.J. Forty years of tropical forest recovery from agriculture: Structure and floristics of secondary and old-growth riparian forests in the Dominican Republic. *Biotropica* **2004**, *36*, 297–317.
74. Yirdaw, E.; Monge, A.M.; Austin, D.; Toure, I. Recovery of floristic diversity, composition and structure of regrowth forests on fallow lands: Implications for conservation and restoration of degraded forest lands in Laos. *New For.* **2019**, *50*, 1007–1026. [CrossRef]
75. Ito, S.; Nakagawa, M.; Buckley, G.P.; Nogami, K. Species richness in sugi (*Cryptomeria japonica* D. DON) plantations in southeastern Kyushu, Japan: The effects of stand type and age on understorey trees and shrubs. *J. Forest. Res.* **2003**, *8*, 49–57. [CrossRef]
76. Devi, L.S.; Yadava, P.S. Above ground biomass and net primary production of semi-evergreen tropical forest of Manipur, north-eastern India. *J. Forest. Res.* **2009**, *20*, 151–155. [CrossRef]
77. Rana, K.; Kumar, M.; Kumar, A. Assessment of annual shoot biomass and carbon storage potential of Grewia optiva: An approach to combat Climate Change in Garhwal Himalaya. *Water Air Soil Pollut.* **2020**, 231. [CrossRef]
78. Zhu, Z.L.; Bouchard, M.; Butman, D.; Hawbaker, T.; Li, Z.; Liu, J.; Liu, S.; McDonald, C.; Reker, R.; Sayler, K.; et al. Baseline and projected future carbon storage and greenhouse-gas fluxes in the Great Plains region of the United States. In *US Geological Survey Professional Paper 1787*; Zhu, Z.L., Ed.; USGS: Reston, VA, USA, 2011.

land

MDPI

Article

Eco-Restoration of Coal Mine Spoil: Biochar Application and Carbon Sequestration for Achieving UN Sustainable Development Goals 13 and 15

Dipita Ghosh and Subodh Kumar Maiti *

Ecological Restoration Laboratory, Department of Environmental Science & Engineering, Indian Institute of Technology (ISM), Dhanbad 826004, India; ghoshdipita@gmail.com
* Correspondence: subodh@iitism.ac.in or skmism1960@gmail.com

Abstract: Open cast coal mining causes complete loss of carbon sink due to the destruction of vegetation and soil structure. In order to offset the destruction and to increase sequestration of carbon, afforestation is widely used to restore these mine spoils. The current field study was conducted to assess the ecosystem status, soil quality and C pool in an 8 years old reclaimed mine spoil (RMS), compared to a reference forest (RF) site and unamended mine spoil (UMS). Biochar (BC) prepared from invasive weed *Calotropis procera* was applied in this 8 year RMS at 30 t ha^{-1} (BC_{30}) and 60 t ha^{-1} (BC_{60}) to study its impact on RMS properties and C pool. Carbon fractionation was also conducted to estimate inorganic, coal and biogenic carbon pools. The C stock of 8 year old RMS was 30.98 Mg C ha^{-1} and sequestered 113.69 Mg C ha^{-1} CO_2. BC_{30} and BC_{60} improved the C-stock of RMS by 31% and 45%, respectively, and increased the recalcitrant carbon by 65% (BC_{30}) and 67% (BC_{60}). Spoil physio-chemical properties such as pH, cation exchange capacity, moisture content and bulk density were improved by biochar application. The total soil carbon at BC_{30} (36.3 g C kg^{-1}) and BC_{60} (40 g C kg^{-1}) was found to be significantly high compared to RMS (21 g C kg^{-1}) and comparable to RF (33 g C kg^{-1}). Thus, eco-restoration of coal mine spoil and biochar application can be effective tools for coal mine reclamation and can help in achieving the UN sustainable development goal 13 (climate action) by increasing carbon sequestration and 15 (biodiversity protection) by promoting ecosystem development.

Keywords: coal mine spoil; reclamation; biochar; carbon sequestration; carbon fractionation

Citation: Ghosh, D.; Maiti, S.K. Eco-Restoration of Coal Mine Spoil: Biochar Application and Carbon Sequestration for Achieving UN Sustainable Development Goals 13 and 15. *Land* **2021**, *10*, 1112. https://doi.org/10.3390/land10111112

Academic Editors: Marina Cabral Pinto, Amit Kumar and Munesh Kumar

Received: 27 September 2021
Accepted: 18 October 2021
Published: 20 October 2021

1. Introduction

The UN sustainable development goal (SDG) 13 stands for climate action and promotes all activities which would ensure successful sequestration of carbon, whereas, SDG 15 safeguards and restores biodiversity protection [1,2]. Burning of fossil fuels is the primary drivers of global warming and climate change, and the extraction of these resources also adds to global concerns regarding the climate crisis [3,4]. Mining activities lead to complete loss of vegetation and the carbon sink in the soil and plants are lost to the atmosphere [5]. Mine spoils are carbon deficit with impoverished soil conditions that cannot support plant and microbial growth. Coal mine restoration can help restore the lost carbon sink by promoting plant growth and enriching the mine spoil, which helps sequester the atmospheric carbon [6,7]. The most common techniques for mine restoration include afforestation, agriculture and grassland development [8,9]. Plantation of hardy species in reclaimed mine spoils (RMS) improves the soil organic carbon (SOC) pool and improves the carbon sequestration potential of the ecosystem [10]. Development of natural forest in mine spoils may take centuries due to the impoverished soil properties and lack of substrate for supporting plant growth. Degraded land can be reclaimed by development of forest cover. Technical reclamation such as leveling and grading of dump, reducing slope length, stabilization of slope by blanketing with coir mat along with grass-legume mixture, application of top soil, fly ash and bio-solids can be used to enhance the vegetation growth [11–13].

Restoration of RMS can potentially enhance soil C sequestration rate and improve soil properties [14–16]. Akala and Lal, [17] reported that the SOC pool of a RMS increased from 14 Mg ha^{-1} to 48.4 Mg ha^{-1} after 21 years of pastureland development in a degraded coal mine site. In another 19 years old revegetated coal mine spoil (Singrauli, India) there was 712% increase in the rate of carbon sequestration [18]. The sequestration of carbon depends on the type of vegetation used for reclamation, age of reclamation and nature of coal mine spoil. A study conducted by Mukhopadhyay et al. [19] in RMS reported that the carbon density was higher for *Dalbergia sissoo* Roxb. and *Acacia auriculiformis* A. Cunn. ex Benth. (39.6–43.7 kg C $tree^{-1}$) and lowest for *Albizzia lebbeck* L. (20.7 kg C $tree^{-1}$). Thus, plantation can be an effective tool for coal mine spoil eco-restoration and enhanced carbon sequestration.

A number of studies reported an increase in the soil carbon stock by biochar application [20,21]. Biochar is a thermal degradation product of biomass produced in a pyrolysis like condition by limiting the supply of oxygen [21]. Pyrolytic conversion of biomass produces aromatic carbon that is resistant to degradation in soil, thus considered an option to address the global CO_2 emission problems by biomass decomposition [22]. Biochar has a high mean residence time and aromaticity, making it highly recalcitrant in nature [11,23]. Thus, carbon that would normally be released as CO_2 through biomass decomposition is converted to biochar which is highly stable and aromatic. The aromaticty of biochar depends on the chemistry of biomass used for biochar production. Mean residence time of biochar depends on feedstock material, pyrolytic method used and the substrate where it is applied [8,24,25]. Fidel et al. [26] reported that biochar has the potential to improve the soil inorganic carbon by 0.023–0.045 mg C kg^{-1} and organic carbon by 0.001–0.0069 mg C kg^{-1}. According to a study conducted by Ghosh and Maiti, [27], *Lantana camara* biochar lowered mine spoil CO_2 flux to 3% (2.60 μmol CO_2 m^{-2} s^{-1}) and 2% (2.85 μmol CO_2 m^{-2} s^{-1}) in comparison to control (4.92 μmol CO_2 m^{-2} s^{-1}). Biochar acts as an amendment and improves physico-chemical; biological and nutritive soil properties [25,28]. An enriched soil supports the growth of vegetation which can facilitate ecosystem development and promote carbon sequestration in vegetation and soil. Thus, it is imperative to investigate the link between the intrinsic characteristics of biochar and mine spoil restoration.

The excessive growth of invasive weeds in RMS during the plantation stage of reclamation causes the problem of allelopathy [28]. These weeds are usually uprooted and left to decompose which adds to the atmospheric CO_2 pool. During the dry tropical summers, they act as fuel and cause even bigger problems of mine fire. One such weed growing abundantly in RMS is *Calotropis procera* (Aiton) W.T.Aiton (family: Apocynaceae). *C. procera* is a hardy shrub with an average height of 2 m and covered with a fissured corky bark which is high in cellulose and lignin. This can be a potential feedstock for biochar production and mine reclamation. Only a few studies have reported on biochar based carbon sequestration in a RMS, and the available data are from laboratory or greenhouse scale experiments [11,27]. The present study was conducted in an 8 year old RMS and the carbon sequestered was calculated in this RMS. A 6 month biochar based field experiment was also conducted to study the effect of biochar as an amendment for reclamation of mine spoil. The study aims to understand how coal mine reclamation along with biochar application can help in achieving UN SDG 13 and 15. Thus, the objectives of the study are: (i) assessment of carbon sequestration in an 8 year old RMS by vegetation, litter and soil carbon stock, (ii) application of *C. procera* biochar in the RMS in a 6 month field based study, (iii) fractionation of carbon in RMS, biochar amended RMS, reserved forest (RF) and unreaclaimed mine spoil (UMS) (iv) calculation of total CO_2 sequestration in each system.

2. Materials and Methods

2.1. Site Description

The study area is located in Damoda colliery, Jharia Coalfield, situated in the Dhanbad district of Jharkhand, India (23°–23°48′ N; 86°11′–86°27′ E). The site map of the study area is presented in Figure 1a,b. Damoda eco-restoration site is an 8 year old backfilled dump site of 4 ha area. The geology of the mine spoil consisted of sandstone, carbonaceous shale, intermixed shale and sandstone, Jhama (heat affected coal) with micaperidotite, subsoil and coal. The area experiences extreme weather conditions with summer temperature of 42 to 46 °C and winter temperature of 22 to 5 °C and received 1900 mm rainfall in the year of the study (2019). Jharia Coalfield is located in a dry tropical region and experiences three main seasons: summer, monsoon and winter. The carbon sequestration study was conducted in the February 2019, 6 months prior to which biochar was incubated in field conditions.

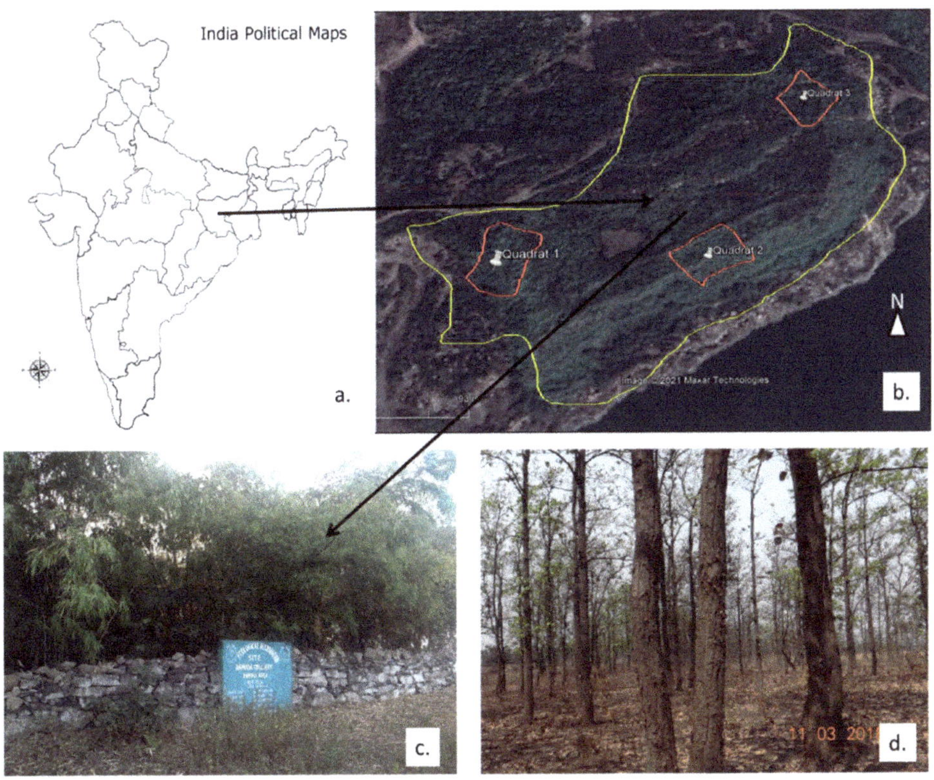

Figure 1. (**a**) Map of India, showing the Jharkhand state (**b**) Location map of Damoda ecorestoration site showing the 3 quadrats in the sampling area (**c**) Outer view of Damoda eco-restoration site showing dense bamboo cumps and stone boundry (**d**) Reserve forest sampling site.

The eco-restored mine dump has a history of shovel–dumper based mining activity. In 2011, plantation of hardy and multipurpose tree saplings was carried out in pits of dimension 30 cm × 30 cm × 30 cm. Grass seeds such as *Pennisetum pedicellatum* Trin. were also spread, which act as pioneer species and develop understory vegetation. Afforested trees such as *A. lebbeck*, *D. sissoo* and *Bambusa arundinacea* (L.) Voss were dominant species with sparse growth of plants such as *Azadirachta indica* A. Juss., *Bauhunia veriegata* (L.) Benth., *Melia azedarach* L., *Psidium guajava* L., *Syzygium cumini* (L.) Skeels., *Terminalia arjuna* (Roxb.) Wight & Arn, and *Zizyphus mauritiana* Lam. Figure 1c shows the Damoda eco-restoration site with stone boundary. RF area near the mining area was used as a positive reference site while UMS was used as a negative reference for the study. The most dominant trees in the RF site were *D. sissoo, A. lebbeck, Butea Monosperma* (Lam.) Taub. and *Shorea robusta* Roth (Figure 1d). UMS was coarser with rock debris, soils and subsoil materials. Since UMS was not revegetated, tree species were absent.

2.2. Biochar Production, Characterization and Field Incubation

C. procera growing in the RMS was collected in bulk, sun-dried, grinded and used for biochar production. Feedstock was pyrolysed in a muffle furnace at 450 °C for 60 min. Biochar characterization was carried out using the methods given in Ghosh et al. [11] and Ghosh and Maiti, [27]. The biochar field experiment was conducted as a completely randomized design in a 2 × 3 factorial scheme, each with 50 cm × 50 cm plots with two biochar application rate of 30 t ha^{-1} (BC_{30}) and 60 t ha^{-1} (BC_{60}), and each with three replications (Figure 2a–c). The carbon sequestration study was done with a 6 months incubation period in natural field conditions.

Figure 2. (**a**) Biochar being applied in RMS, (**b**) Plots showing biochar application (50 cm × 50 cm × 10 cm), (**c**) Ecological restoration project site of Damoda, showing boundary wall and sign board, (**d**) Soil sampling being done by a soil corer (80 cm × 20 cm), (**e**,**f**) Collection of litter from a metal quadrat (50 cm × 50 cm).

2.3. Soil Sampling

Soil samples were collected from the rhizospheric region (0–15 cm) of different tree species, 8 samples collected from each quadrat with a metallic corer (80 cm × 20 cm) after removing the litter (Figure 2d). A total of 24 samples (8 samples × 3 quadrats) were collected from each RMS and RF. In the biochar incubation sites, 4 samples were collected from each plot making the total number of samples 24 (2 application level × 3 replicates × 4 samples per plot). Samples were placed in zip-lock packets and brought back to the laboratory for physico-chemical analysis. Samples were air-dried inside the laboratory for a week at room temperature.

2.4. Plant Biodiversity and Vegetation Analysis

Vegetation carbon stock was only analysed for RMS, RF and UMS, and no observable changes were observed with 6-month biochar application on the tree stocks, hence this data were not included. Three random quadrats of size 10 m × 10 m, covering total area of 300 m^2 were laid down for relative density [3]. Details on the density of the plants, IVI values and total number of species in RMS are provided in the Supplementary Materials. The density of species present in each site was expressed as number of individual species present per hectare area [3].

$$\text{Relative density } (\%) = \frac{\text{Number of individual plant species}}{\text{Total number of species in a quadrat}} \times 100 \tag{1}$$

Circumference of large tree species in the quadrats were measured at 1.37 m for the measurement of diameter at breast height (DBH), and the height of the tree was measured with a Distometer (Bosch GLM 40, India), while smaller vegetation (<3 m height) was measured using a digital Vernier caliper (Precise®, India). Specific gravity of the wood was measured by water displacement method. The aboveground biomass (AGB) was estimated by the regression model developed by Chave et al. [29], which showed the best-fit for tropical forests. The model estimated tree AGB by the following equation:

$$\text{AGB} = 0.0673 \times \left(\rho D^2 H\right)^{0.976} \tag{2}$$

where, AGB = above ground biomass (kg), ρ = wood specific gravity (g cm^{-3}), D = DBH (cm) and H = tree height (cm).

Root biomass (RB) was calculated by multiplying AGB by a factor of 2.25 [30]:

$$\text{RB}\left(\text{Mg ha}^{-1}\right) = \text{AGB}\left(\text{Mg ha}^{-1}\right) \times 2.25 \tag{3}$$

Tree carbon stock was calculated multiplying a factor of 0.5 by total tree biomass [31].

$$\text{Tree C stock}\left(\text{Mg ha}^{-1}\right) = \text{Total tree biomass}\left(\text{Mg ha}^{-1}\right) \times 2.25 \tag{4}$$

The CO_2 sequestered by the plantation stock is calculated by relation given by IPCC [30].

$$\text{CO}_2\text{sequestered}\left(\text{Mg ha}^{-1}\right) = \text{Tree C stock}\left(\text{Mg ha}^{-1}\right) \times 3.67 \tag{5}$$

The AGB of bamboo clumps were calculated by the allometric relationship given by Nath et al. [32] and Mazumder et al. [33]. This equation was primarily developed to establish a relationship between culm height, density and AGB in thick walled bamboo. The equation is as follows:

$$\text{AGB} = 7.5 \times \left(D^2 H\right)^{0.91} \tag{6}$$

where H is total height of the bamboo culm, and D is DBH of the bamboo culm. 47% of the total biomass stock was considered as total carbon stock [32].

2.5. Herbaceous Biomass and Litter Analysis

The herbaceous biomass and litter present in the respective quadrats were measured by placing three litter traps (50 cm × 50 cm) under the tree canopy per quadrat [34] as shown in Figure 2e,f. The collected biomass was then dried in a hot air oven at 65 °C for 48 h. The dry weight of the litter obtained was converted to kg m^{-2} by dividing it by quadrat area (50 cm × 50 cm) and then converting to kg ha^{-1}. Litter was assumed to have 40% carbon; hence C stock was calculated by a conversion factor of 0.4.

2.6. Soil Characterization

2.6.1. Soil Carbon Fractionation

Soil fractionation for the determination of inorganic, biogenic (labile and stable) and coal carbon present in the mine spoil was determined by the sequence of steps given by Ussiri and Lal. [10] The steps followed for sequential extraction of different forms of soil organic carbon, coal carbon and inorganic carbon in RMS, BC_{30}, BC_{60} and RF are given in Figure 3.

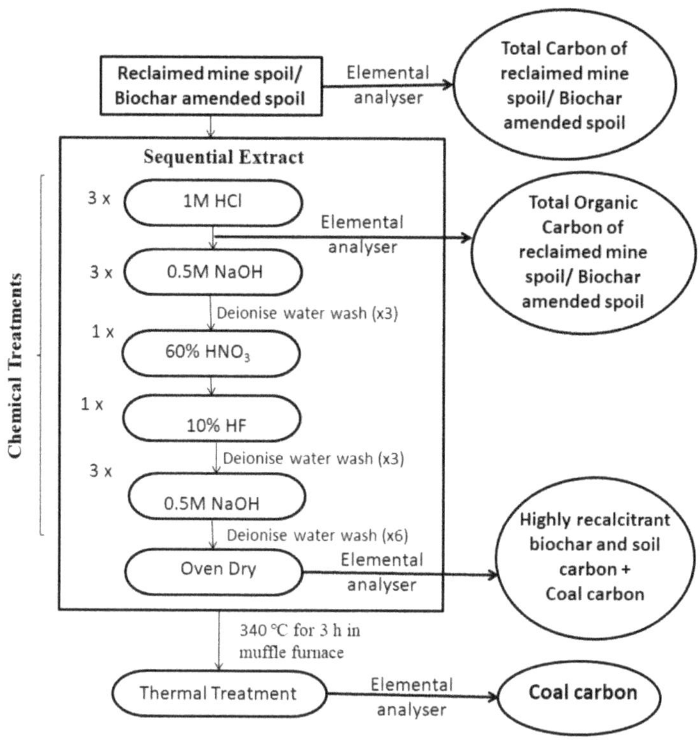

Figure 3. Sequential methods for the determination of total carbon, inorganic carbon and biogenic carbon pools in RMS, BC_{30}, BC_{60} and RF (n x indicated the number of times the step was repeated).

2.6.2. Soil Physico-Chemical Properties

The soil samples were air dried and sieved by a 2-mm sieve to remove the coarse fraction from the fine earth fraction (<2 mm). pH and EC were determined in a soil and water slurry (spoil: water, 1:2.5, *w/v*) by a multiparameter probe (HI-2020, Hanna Instruments, India). Cation exchange capacity (CEC) was calculated by the ammonium acetate extraction method [35]. Available-N was determined by a Kjeldahl distillation unit (KJELODIST-EAS VA, Pelican Equipment Inc. India). Available-P was extracted by $NaHCO_3$ (pH 8.5) and measured by a UV-VIS Spectrophotometer (Shimadzu Corporation,

Japan) [36]. Available-K was calculated by 1 N ammonium acetate method by using flame photometer [5]. The C-stock is often underestimated due to the coarse fraction in mine soils, hence only soil fraction (<2-mm particle size) was considered for bulk density calculation. The bulk density was corrected by the equation given by Ahirwal et al. [6]:

$$\text{Corrected Bulk Density}\left(\text{Mg m}^{-3}\right) = \frac{\text{Sample weight (Mg)} \times \text{Fine earth fraction (\%)}}{\text{Volume of corer }(\text{m}^3) \times 100} \quad (7)$$

Soil organic carbon (SOC) of the study sites were calculated by the relation [6]:

$$\text{SOC stock}\left(\text{Mg ha}^{-1}\right) = \frac{\text{Biogenic carbon pool(\%)} \times \text{BD}\left(\text{Mg m}^{-3}\right) \times \text{T (m)} \times 10^4\ (\text{m}^2\ \text{ha}^{-1})}{100} \quad (8)$$

where SOC = Soil organic carbon; BD = corrected bulk density; and T = depth of the soil layer.

2.7. Carbon Sequestration Study

The total C sequestration pool of an ecosystem is calculated by adding the C-stock associated with (i) AGB and RB, (ii) understory vegetation and litter layer, and (iii) SOC stock. Carbon is accumulated in vegetative parts such as leaf, twigs, and logs, live and dead roots, and soil organic matter. The biomass carbon pool varies from plant to plant and also by age of vegetation. Thus, the total ecosystem carbon pool can be assessed by adding (i) AGB and RB carbon-pool, (ii) understory and litter C stock and (iii) biogenic carbon stock at (0 to 15 cm). To determine the Carbon sequestration rate (Mg C ha^{-1} $year^{-1}$), total C stock (Mg C ha^{-1}) was divided with by age of reclamation.

2.8. Statistics

One-way ANOVA was used to compare the means of data obtained from RMS, BC_{30}, BC_{60} and RF. Post hoc Duncan's multiple range tests at $p < 0.05$ significance level was used to test the significant difference in the C-stock in each level of analysis. SPSS 23 was used for statistical studies, and software such as MS-EXCEL and ORIGIN Pro-8 was used for graphical representation.

3. Results and Discussions

3.1. Biochar Characteristics

The general characteristics *C. procera* biochar are presented in Table 1. Biochar yield obtained from *C. procera* feedstock was 51.87%. Biochar obtained was alkaline in nature with a pH of 7.75 and EC of 4.7 mS cm^{-1}. The total elemental C, H and N were 68.25%, 35.39% and 13.62% respectively. The C/N ratio of 5.01 indicates that the biochar is rich in labile carbon, providing substrate for microbial action in the mine spoil, while H/C ratio of 0.51 for *C. procera* biochar represents its high degree of aromatization. *C. procera* has a high organic carbon content of 42.24%, porosity of 78% and low bulk density of 0.25 g cm^{-3}.

The high surface area and the porous morphology of the biochar surface can be seen in the FE-SEM image of *C. procera* biochar given in Figure 4a,b. The porous structure provides an enlarged surface area and substrate for microbial action. Several spectral peaks representing various functional groups were observed on the *C. procera* biochar surface (Figure 4c). At transmittance of 3391 cm^{-1} an O-H bond is prominent due to the breaking of hydrogen bonded hydroxyl groups. Other bonds such as $-CH_3$ (2924 cm^{-1}), $-CH_2$ (2870 cm^{-1}), C = O (1600–1700 cm^{-1}), due to cellulose of the feedstock, are also present. The peaks at 500–600 cm^{-1} represent the aromatic carbons in the biochar surface.

Table 1. Physio-chemical properties of *C. procera* biochar ($n = 5$, mean ± standard deviation).

Characteristics	Values
Yield (%)	51.87 ± 2.27
pH Water (1:5; *w*/*v*)	7.75 ± 1.62
EC $_{Water}$ (1:5; *w*/*v*) (mS cm^{-1})	4.70 ± 0.12
C (%)	68.25 ± 4.58
H (%)	35.39 ± 5.22
N (%)	13.62 ± 2.40
H/C	0.51 ± 0.22
C/N	5.01 ± 1.28
OC (%)	42.24 ± 0.89
Porosity (%)	78 ± 4.00
Bulk density (g cm^{-3})	0.25 ± 0.01

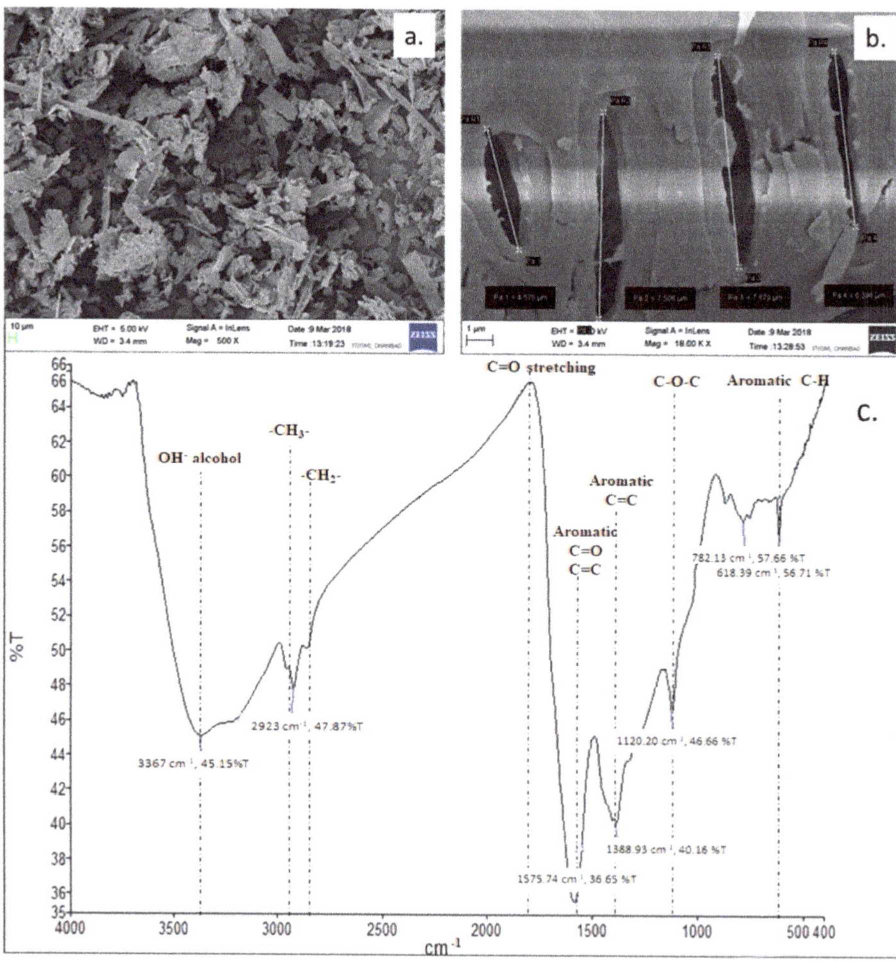

Figure 4. (**a**) FE-SEM image of *C. procera* biochar at 500× magnification, (**b**) FE-SEM image of *C. procera* biochar at 1800× magnification showing the pore sizes, (**c**) FTIR spectra of *C. procera* biochar showing the surface functional groups.

3.2. Plant Biodiversity and Vegetation Analysis

B. arundinacea clump density was 4033 clumps ha^{-1}, whereas the tree density was 3233 trees ha^{-1}. Relative distribution of the species in the reclaimed site is shown in Figure 5. *B. arundinacea* clumps are most abundant (56%), followed by *Albizzia* spp. (18%), *D. sissoo* (10.5%), and *Z. mauritiana* (5%). Das and Maiti [37] reported that the same reclaimed mine spoil at 4 years old had a bamboo clumps density of 3033 clumps ha^{-1}, whereas the tree density was 2500 trees ha^{-1}.

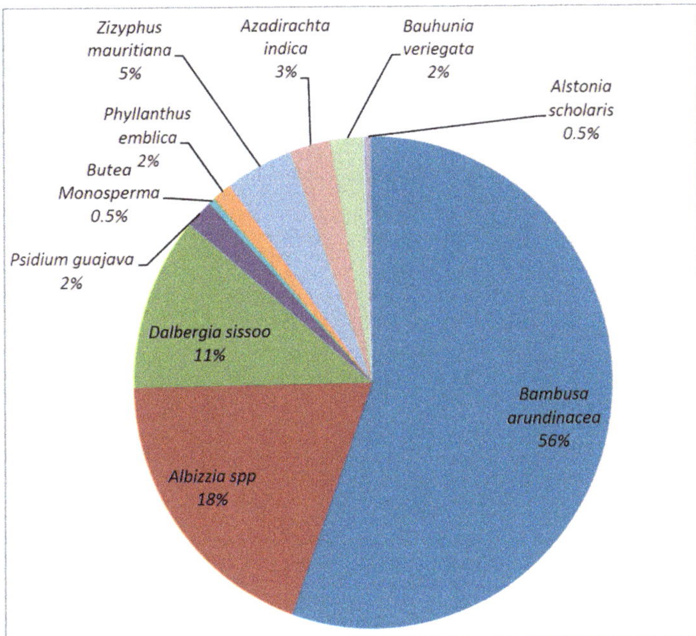

Figure 5. Relative density of each species in the RMS showing the percentage of each species in the study area.

3.3. Estimation of Biomass Carbon Stock

Plant biomass and C pool associated with RMS and RF are summarized in Table 2. In RMS, *Albizzia* spp. (4.34 Mg C ha^{-1}) had the highest total tree carbon stock, followed by *B. arundinacea* (2.61 Mg C ha^{-1}) and *D. sissoo* (2.91 Mg C ha^{-1}). The total carbon stock from the plant biomass was 12.59 Mg C ha^{-1} and CO_2 sequestered was 48.17 Mg ha^{-1}. Tree biomass of the RF was three times (30.63 Mg C ha^{-1}) higher than the RMS (12.59 Mg C ha^{-1}) and UMS (3.52 Mg C ha^{-1}). Čížková et al. [38] reported 1.6 t ha y^{-1} potential for carbon sequestration in reclaimed grasslands in a reclaimed lignite mine. Ahirwal et al. [39] reported the effect of fast-growing trees on soil properties and the ecosystem carbon pool after eight years of afforestation. The study reported greater carbon storage in *D. sissoo* (39 Mg C ha^{-1}) compared to *A. lebbeck* (34 Mg C ha^{-1}) and *A. procera* (26 Mg C ha^{-1}). In a 16 year reclaimed coal mine site, Ahirwal and Maiti, [40] reported that the tree carbon stock was 75% of the reference forest site, plantation of multipurpose tree species improved mine spoil fertility, facilitates natural growth of indigenous tree species. Due to the plant soil-interaction in a coal mine spoil, the roots of the re-vegetated plants alter soil structure and function [8]. Although carbon stock associated with RMS was less than that of RF, yet the growth of native species proves that proper reclamation technology can influence ecosystem development Thus, increase in C-stock proves that plantation is a successful means of mine reclamation which certainly helps in achieving SDG 13 and 15.

Table 2. Total biomass, tree carbon stock and CO_2 sequestered in various coal mine spoils of the world and their values in the current study (RMS: reclaimed mine spoil and RF: reserve forest).

Reclamation Type	Location	Age	Total Biomassk (Mg C ha^{-1})	Tree Carbon Stock (Mg C ha^{-1})	CO_2 Sequestered * (Mg ha^{-1})	References
Albizia lebbeck *Albizia procera* *Tectona grandis* *Dendrocalamus strictus*	Singrauli, India	4–5	17.28 8.48 7.34 15.02	8.64 4.24 3.67 7.51	31.70 15.56 12.36 27.56	[41]
Mixed Plantation	Singrauli, India	10	15.64	13.68	50.20	[18]
Reclaimed dump Reference forest	Jharia, India	5	18.07 55.25	9.03 27.63	33.14 101.40	[42]
Albizia lebbeck *Albizia procera* *Dalbergia sisso* Natural forest	Jharia, India	8	73.66 55.76 83.28 250.28	34.62 10.89 26.21 117.63	127.05 39.96 96.19 431.702	[39]
RMS RF UMS	Damoda eco-restoration, Jharia, India	8	25.18 61.26 7.05	12.59 30.63 3.52	46.17 112.41 12.93	Present Study

* CO_2 sequestered (Mg C ha^{-1}) = Tree carbon stock (Mg C ha^{-1}) × 44/12.

3.4. Herbaceous Biomass and Litter Analysis C-Stock

Litter consists of twigs, plant debris, foliage and branches which possess high nutrient content. It is one of the most important sources of organic matter in the mine spoils, and litter decomposition contributes to nutrients recycling and improvement in soil fertility [36]. During reclamation of a degraded mine spoil, increase in litter C improves the SOC of the ecosystem and can be an indicator of restoration success. Litter and understory carbon stock in RMS, BC_{30}, BC_{60}, RF and UMS are presented in Figure 6. The total carbon stock in RMS was 1.24 Mg ha^{-1}, while higher values of 1.64 Mg ha^{-1} and 1.73 Mg ha^{-1} were observed for BC_{30} and BC_{60} respectively. Biochar application had carbon stock comparable to the RF (1.79 Mg ha^{-1}) and was significantly higher ($p < 0.05$) than UMS (0.5 Mg ha^{-1}).

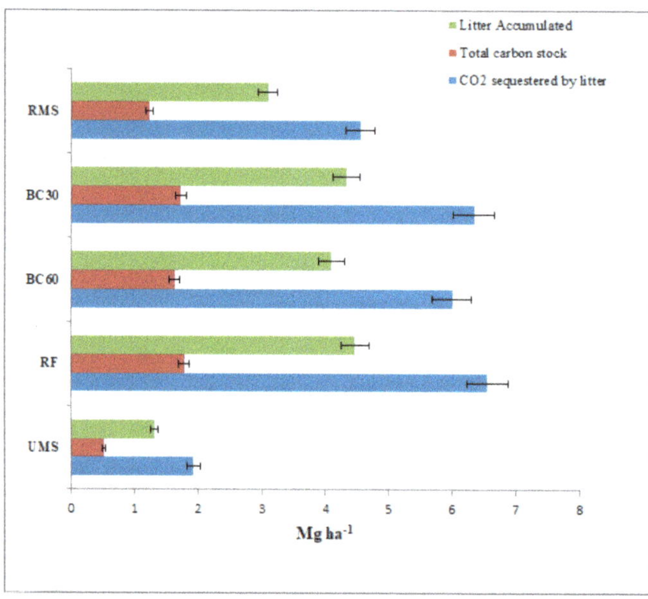

Figure 6. Litter and understory carbon stock in reclaimed mine spoil (RMS), biochar treatment at 30 t ha^{-1} (BC_{30}) and 60 t ha^{-1} (BC_{60}), reserved forest (RF) and unreclaimed mine spoil (UMS) ($n = 24$).

3.5. Mine Spoil Properties

3.5.1. Inorganic, Biogenic and Coal Carbon Estimation

Carbon fractions in RMS, BC_{30}, BC_{60}, RF and UMS are presented in Figure 7a. The total soil carbon at BC_{30} (36.3 g C kg^{-1}) and BC_{60} (40 g C kg^{-1}) was found to be significantly high ($p < 0.05$) compared to RMS (21 g C kg^{-1}), comparable to RF (33 g C kg^{-1}), and low in UMS (12 g C kg^{-1}). Although there was no significant difference between the soil inorganic carbon of RMS (1.9 g C kg^{-1}), RF (2 g C kg^{-1}) and UMS (0.7 g C kg^{-1}), the inorganic fraction was found to increase to 3.5 g C kg^{-1} and 4.5 g C kg^{-1} at BC_{30} and BC_{60}, respectively. Average coal carbon was higher in RMS, BC_{30}, BC_{60} and UMS compared to RF. A greenhouse experiment conducted by Rodríguez-Vila et al. [43] on copper mine spoils reported a range of 20–207 g C kg^{-1} for total soil carbon and 3–27 g C kg^{-1} for inorganic carbon by biochar application rate of 20–100%.

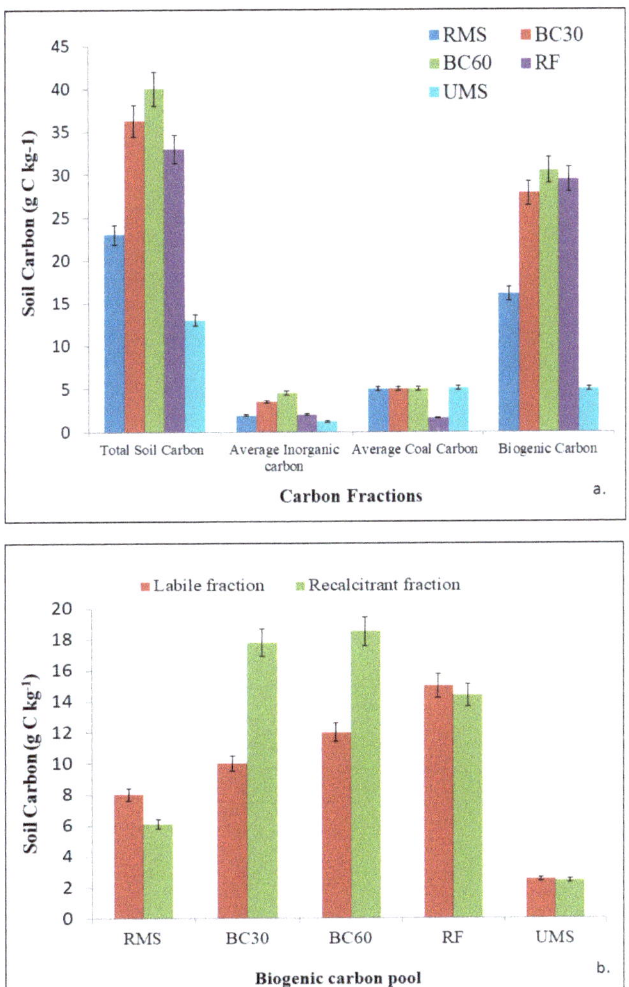

Figure 7. (**a**) Comparison of different carbon fractions in reclaimed mine spoil (RMS), biochar treatment at 30 t ha^{-1} (BC_{30}) and 60 t ha^{-1} (BC_{60}), reserved forest (RF) and unreclaimed mine spoil (UMS). (**b**) Comparison of labile and recalcitrant carbon fractions in RMS, BC_{30}, BC_{60},RF and UMS.

Biogenic carbon fraction is a complex pool which can be broadly divided into labile and recalcitrant carbon pool. Labile pool has a residence time of years to a few decades while the recalcitrant carbon pool can remain in the soil for hundreds to thousands of years [37]. Figure 7b shows the labile and the recalcitrant carbon present in RMS, BC_{30}, BC_{60} and RF. Labile fraction was found to be 56% and 51% of biogenic carbon pool for RMS and RF, respectively. The labile carbon trend was of the order UMS > RMS > RF > BC_{30} > BC_{60}. Recalcitrant fraction was found to be 44% and 49% of the biogenic carbon pool for BC_{30} and BC_{60}, respectively. The recalcitrant carbon trend was of the order BC_{60} > BC_{30} > RF > RMS > UMS. Labile C pools can have a large impact on soil-biochar interactions in the short term (~6 months), whereas recalcitrant carbon pool is influential in soil properties and soil function for a longer period of time [24,26]. Sub-fractions of soil organic carbon are indicators of soil fertility and are instrumental in understanding the influence of a management practice. In a study conducted on a 10 year old reclaimed coal mine spoil, Das and Maiti. [44] reported that the biogenic C constituted 45–66% of total soil carbon in RMS. Fidel et al. [26] reported that biochar improves the soil inorganic carbon by 0.023–0.045 mg C kg^{-1} and organic carbon by 0.001–0.0069 mg C kg^{-1} in eroded soil. The study also reported that labile biochar pools are stabilized by the recalcitrant C pool. Thus, because of the higher recalcitrant fraction, biochar application will play an instrumental role in increasing the C-stock and help achieve the UN SDG 13 and 15.

3.5.2. Other Physio-Chemical Properties

RMS, BC_{30} and BC_{60}, RF and UMS properties are summarized in Table 3. Soil fraction was significantly ($p < 0.05$) high in the reference forest compared to the 8 year old RMS and UMS, biochar application had no effect in the soil fractions. pH of RMS was neutral, biochar application resulted in an alkaline mine spoil, while it was slightly acidic in the RF and UMS. EC of the RMS was 0.16 dS cm^{-1} compared to 0.11 dS cm^{-1} in RF soil, 0.17 dS cm^{-1} in UMS, 0.09 dS cm^{-1} in BC_{30} and 0.1 dS cm^{-1} in BC_{60}. CEC was higher in RF (13.1 C mol kg^{-1}) compared to RMS (8.22 C mol kg^{-1}). Moisture content in RF was 26% higher that of RMS, while BC_{60} and BC_{30} improved the moisture content by 33% and 55% respectively ($p < 0.05$). Available N and P showed significant difference ($p < 0.05$) in the RMS and RF; available N and P in RF were 26% and 47% higher, respectively. The exchangeable potassium was also found to be higher in RF (55.22 mg kg^{-1}) compared to the 8 year old RMS (102.1 mg kg^{-1}). BC_{30} and BC_{60} improved the NPK values in the mine spoil significantly ($p < 0.05$) which helps in vegetation growth. The corrected bulk density was higher in the RF site (1.34 Mg m^{-3}) compared to the RMS (0.71 Mg m^{-3}), BC_{30} (0.65 Mg m^{-3}) and BC_{60} (0.63 Mg m^{-3}). Ghosh et al. [11] reported an increase in organic carbon by threefold, CEC by twofold, with a decrease in bulk density to half, by *Lantana* biochar application in a coal mine spoil. Thus, biochar application can effectively improve the physio-chemical properties of the mine spoil which will help accelerate the process of soil formation in a RMS and increase C-stock to near RF level. The application of biochar improves the soil physico-chemical properties and increases the carbon stock in the soil. This ameliorated mine spoil will support plant growth and help in ecosystem development. This in the long run will help in achieving the UN SDGs 13 and 15.

3.6. Total C-Pool

The total C stock, CO_2 sequestered and rate of C sequestration in RMS, BC_{30}, BC_{60}, RF and UMS are presented in Table 4. The C stock of RF (72.11 Mg C ha^{-1}) was almost thrice the RMS (30.98 Mg C ha^{-1}) and 5 times the UMS (13.92 Mg C ha^{-1}). Application of biochar @ 30 t ha^{-1} improved the C-stock of RMS by 33%, but was 42% lower than the RF. Similarly, biochar @ 60 t ha^{-1} improved the C-stock of RMS by 47%, but was 36% lower than the RF C-stock. CO_2 sequestered had the sequence, RF (264.64 Mg C ha^{-1}) > BC_{60} (168.22 Mg C ha^{-1}) > BC_{30} (151.70 Mg C ha^{-1}) > RMS (113.69 Mg C ha^{-1}) > USM (48.37 Mg C ha^{-1}). Mukhopadhyay and Masto [45] reported yard waste biochar improved the stable carbon pool in biochar amended mine spoil to 0.873 g CO_2–C kg^{-1} compared to 0.03 g CO_2–C kg^{-1} in mine spoil. Xu et al. [46] reported that the application of bamboo leaf biochar at 5 and 15 Mg ha^{-1} application rate increased the ecosystem carbon stock by 1486.31% and 252.98%, respectively. This increase could be due to the increase in vegetative cover with time. Plant growth can play an important role in decomposition and humus layer formation, which later turned into soil organic matter and increases the organic carbon pool of the soil and provides nutrients to the reclaimed vegetation.

Table 3. Characteristics of RMS (reclaimed mine soil), BC_{30} (biochar @ 30 t ha^{-1}), BC_{60} (biochar @ 60 t ha^{-1}), RF (reference forest) soil and unreclaimed mine spoil (UMS) (n = 24, mean ± Standard deviation, within each row, values with same letter are not significantly different, $p < 0.05$ with Duncan's multiple range test).

Soil Parameters	RMS	BC_{30}	BC_{60}	RF	UMS
Soil Fraction (<2 mm size)%	63.10 ± 12.06 b	63.10 ± 12.06 b	63.10 ± 12.06 b	88.12 ± 6.57 a	25.6 ± 3.4 c
Non- Soil Fraction (>2 mm size)%	36.89 ± 12.04 b	36.89 ± 12.04 b	36.89 ± 12.04 b	11.88 ± 3.20 c	74.4 ± 3.3 a
pH(water, 1:2.5, w/v)	7.15 ± 0.41 b	9.66 ± 0.2 a	9.53 ± 0.57 a	6.12 ± 1.22 c	6.3 ± 0.4 c
EC (water, 1:2.5, w/v) dS/m	0.16 ± 0.27 a	0.09 ± 0.3 c	0.1 ± 0.56 b	0.11 ± 0.08 b	0.17 ± 0.06 ab
CEC (C mol kg^{-1})	8.22 ± 1.55 c	13.25 ± 4.22 b	18.39 ± 0.38 a	13.1 ± 0.27 b	5.22 ± 1.54 d
Moisture Content (%)	5.37 ± 5.35 d	8.14 ± 0.12 b	10.77 ± 0.28 a	7.29 ± 2.18 c	4.26 ± 2.6 d
Available-N (mg kg^{-1})	96 ± 8.34 cd	102 ± 5.87 b	105 ± 6.27 b	130 ± 5.22 a	58.72 ± 4.2 d
Available-P (mg kg^{-1})	3.82 ± 0.84 d	8.91 ± 1.12 b	10.18 ± 0.4 a	7.26 ± 0.87 bc	3.24 ± 1.3 d
Exchangeable K (mg kg^{-1})	55.22 ± 3.57 c	423.3 ± 35.11 a	456.66 ± 7.4 a	102.1 ± 5.22 b	30.56 ± 3.2 d
Corrected bulk density (Mg m^{-3})	0.71 ± 0.51 b	0.65 ± 0.58 c	0.63 ± 0.91 c	0.91 ± 0.85 ab	1.05 ± 0.10 a
SOC (Mg ha^{-1})	31.33 ± 0.75 c	41.29 ± 1.22 b	45.7 ± 1.89 b	72.11 ± 5.22 a	12.6 ± 0.32 d

Table 4. Effect of biochar application on carbon stocks of different landforms. Comparison of total C stock, CO_2 sequestered, Rate of C accumulation in 8 year old RMS (reclaimed mine spoil), BC_{30} (biochar @ 30 t ha^{-1}), BC_{60} (biochar @ 60 t ha^{-1}), RF (reserve forest) and unreclaimed mine spoil (UMS).

Land use	Country/Location	Biochar Feedstock	Results		References
Sub-urban red soil	Hangzhou, China	Oak wood, bamboo	-Lability index increased by 4 and 6%, respectively, -The carbon management index (CMI) increased by 50 to 286%.		[24]
Agricultural Soil	Pottawattamie County, USA	Corn stover	Increase in soil inorganic carbon by 0.023–0.045 mg C kg^{-1} and organic carbon by 0.001–0.0069 mg C kg^{-1}.		[26]
Moso bamboo forest	Zhejiang, China	Bamboo leaf	5 and 15 Mg ha^{-1} increased the ecosystem carbon stock by 1486.31% and 252.98%, respectively.		[46]
Agricultural soil	Atlantic, USA	Wood	C stocks nearly twice (14.07 Mg soil C ha^{-1}) the amount of C added with biochar 6 years earlier (7.25 Mg biochar C ha^{-1})		[47]
Fresh Coal Mine spoil	Jharia, India	Yard waste	Stable carbon pool in biochar amended mine spoil was 0.873 g CO_2–C kg^{-1} compared to 0.03 g CO_2–C kg^{-1} in mine spoil		[45]
Copper Mine spoil	Touro, Spain	Holm oak wood	20–207 g C kg^{-1} for total soil carbon and 3–27 g C kg^{-1} for inorganic carbon by biochar application rate of 20–100%		[43]
			C stock (Mg C ha^{-1})	CO_2 sequestered (Mg ha^{-1})	
RMS	Damoda, Eco-restoration, Jharia, India	*Calotropis procera*	30.98 ± 1.25	113.69 ± 3.5	Present Study
BC30			41.34 ± 1.3	151.70 ± 5.31	
BC60			45.84 ± 1.5	168.22 ± 4.25	
RF			72.11 ± 3.2	264.64 ± 5.65	
UMS			13.18 ± 0.87	48.37 ± 1.27	

Application of biochar increased the C-stock of RMS by 33%, and 47% at application rates of 30 t ha^{-1} and 60 t ha^{-1}, respectively. As mentioned earlier, recalcitrant fraction of this C-stock is 44% and 49% of the biogenic carbon pool for BC_{30} and BC_{60}, respectively. Thus, it can be concluded that the increase in the C-pool in the 6 months of the study was due to the labile fraction. The remaining 44% and 49% will remain in the form of recalcitrant carbon in the RMS, and thus, will aid in achieving the climate action goals of the SDG.

4. Future Recommendations

Through the course of this study, future goals and recommendations for continuing research in this field are as follows:

(i) A biochar based chrono-sequence study in the RMS to study the trends in carbon stock for a prolonged period of time.
(ii) Field based long term studies are needed to understand the behaviour of carbon that is fixed in the soil by biochar application. Studies need to be done to ensure that the carbon in the biochar is fixed in soil for a long period of time and not emitted into the atmosphere.
(iii) Effect of biochar application on the existing humus of the RMS.
(iv) Conducting life cycle assessment (LCA) to confirm that net soil-ecosystem C pools is increased by biochar application.
(v) Environmental cost benefit by biochar application should be carried out for the applicability of biochar.
(vi) Developing techniques for the large-scale production of biochar in the field itself will help reduce the transportation cost.

5. Conclusions

Plantation of hardy species in coal mine spoil restores the derelict ecosystem which promotes natural colonization of indigenous species. In the current study, 8 year old RMS effectively increased the biomass, litter and biogenic carbon stock in the soil. The rate of C accumulation for 8 year old RMS was calculated to be 3.92 Mg C ha^{-1} year^{-1}. Application of *C. procera* biochar in the RMS improved the soil physio-chemical properties. The inorganic and biogenic carbon pool, especially the recalcitrant pool, was improved by biochar application, suggesting that biochar can be an effective mode of enhanced carbon fixation in the spoil, along with plantation activities. A mere 6 month application period increased the C-stock by 36–42%, thus its recalcitrant carbon content can be fixed in the mine spoil for a longer period of time. This proves that biochar has tremendous potential in fixing carbon, along with forestry based reclamation of coal mine spoil. Thus, carbon stock increases with age of reclamation, and biochar application can increase the carbon stock close to reference forest site level. As the biochar-plantation synergy can both sequester carbon and also promote biodiversity, it can be an effective tool for achieving United Nations SDG 13 and 15.

Supplementary Materials: The following are available online at https://www.mdpi.com/article/10.3390/land10111112/s1, Table S1. List of species recorded in quadrats of RMS sites showing some biodiversity parameters.

Author Contributions: Both authors contributed to the study conception and design. Material preparation, data collection and analysis were performed by D.G., S.K.M. was the overall supervisor of the work. All authors have read and agreed to the published version of the manuscript.

Funding: This research received no external funding.

Institutional Review Board Statement: Not applicable.

Informed Consent Statement: Not applicable.

Data Availability Statement: Not applicable.

Acknowledgments: Sincere thanks to Indian Institute of Technology (Indian School of Mines), Dhanbad and MHRD, GOI for providing fellowship to the first author (17DR000426). We would also like to thank the reviewers and the editors for their insightful comments during the review of the manuscript.

Conflicts of Interest: The authors declare no conflict of interest.

References

1. Dubey, P.; Singh, A.; Raghubanshi, A.; Abhilash, P.C. Steering the restoration of degraded agroecosystems during the United Nations Decade on Ecosystem Restoration. *J. Environ. Manag.* **2020**, *280*, 111798. [CrossRef] [PubMed]
2. Ghosh, D.; Maiti, S.K. Biochar-assisted eco-restoration of coal mine degraded land to meet United Nation Sustainable Development Goals. *Land Degrad. Dev.* **2021**, *32*, 4494–4508. [CrossRef]
3. Angulo-Mosquera, L.S.; Alvarado-Alvarado, A.A.; Rivas-Arrieta, M.J.; Cattaneo, C.R.; Rene, E.R.; García-Depraect, O. Production of solid biofuels from organic waste in developing countries: A review from sustainability and economic feasibility perspectives. *Sci. Total Environ.* **2021**, *795*, 148816. [CrossRef]
4. Shahid, M.K.; Batool, A.; Kashif, A.; Nawaz, M.H.; Aslam, M.; Iqbal, N.; Choi, Y. Biofuels and biorefineries: Development, application and future perspectives emphasizing the environmental and economic aspects. *J. Environ. Manag.* **2021**, *297*, 113268. [CrossRef] [PubMed]
5. Maiti, S.K. *Ecorestoration of Coal Mine Degraded Lands*; Springer: New Delhi, India, 2013. [CrossRef]
6. Ahirwal, J.; Maiti, S.K.; Singh, A.K. Changes in ecosystem carbon pool and soil CO_2 flux following post-mine reclamation in dry tropical environment, India. *Sci. Total Environ.* **2017**, *583*, 153–162. [CrossRef]
7. Frouz, J.; Hüblov, L. Contrasting effect of coniferous and broadleaf trees on soil carbon storage during reforestation of forest soils and afforestation of agricultural and post-mining soils. *J. Environ. Manag.* **2021**, *290*, 112567. [CrossRef]
8. Maiti, S.K.; Ghosh, D. Plant–soil interactions as a restoration tool. In *Climate Change and Soil Interactions*; Prasad, M.N.V., Pietrzykowski, M., Eds.; Elsevier: Amsterdam, The Netherlands, 2020; pp. 689–730.
9. Chen, Y.; Liu, Z.; Rao, X.; Wang, X.; Liang, C.; Lin, Y.; Zhou, L.; Cai, X.A.; Fu, S. Carbon storage and allocation pattern in plant biomass among different forest plantation stands in Guangdong, China. *Forests* **2015**, *6*, 794–808. [CrossRef]
10. Ussiri, A.N.; Lal, R. Method for Determining Coal Carbon in the Reclaimed Minesoils Contaminated with Coal. *Soil Sci. Soc. Am. J.* **2008**, *72*, 231–237. [CrossRef]
11. Ghosh, D.; Masto, R.E.; Maiti, S.K. Ameliorative effect of Lantana camara biochar on coal mine spoil and growth of maize (*Zea mays*). *Soil Use Manag.* **2020**, *36*, 726–739. [CrossRef]
12. Kumari, S.; Maiti, S.K. Reclamation of coalmine spoils with topsoil, grass, and legume: A case study from India. *Environ. Earth Sci.* **2019**, *78*, 1–14. [CrossRef]
13. Raj, D.; Kumar, A.; Maiti, S.K. Mercury remediation potential of *Brassica juncea* (L.) Czern. for clean-up of flyash contaminated sites. *Chemosphere* **2020**, *248*, 125857. [CrossRef] [PubMed]
14. Frouz, J.; Kuraz, M. Changes in Some Physical Properties of Soils in the Chronosequence of Self Overgrown Dumps of the Sokolov Quarry-Dump Complex Czechia. *Eurasian Soil Sci.* **2012**, *45*, 266–272. [CrossRef]
15. Józefowska, A.; Pietrzykowski, M.; Frouz, J. Geoderma The effects of tree species and substrate on carbon sequestration and chemical and biological properties in reforested post-mining soils. *Geoderma* **2017**, *292*, 9–16. [CrossRef]
16. Shukla, M.K.; Lal, R. Soil organic carbon stock for reclaimed minesoils in northeastern Ohio. *L. Degrad. Dev.* **2005**, *16*, 377–386. [CrossRef]
17. Akala, V.A.; Lal, R. Soil organic carbon pools and sequestration rates in reclaimed minesoils in Ohio. *J. Environ. Qual.* **2001**, *30*, 2098–2104. [CrossRef]
18. Tripathi, N.; Shekhar, R.; Nathanail, C.P. Mine spoil acts as a sink of carbon dioxide in Indian dry tropical environment. *Sci. Total Environ.* **2014**, *468–469*, 1162–1171. [CrossRef]
19. Mukhopadhyay, S.; Masto, R.E.; Cerdà, A.; Ram, L.C. Rhizosphere soil indicators for carbon sequestration in a reclaimed coal mine spoil. *Catena* **2016**, *141*, 100–108. [CrossRef]
20. Ahmad, M.; Rajapaksha, A.U.; Lim, J.E.; Zhang, M.; Bolan, N.; Mohan, D.; Vithanage, M.; Lee, S.S.; Ok, Y.S. Biochar as a sorbent for contaminant management in soil and water: A review. *Chemosphere* **2014**, *99*, 19–33. [CrossRef]
21. Novak, J.M.; Ippolito, J.A.; Ducey, T.F.; Watts, D.W.; Spokas, K.A.; Trippe, K.M.; Sigua, G.C.; Johnson, M.G. Remediation of an acidic mine spoil: *Miscanthus* biochar and lime amendment affects metal availability, plant growth, and soil enzyme activity. *Chemosphere* **2018**, *205*, 709–718. [CrossRef]
22. Lal, R. Biochar and Soil Carbon Sequestration. *Agric. Environ. Appl. Biochar Adv. Barriers* **2015**, *43210*, 175–197. [CrossRef]
23. Wang, T.; Camps-Arbestain, M.; Hedley, M. Predicting C aromaticity of biochars based on their elemental composition. *Org. Geochem.* **2013**, *62*, 1–6. [CrossRef]
24. Demisie, W.; Liu, Z.; Zhang, M. Effect of biochar on carbon fractions and enzyme activity of red soil. *Catena* **2014**, *121*, 214–221. [CrossRef]
25. Masto, R.E.; Kumar, S.; Rout, T.K.; Sarkar, P.; George, J.; Ram, L.C. Biochar from water hyacinth (*Eichornia crassipes*) and its impact on soil biological activity. *Catena* **2013**, *111*, 64–71. [CrossRef]

26. Fidel, R.B.; Laird, D.A.; Parkin, T.B. Impact of Biochar Organic and Inorganic Carbon on Soil CO_2 and N_2O Emissions. *J. Environ. Qual.* **2017**, *46*, 505–513. [CrossRef] [PubMed]
27. Ghosh, D.; Maiti, S.K. Effect of invasive weed biochar amendment on soil enzymatic activity and respiration of coal mine spoil: A laboratory experiment study. *Biochar* **2021**. [CrossRef]
28. Ghosh, D.; Maiti, S.K. Can biochar reclaim coal mine spoil? *J. Environ. Manag.* **2020**, *272*, 111097. [CrossRef]
29. Chave, J.; Réjou-Méchain, M.; Búrquez, A.; Chidumayo, E.; Colgan, M.S.; Delitti, W.B.; Duque, A.; Eid, T.; Fearnside, P.M.; Goodman, R.C.; et al. Improved allometric models to estimate the aboveground biomass of tropical trees. *Glob. Chang. Biol.* **2014**, *20*, 3177–3190. [CrossRef]
30. IPCC. *IPCC Guidelines for National Greenhouse Gas Inventories. Intergovernmental Panel on Climate Change*; IGES: Kanagawa, Japan, 2006.
31. FAO, Food and agriculture organization (FAO). Global Forest Resources Assessment 2005. Global Assessment of Growing Stock, Biomass and Carbon Stock. 2005. Available online: www.fao.org/docrep/013/i1757e/i1757e.pdf (accessed on 15 September 2021).
32. Nath, J.A.; Sileshi, G.W.; Kumar, A. Land Use Policy Bamboo based family forests offer opportunities for biomass production and carbon farming in North East India. *Land Use Policy* **2018**, *75*, 191–200. [CrossRef]
33. Majumder, A.F.; Das, A.K.; Nath, A.J. Biomass storage and carbon sequestration in priority bamboo species in relation to village physiography. *Int. J. Ecol. Environ. Sci.* **2019**, *45*, 85–95.
34. Thakur, T.K.; Swamy, S.L.; Bijalwan, A.; Dobriyal, M.J. Assessment of biomass and net primary productivity of a dry tropical forest using geospatial technology. *J. For. Res.* **2019**, *30*, 157–170. [CrossRef]
35. Jackson, M.L. *Soil Chemical Analysis*; PHI Pvt. Ltd.: New Delhi, India, 1973.
36. Ahirwal, J.; Maiti, S.K. Assessment of carbon sequestration potential of revegetated coal mine overburden dumps: A chronosequence study from dry tropical climate. *J. Environ. Manag.* **2017**, *201*, 369–377. [CrossRef]
37. Das, R.; Maiti, S.K. Estimation of carbon sequestration in reclaimed coalmine degraded land dominated by *Albizia lebbeck, Dalbergia sissoo* and *Bambusa arundinacea* plantation: A case study from Jharia Coalfields, India. *Int. J. Coal Sci. Technol.* **2016**, *3*, 246–266. [CrossRef]
38. Čížková, B.; Woś, B.; Pietrzykowski, M.; Frouz, J. Development of soil chemical and microbial properties in reclaimed and unreclaimed grasslands in heaps after opencast lignite mining. *Ecol. Eng.* **2018**, *123*, 103–111. [CrossRef]
39. Ahirwal, J.; Kumar, A.; Maiti, S.K. Effect of Fast-Growing Trees on Soil Properties and Carbon Storage in an Afforested Coal Mine. *Minerals* **2020**, *10*, 840. [CrossRef]
40. Ahirwal, J.; Maiti, S.K. Development of Technosol properties and recovery of carbon stock after 16 years of revegetation on coal mine degraded lands, India. *Catena* **2018**, *166*, 114–123. [CrossRef]
41. Singh, A.N.; Zeng, D.H.; Chen, F.S. Effect of young woody plantations on carbon and nutrient accretion rates in a redeveloping soil on coalmine spoil in a dry tropical environment, India. *Land Degrad. Dev.* **2006**, *17*, 13–21. [CrossRef]
42. Ahirwal, J.; Maiti, S.K. Assessment of soil carbon pool, carbon sequestration and soil CO_2 flux in unreclaimed and reclaimed coal mine spoils. *Environ. Earth Sci.* **2018**, *77*, 9. [CrossRef]
43. Rodríguez-vila, A.; Asensio, V.; Forján, R.; Covelo, E.F. Carbon fractionation in a mine soil amended with compost and biochar and vegetated with *Brassica juncea* L. *J. Geochem. Explor.* **2016**, *169*, 137–143. [CrossRef]
44. Das, R.; Maiti, S.K. Importance of carbon fractionation for the estimation of carbon sequestration in reclaimed coalmine soils-A case study from Jharia coalfields, Jharkhand, India. *Ecol. Eng.* **2016**, *90*, 135–140. [CrossRef]
45. Mukhopadhyay, S.; Masto, R.E. Effect of fly ash on carbon mineralization of biochar and organic manures added to mine spoil. *SN Appl. Sci.* **2019**, *1*, 1–10. [CrossRef]
46. Xu, L.; Fang, H.; Deng, X.; Ying, J.; Lv, W.; Shi, Y.; Zhou, G. Biochar application increased ecosystem carbon sequestration capacity in a Moso bamboo forest. *For. Ecol. Manag.* **2020**, *475*, 118447. [CrossRef]
47. Blanco-canqui, H.; Acharya, B.S.; Laird, D.A.; Heaton, E.A. Soil carbon increased by twice the amount of biochar carbon applied after 6 years: Field evidence of negative priming. *GCB Bioenergy* **2020**, *12*, 240–251. [CrossRef]

land

Article

The Prediction of Carbon Emission Information in Yangtze River Economic Zone by Deep Learning

Huafang Huang [1,2,3], Xiaomao Wu [4,5] and Xianfu Cheng [1,3,*]

1 School of Geography and Tourism, Anhui Normal University, Wuhu 241002, China; huanghf@ahnu.edu.cn
2 College of Economics & Management, Hefei Normal University, Hefei 230601, China
3 Anhui Key Laboratory of Natural Disaster Process and Prevention, Wuhu 241002, China
4 Department of Chemical and Biological Engineering, University of Sheffield, Sheffield S10 2TN, UK; wxm19911118@gmail.com
5 Tongling Nonferrous Design and Research Institute Company Limited Hefei Branch, Hefei 230000, China
* Correspondence: xf990112@mail.ahnu.edu.cn

Abstract: This study aimed to respond to the national "carbon peak" mid-and long-term policy plan, comprehensively promote energy conservation and emission reduction, and accurately manage and predict carbon emissions. Firstly, the proposed method analyzes the Yangtze River Economic Belt as well as its "carbon peak" and carbon emissions. Secondly, a support vector regression (SVR) machine prediction model is proposed for the carbon emission information prediction of the Yangtze River Economic Zone. This experiment uses a long short-term memory neural network (LSTM) to train the model and realize the experiment's prediction of carbon emissions. Finally, this study obtained the fitting results of the prediction model and the training model, as well as the prediction results of the prediction model. Information indicators such as the scale of industry investment, labor efficiency output, and carbon emission intensity that affect carbon emissions in the "Yangtze River Economic Belt" basin can be used to accurately predict the carbon emissions information under this model. Therefore, the experiment shows that the SVR model for solving complex nonlinear problems can achieve a relatively excellent prediction effect under the training of LSTM. The deep learning model adopted herein realized the accurate prediction of carbon emission information in the Yangtze River Economic Zone and expanded the application space of deep learning. It provides a reference for the model in related fields of carbon emission information prediction, which has certain reference significance.

Keywords: carbon emission; SVR; LSTM neural network; carbon emission prediction

check for updates

Citation: Huang, H.; Wu, X.; Cheng, X. The Prediction of Carbon Emission Information in Yangtze River Economic Zone by Deep Learning. *Land* **2021**, *10*, 1380. https://doi.org/10.3390/land10121380

Academic Editors: Marina Cabral Pinto, Amit Kumar and Munesh Kumar

Received: 15 November 2021
Accepted: 12 December 2021
Published: 13 December 2021

1. Introduction

In recent years, with the frequent occurrence of extremely severe weather due to global warming, countries around the world have begun to pay attention to the imbalance of carbon emissions caused by the emissions of greenhouse gases such as carbon dioxide (CO_2) [1]. Excessive CO_2 and other greenhouse gas emissions have caused irreparable damage to the environment [2]. Meanwhile, the process of social development cannot avoid the problem of carbon emissions, so the real-time prediction and monitoring of carbon emissions information has become extremely crucial [3]. Many scholars have performed a lot of research in the field of carbon emissions.

The earliest representative studies abroad mainly used the factor decomposition method, index decomposition method, input–output method, and combination model forecasting. Scholars have proposed an improved cuckoo optimization algorithm neural network (COANN) artificial neural network structure, which is optimized by the cuckoo algorithm (COA). The performance of COANN is evaluated by the mean square error (MSE), root mean square error (RMSE), mean absolute error (MAE), and correlation coefficient (CC) between the model output and the actual data set. The COANN prediction model can predict the world's CO_2 emissions by 2050 [4]. Some researchers proposed a multi-objective predictive energy management strategy for residential grid-connected hybrid energy systems via machine learning technology. The strategy proposed includes three levels of control: 1—the logical level of management of computational load and accuracy; 2—on the dual prediction model of the residual causal expansion convolutional network, it is used for energy production and system power load; and 3—as well as multi-objective optimization for effective transactions, to provide energy for the public grid through battery charging scheduling [5]. It was proposed that energy efficiency on non-intrusive load monitoring (NILM) can save electricity by improving the awareness of behavior changes and reducing carbon dioxide emissions into the environment. In the data published in Malaysia from 1996 to 2018, a predictive model was established and scenario simulations were carried out. Malaysia's public database was also sued for predicting the impact of CO_2 emissions and NILM on environmental degradation from 2019 to 2030 [6]. Some scholars have used Eviews software to analyze the carbon emission data of Beijing, Henan, Guangdong, and Zhejiang from 1997 to 2017. They also used differential stationary processing, moving average, and substituting strong impact points for data preprocessing. Through model identification, parameter estimation, and model testing, they established an integrated moving average autoregressive (ARIMA) models to predict carbon emissions in four regions [7]. It is believed that it is important to objectively evaluate the impact of relevant factors on carbon emissions. They proposed a modified production theory decomposition analysis (PDA) model under the semi-disposable hypothesis, and correspondingly decomposed the carbon emission changes of China's thermal power generation industry [8]. Some scholars have used the Lasso regression model to screen out eight significant factors affecting carbon emissions, and used the BP neural network model to predict the carbon emissions of Jiangsu Province from 2019 to 2030. They used artificial neural networks (ANN) to develop carbon emission intensity prediction models for Australia, Brazil, China, India, and the United States. Nine parameters that play an important role in the intensity of carbon emissions were selected as input variables. After many iterations, the best model was selected for each country by predefined criteria. They used a 9–5–1 multilayer perceptron with a backpropagation algorithm to build, validate, and train the model. The results of the verification model show that the error between the predicted value and the actual value is approximately 0, and the proposed ANN model can accurately predict carbon emissions [9,10]. Wang et al. (2021) used the random forest (RF) machine learning algorithm to analyze the relationship between urban factors and carbon emissions using real data from Chinese cities [11]. Yan et al. (2021) proposed a new integrated inversion model. This model was used for the intelligent assessment and prediction of water, carbon, and ecological footprint based on integrated multi-task machine learning (MML) and multi-model stack (MMS) algorithms. The accuracy and generalization ability of the model is further explained through the three largest urban agglomerations in the middle reaches of the Yangtze River [12]. Huang et al. (2021) proposed a new method to simulate the dual relationship between emission inventory and pollution concentration for emission inventory estimation [13].

Deep learning has become the latest method and means of studying carbon emissions. At present, most studies use one algorithm to study carbon emissions, and do not consider the combination of multiple algorithms to study carbon emissions. Therefore, the prediction accuracy of the established model is key to measuring whether the algorithm is suitable for carbon emission prediction. Because the long short-term memory (LSTM) neural network is very effective in predicting time-dependent problems, the time factor has a greater impact on carbon emissions. In addition, the problem of carbon emissions is a complex non-linear problem. Additionally, a large amount of data need to be classified and processed. Due to the large error term in the LSTM prediction, slack variables are introduced into the SVR model. Slack variables are introduced to correct the larger prediction errors in the LSTM model, the LSTM-SVR hybrid model is established, and a better prediction effect is achieved. The innovation lies in the indexing of "carbon emissions" information and expanding the area of data availability. Meanwhile, the LSTM is revised by introducing slack variables into the SVR model. This research has a certain reference value for energy-saving development entities in the Yangtze River Economic Zone to implement energy-saving emission reduction and carbon emission refined control forecasts.

2. Related Concepts and Algorithm Analysis

2.1. Yangtze River Economic Belt

The Yangtze River Economic Belt comprises Guizhou, Sichuan, Yunnan, and Chongqing in the western region, and Hubei, Hunan, and Jiangxi provinces in the central region, and Anhui, Jiangsu, Zhejiang and Shanghai in the eastern region, with 11 provinces and cities. The Yangtze River Economic Belt covers an area of approximately 2.05 million square kilometers, has a total population of approximately 600 million, and represents 40% of the country's GDP. It is regarded as a dynamic economic belt, second only to the coastal economic belt [14]. Its rapid economic development has been accompanied by increasingly prominent ecological and environmental problems in the Yangtze River Basin, such as soil erosion, floods, and ecological imbalances along the river. Due to the increasingly prominent environmental resource problems of the Yangtze River, the protection of its natural ecology has become ineffective, restricting the growth rate of the Yangtze River Economic Belt [15]. Figure 1 shows a sketch of the regional locations of provinces and cities in the Yangtze River Economic Belt.

Figure 1. Sketch map of the geographic regions of provinces and cities in the Yangtze River Economic Belt.

2.2. Definition and Analysis of Carbon Sources

2.2.1. Definition

The "source" of greenhouse gases, in layman's terms, is the activity of emitting gases into the atmosphere. "Greenhouse gases" can make the Earth's surface temperature higher, generally by absorbing and re-emitting infrared radiation to play a role [16]. Greenhouse gases mainly include CO_2, ozone (O_3), and methane (CH_4). In addition, this also includes man-made greenhouse gases such as hydrofluorocarbons (HFCs) [17], of which CO_2 has the most obvious warming effect. The warming effects and life cycles of some greenhouse gases are shown in Figure 2.

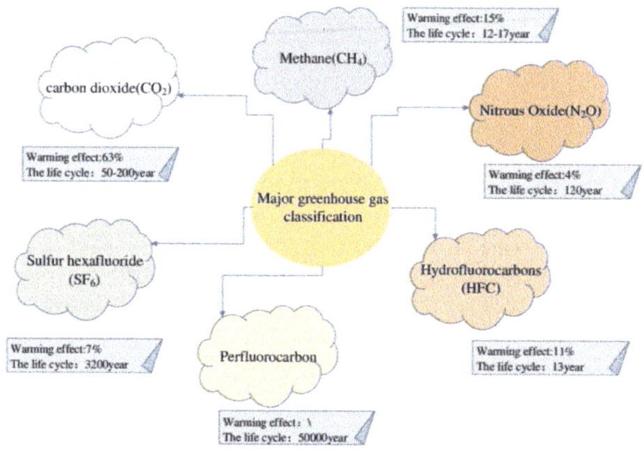

Figure 2. Diagram of the warming effects and life cycles of some greenhouse gases.

2.2.2. Analysis

"Carbon source", as the name implies, is a gas component that increases the amount of CO_2 in the Earth's atmosphere [18]. These gas components enter the atmosphere from the surface of the earth or are formed by the chemical conversion of CO_2 in the atmosphere. Corresponding to the "carbon source" is the "carbon sink". Simply put, the "carbon sink" refers to the mechanism of removing greenhouse gases from the atmosphere, such as by means of the photosynthesis process of plants. From this point of view, the reduction in "carbon sinks" will also lead to an increase in carbon emissions [19]. The specific classification of "carbon sources" is shown in Figure 3.

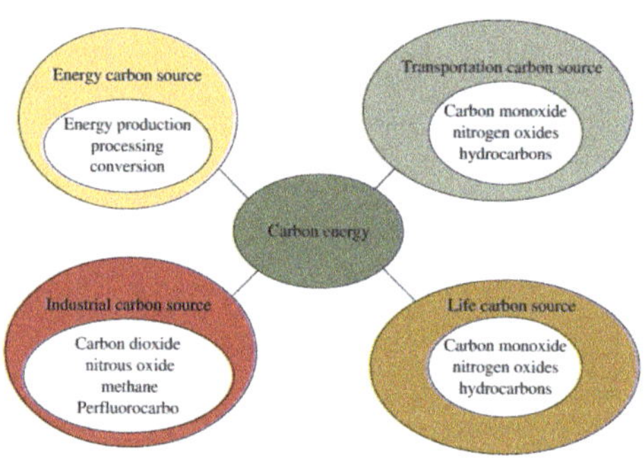

Figure 3. Schematic diagram of the classification of "carbon sources".

2.2.3. Causes of Carbon Emissions

Carbon Emissions in the Process of Urbanization

The factors inducing high carbon emissions in the process of China's urbanization can be divided into two categories. The first is that of economic factors such as the expansion of infrastructure construction, the growth of residents' consumption, and the transformation of land use patterns. The other is the policy incentives that lead to phenomena such as short-lived construction, major demolition, and construction, and low-density urban sprawl. On the one hand, the new construction in the process of urbanization development constitutes an incremental part of carbon emissions. On the other hand, repeated construction and wasted building energy have aggravated high energy consumption and high carbon emissions in the process of urbanization. The combination of economic factors and policy factors in China's urbanization has led to the phenomenon of high carbonization becoming more and more obvious. Firstly, industrial production has brought about an increase in carbon emissions. Rapid industrial development is the main driving force for China's carbon emissions growth. In addition, the embodied carbon in China's export trade plays an important role in the rise of China's carbon emissions. China's exports are dominated by processing trade, with high energy consumption, which is also one of the important factors that constitute China's energy demand growth. Secondly, carbon emissions from the construction industry rapidly increased, and the increase in building areas also brought more carbon emissions. Finally, transportation carbon emissions have rapidly increased, and the increase in transportation demand has led to an upward trend in the total energy consumption of transportation and its share. In recent years, there have been significant changes in China's transportation, road transportation infrastructure, and residents' travel. With the acceleration of urban logistics circulation, the freight capacity of cities and towns has gradually strengthened. The urban expansion in the center will increase the distance traveled by residents, and the level of urban motorization will thus rapidly increase.

Carbon Emissions from Animals and Plants

The respiration of plants and animals produces CO_2. CO_2 generates organic matter and oxygen through photosynthesis. Carbon exists in the form of CO_2 in nature, and plant straw is a biological resource produced in the process of crop production. Burning plant straws will bring about a lot of carbon emissions.

2.3. *Algorithmic Analysis of Carbon Emissions*

Through the classification of "carbon sources", it can be seen that CO_2 emissions involve a wide range and many uncertain factors. The amount of CO_2 cannot be directly obtained through the monitoring instrument. It needs to be calculated by methods. Common methods are subsequently explicated.

2.3.1. Oak Ridge National Laboratory (ORNL)

The Oak Ridge National Laboratory was established by the US Department of Energy in 1943 and is the world's largest scientific energy research laboratory. In 1990, members of the laboratory proposed a method for CO_2 emissions from fossil fuel combustion:

$$\text{Carbon emissions from coal} = \text{coal consumption} \times 0.982 \times 0.733$$
$$\text{Fuel oil carbon emissions} = \text{standard coal equivalent} \times 0.982 \times 0.733 \times 0.813$$
$$\text{Gas carbon emissions} = \text{standard coal equivalent} \times 0.982 \times 0.733 \times 0.561$$

In the equation, 0.982 is the effective oxidation fraction, 0.733 is the carbon content per ton of standard coal, 0.813 means that under the premise of obtaining the same heat energy, the CO_2 released by fuel oil is 0.813 times the CO_2 released by coal, and 0.561 represents that when the same heat energy is obtained. The CO_2 released by natural gas is 0.561 times that of the CO_2 released by coal [20].

2.3.2. Logistic Model

Most economic indicators are increasing functions that change over time. Conditions such as the environment restrict their growth rate and will gradually slow down their growth rate. Most economic indicators show changes in graphs that resemble a flattened S-shaped curve (logistic curve). The relationship between carbon emissions and time is closely resembles an S-shaped curve [21].

2.3.3. System Dynamics Model

The system dynamics Stella software is used to construct the energy consumption model. The energy consumption model constructed by Stella software can obtain a simulation estimation model of energy consumption by inputting conditions such as GDP, population, and the proportion of output value of each industry. This model can effectively overcome errors caused by missing data [22].

The model obtains carbon emissions according to the following equation: carbon emissions = energy consumption x carbon emission factor × (1 − carbon sequestration rate) × oxidation rate.

2.3.4. Input–Output Analysis (IOA)

IOA combines input–output tables and uses mathematical methods to build models. IOA calculates carbon emissions by establishing relationships through models [23]. The principle is as follows:

First: Calculate the carbon emissions of energy consumption in various sectors.

1. Calculate the direct carbon emission coefficient: $ce_j(\text{direct}) = f_j{\cdot}e_j$, where the energy carbon emission coefficient of the j sector is represented by f_j as $\frac{tCO_2}{tce}$, that is, the CO_2 emission per ton of energy, and e_j is the energy consumption intensity of the j sector (10,000 tons of standard coal/CNY 10,000).
2. Calculate the indirect carbon emission coefficient: $ce_j(\text{indirect}) = f_j\left(\sum_{i=1}^{n} a_j b_{ij}\right)$, where the complete consumption of the i-sector product caused by the unit product is represented by b_{ij}. $\sum_{i=1}^{n} a_j b_{ij}$ is the total indirect energy consumption of all n products caused by the unit product of the j sector. The complete carbon emission coefficient of energy consumption is equal to the sum of direct and indirect carbon emission coefficients, expressed as Equation (1):

$$ce_j = f_j{\cdot}e_j + f_j{\cdot}\left(\sum_{i=1}^{n} a_j b_{ij}\right) \tag{1}$$

Second: Calculate the embodied carbon emission coefficient in the production process.

First of all, a prerequisite must be met: the i-th sector's products will produce CO_2 during the production process; then, since the j-th sector consumes the i-th sector's products, it will cause the j-th sector to produce implicit carbon emissions during the production process, which holds Equation (2):

$$\begin{cases} i=j, ce_j=g_j+g_j{\cdot}b_{ij} \\ i\neq j, ce_j=g_j{\cdot}b_{ij} \end{cases} \tag{2}$$

In $g_j = \frac{\omega_i{\cdot}Q_i}{X_i}$, ω_i represents the carbon emissions from the industrial production process of the unit physical quantity product in the i sector (tons of CO_2/ton of product), and Q_i represents the product output of the i sector (10,000 tons); X_i represents the total output of the i sector (CNY 10,000). Combining the two equations yields $ce_j = g_j{\cdot}c_{ij}$.

Therefore, Equation (3) is the complete carbon emission coefficient of each sector:

$$ce_j(\text{completely}) = f_j{\cdot}\left(\sum_{i=1}^{n} e_i{\cdot}b_{ij}\right) + g_j{\cdot}c_{ij} \tag{3}$$

On Equation (3), the total carbon emissions in the air can be calculated, as shown in Equation (4):

$$CE = \sum_{j=1}^{n} ce_j{\cdot}Y_j \tag{4}$$

(Y_j is the final use of the department (CNY 10,000).)

The CO_2 estimate provided in the IPCC National Greenhouse Gas Inventory Guidelines for energy-based CO_2 is shown in Equation (5):

$$CO_2 = \sum_{i=1}^{n} CO_2 = \sum_{i=1}^{n} E_i \times NCV_i \times CEF_i \times COF_i \times (44/12) \tag{5}$$

In Equation (5), CO_2 is the estimated amount of CO_2 emissions; i represents the estimated i-th energy; E_i represents each energy consumption; NCV_i is the average low calorific value of each energy source; CEF_i represents the carbon emission factor per calorific value provided by the IPCC Greenhouse Gas Inventory; COF_i stands for carbon oxidation factor. The molecular coefficient is 44/12 units of CO_2, which represents the amount of CO_2 that 1 unit of carbon element can be converted into [24].

2.4. Temporal and Spatial Characteristics of Carbon Emissions

According to relevant data, the industrial structure, population size, economic development level, opening-up level, carbon emission intensity, etc. are all factors that affect the scale of carbon emission [25]. Carbon emissions have formed a time–space effect along with changes in time series and spatial sequences. When studying carbon emissions, it is necessary to consider the following: 1—spatial connections and changes in different economic regions; and 2—carbon emission prediction requires analysis and research on the spatial characteristics of carbon emission.

2.5. Analysis of Deep Learning Algorithms

(1) Recurrent Neural Network (RNN)

RNN is a neural network with cyclic characteristics. It can perform calculations on the characteristics of carbon emission time series data because it can continuously circulate information and has a short-term information memory function [26]. The most basic RNN network is composed of multiple neuron nodes, and each node has an activation function with time as a variable to enhance adaptability. Meanwhile, all function parameters of the node can be adjusted in real time. The expanded diagram of the RNN network structure is shown in Figure 4.

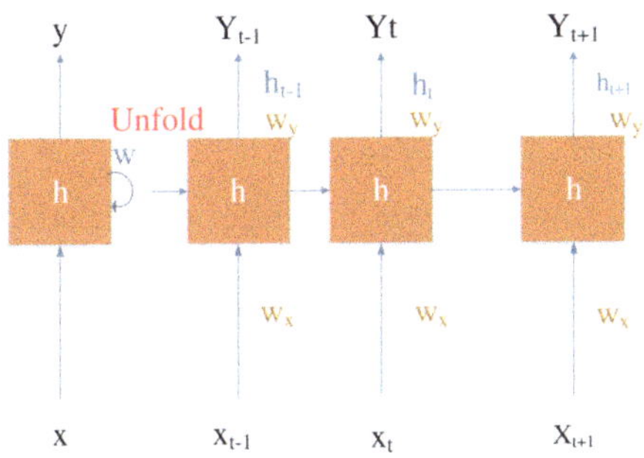

Figure 4. Expanded schematic diagram of an RNN network structure.

In Figure 4, h_t represents the output of the hidden layer node at time t; h_{t-1} represents the output at the previous moment; y_t represents the output vector; x_t represents the input vector; h_w, x_w, and y_w all represent the weight vector of the hidden layer neuron node and the next hidden layer neuron node, respectively.

(2) Long Short-Term Memory (LSTM)

LSTM is a classic variant of RNN. LSTM has powerful classification and prediction capabilities and can handle operations with relatively long-time intervals and delays. The important thing is that LSTM can solve the problem of gradient disappearance and gradient explosion during long-sequence training [27].

The LSTM neural network structure uses a gate control unit. The neuron of each cell contains a forget gate, input gate, and output gate to strengthen the network structure, as shown in Figure 5.

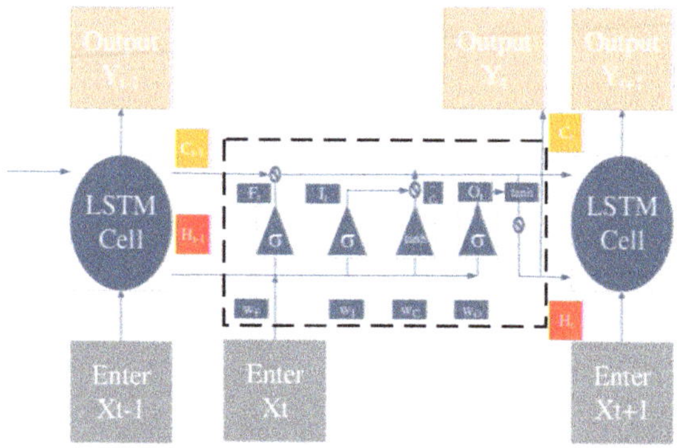

Figure 5. Schematic diagram of an LSTM neural network structure.

In the network structure diagram of LSTM, the cell state similar to the conveyor belt runs on the entire chain. If the cell state undergoes a slight linear operation, the information flows through the entire chain and remains stable. The gate is a weight composed of a sigmoid function, a tanh function, and a point multiplication operation to realize the selection of information.

Demonstrate the operation according to the structure of Figure 5 LSTM.

1. Forgetting door calculation:

The forgetting door decides whether to leave the information, as shown in Equation (6). In the information entry gate, the information status is read under sigmoid, and the output is a value between 0 and 1—where 1 means complete retention and 0 means complete deletion, as shown in Equation (7):

$$F_t = \sigma(w_F[h_{t-1}, x_t] + b_F) \tag{6}$$

$$S(t) = \frac{1}{1+e^{-t}} \tag{7}$$

2. Input gate calculation:

The input gate determines how much new information enters, as shown in Equation (8). There are two steps: the first step is to enter the gate to determine the new information that is allowed to enter the cell; the second step is to obtain the candidate information that needs to be remembered at tanh, as shown in Equation (9):

$$I_t = \sigma(\mathrm{w}_I \cdot [h_{t-1}, x_t] + b_I) \tag{8}$$

$$\widetilde{C}_t = \tanh(\mathrm{w}_c \cdot [h_{t-1}, x_t]) \tag{9}$$

3. Cell status update:

The cell state that C_{t-1} is updated to C_t is to multiply the last state value C_{t-1} by F_t, discard the unnecessary part, and add the value that allows it to be remembered and multiplied by it. Finally, the information C_t that the update wants to add to the unit state is obtained, as shown in Equation (10):

$$\underline{C}_t = F_t * C_{t-1} + I_t{}^* \widetilde{C}_t \tag{10}$$

4. Output gate

The output gate determines the information to be output from the cell state, as shown in Equation (11). By activating the sigmoid function, the cell state information output is determined. The tanh function is used to process the final output information h_t of the cell state at the last current moment, as shown in Equation (12):

$$O_t = \sigma(w_O \cdot [h_{t-1}, x_t] + b_O) \tag{11}$$

$$h_t = O_t * \tanh(C_t) \tag{12}$$

In Equations (11) and (12), F_t, I_t, and O_t represent the calculation of the forget gate, input gate, and output gate at time t, respectively; C_{t-1} represents the cell state at the previous moment; C_t represents the cell state at time t; x_t represents input information; h_{t-1} represents the output at the previous moment; w_F, w_I, and w_O represent the weight vectors of the forget gate, input gate, and output gate, respectively; b_F, b_I, and b_O represent the bias vectors of the forget gate, input gate, and output gate, respectively; σ (•) represents the activation function sigmoid function. The entire forward calculation of the LSTM unit cell is completed through three gates and one cell state.

3. Indicator Creation and Model Design

3.1. Creation of Carbon Emission Information Indicators

After analyzing the industry characteristics of the Yangtze River Economic Belt, the proposed method derives key factors such as industry investment scale and labor output efficiency, which affect carbon emissions in the "economic belt". Then, the extensible random environmental impact assessment (STIRPAT) model is optimized to determine the information indicators:

(1) ZB1 is used to represent the total carbon emissions in the "Yangtze River Economic Belt" basin, with a unit of 10,000 tons;

(2) ZB2 is used to represent the investment scale of the industry, combined with the research, the sum of fixed assets, and the current assets of enterprises above the designated size used as measurements, with a unit of CNY 100 million;

(3) ZB3 is used to represent industrial economic efficiency, and labor efficiency is used as output as a measure;

(4) ZB4 is used to represent carbon emission intensity, that is, CO_2 emissions per unit of industrial added value;

(5) ZB5 is used to represent the scale of opening up to the outside world, and it is measured by the proportion of the sum of investments from Hong Kong, Macao, and Taiwan in addition to foreign investment in the industrial added value, and the unit is %;

(6) ZB6 is used to represent the intensity of environmental protection, and the investment in environmental pollution control is expressed as the proportion of industrial GDP, for which the unit is % [28].

The expanded model indicator variable list is shown in Table 1.

Table 1. Carbon emission information forecast and expansion model indicators.

Indicator Definition	Representative Symbol	Index Quantification Unit
Total carbon emissions of provinces and cities in the Yangtze River Economic Belt	ZB1	Ten thousand tons
Industry investment scale	ZB2	CNY 100 million
Labor efficiency output	ZB3	CNY 100 million /10,000 people
Carbon intensity	ZB4	10,000 tons/10,000 people
Openness to the outside world	ZB5	%
Environmental protection	ZB6	%

3.2. Carbon Emission Modeling Process

The problem of carbon emission prediction is actually predicting carbon emission information in the future based on the information of carbon emission indicators in the past.

Therefore, this study is actually on the relationship between a set of time series containing characteristic data to perform regression prediction on the target value.

Regardless of whether it is for regression algorithms or classification algorithms, data need to be preprocessed before the model is built, especially for practical research on multi-dimensional data. In response to the problem to be dealt with, in the selection of data features, the six indicators in Table 1 are used to predict carbon emission information. Because of the various characteristics of the original data in different dimensions, it needs to perform standardized preprocessing operations on the data.

(1) Data normalization method:

The min–max standardization is also called the maximum value normalization, which is a linear transformation of the original data, and the result value is mapped to between [0,1]. The expression Equation of the normalization method of the data is as in Equation (13):

$$x_{\text{scale}} = \frac{x - x_{\text{min}}}{x_{\text{max}} - x_{\text{min}}} \tag{13}$$

Maximum normalization is to calculate the maximum and minimum values of each dimension data and then convert the original data. However, maximum normalization has its limitations. If the data change, their maximum and minimum values need to be recalculated. In addition, the maximum normalization is extremely susceptible to extreme values. Therefore, maximum normalization is more suitable for processing boundary data [29].

(2) Normalization of mean variance:

$$x_{\text{scale}} = \frac{x - x_{\text{mean}}}{\sigma} \tag{14}$$

Mean variance normalization is a common method for preprocessing data. The essence of mean variance normalization is to calculate the mean and standard deviation of the data, so that the original data obey a normal distribution with a mean value of 0 and a standard deviation of 1 [30].

The focus of the evaluation model is to divide the data set that the research has collected. The data can be divided into three sets: training set, validation set, and test set [31]. The training set is passed into the model for model fitting, and the model parameters are optimized on the validation set, and then the model is evaluated. When the model works well, this means that the experiment has found the best model parameters, and then uses the test set for model testing.

Common indicators for evaluating the prediction accuracy of regression models are as follows:

(a) The MSE is the sum of squares of the difference between the results of the original feature data predicted by the model and the real results, but the sum of squares will continue to accumulate as the number of samples increases. In order to eliminate the influence of the number of samples, the mean value of the square error is calculated, and the MSE is obtained, as shown in Equation (15):

$$MSE = \frac{1}{N}\sum_{i=1}^{N}|y_i - y_i'|^2 \tag{15}$$

(b) The average absolute error (MAE) is the average of the absolute value of the difference between the predicted value and the true value, as shown in Equation (16):

$$MAE = \frac{1}{N}\sum_{i=1}^{N}|y_i - y_i'|^2 \tag{16}$$

(c) RMSE is shown Equation (17):

$$RMSE = \sqrt{MSE} = \sqrt{\frac{1}{N}\sum_{i=1}^{N}\left|y_i - y_i'\right|^2} \tag{17}$$

(d) The absolute error of the median, the absolute value of the difference between the predicted value, and the true value are not averaged, but the median is taken, which is MedAE, as shown in Equation (18):

$$MedAE = median_{i=1,\ldots,\mathrm{N}}\left|y_i - y_i'\right| \tag{18}$$

For the median absolute error index, because the expression contains the absolute value, it is necessary to derive the loss function of the model, and the absolute value index usually fails [32].

3.3. Data Source

The data of this experiment were obtained according to the online data query menu system on the official website of the National Bureau of Statistics (https://data.stats.gov.cn) (accessed on 11 December 2021). The database of the National Bureau of Statistics collected data from 2018 to 2020, and the monitoring interval was monthly. The relevant index data obtained are mainly data on the 11 provinces and cities in the Yangtze River Economic Belt. The data collected from the official website of the National Bureau of Statistics were used as the training set and test set of the carbon emission prediction model proposed; the data collected by the official statistical bureaus of local provinces and cities were used as the verification data set for demonstration applications.

Table 2 shows the descriptive data of carbon emissions in the Guizhou and Jiangsu provinces from 2000 to 2020.

Table 2. Carbon emissions in the Jiangsu and Guizhou provinces during the period 2000–2020.

Guizhou Province		Jiangsu Province	
Year	**Carbon Emissions ($\times 10^8$ t) Year**	**Year**	**Carbon Emissions ($\times 10^8$ t)**
2000	1.845820987	2000	5.521659376
2001	1.50698462	2001	5.864446639
2002	1.622689431	2002	5.866892432
2003	2.079300127	2003	7.11895019
2004	2.422275528	2004	7.12083157
2005	2.765439067	2005	7.123089224
2006	2.767320446	2006	7.352053055
2007	2.767320446	2007	7.581581299
2008	2.656319082	2008	8.038003857
2009	3.226565072	2009	8.494990828
2010	3.455905178	2010	9.065236819
2011	3.45759842	2011	9.521471238
2012	3.800573821	2012	9.410469874
2013	4.257184516	2013	8.957621937
2014	4.259254033	2014	8.846244297
2015	4.261135412	2015	9.18903156
2016	4.3772165	2016	9.532195099
2017	4.492545036	2017	9.8751705
2018	4.492921311	2018	11.467946
2019	4.608249847	2019	13.51563896
2020	4.950848972	2020	14.54004986

3.4. SVR Machine Model Creation

Support vector machine (SVM) is a powerful machine learning algorithm [33]. SVM can solve classification and regression problems at the same time. In addition, SVM can handle both supervised learning target variables and unsupervised learning without target variables. Additionally, its application scenarios are very rich, which can be used for binary classification problems, multi-classification problems, linear and nonlinear problems, etc.

The problem being studied is essentially to predict the regression problem. The application of SVM to regression is also called support vector regression (SVR) [34]. SVR is achieved by adding an insensitive loss function to SVM. It extends the classification problem to the regression problem, finds an error, and makes all the sample points as far as possible within this error to achieve a prediction of the data.

SVR is an application of SVM in the field of regression. Its principle is to obtain a regression model (Equation (20)) on the known sample set (Equation (19)):

$$D = \{(x_1, y_1), (x_2, y_2), \ldots, (x_k, y_k), x_i \in R^n, y_i \in R\} \tag{19}$$

$$f(x) = \omega^T x + b \tag{20}$$

Make $f(x)$ and y as close as possible. Among them, w and b are the parameters to be determined in the model; w is the normal vector of the hyperplane; and b is the displacement term. For general regression problems, only when $f(x)$ and y are exactly equal will the loss of the model be zero. In the SVR model, a certain degree of tolerance deviation ε is given, so that if and only when the absolute value of the difference between $f(x)$ and y is greater than the tolerance deviation ε, it is considered as a loss. At this time, it is equivalent to taking $f(x)$ as the center to construct an isolation band with a width of 2ε, as shown in Figure 6.

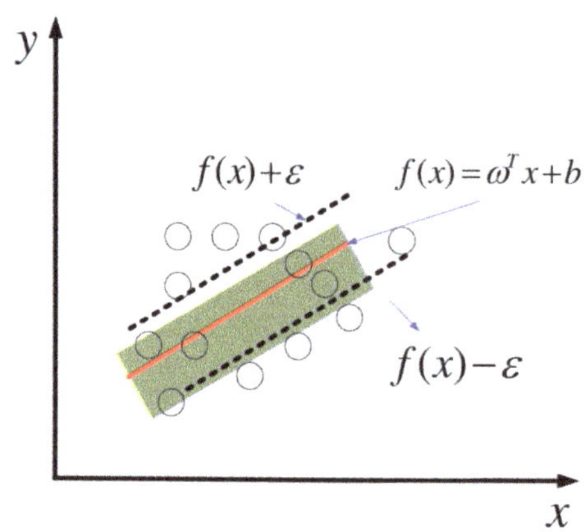

Figure 6. Schematic diagram of SVR.

The problem of the SVR model can be transformed into:

$$\min_{w,b} \frac{1}{2} \parallel \omega \parallel^2 + C \sum_{i=1}^{m} l_s(f(x_i) - y_i) \tag{21}$$

$$l_\varepsilon = \begin{cases} 0, & \text{if} |z| < \varepsilon \\ |z| - \varepsilon, & \text{other} \end{cases} \tag{22}$$

where C is the regularization constant and $l\varepsilon$ is the ε-insensitive loss function in Figure 6. With the introduction of relaxation factors $\xi_i^{\vee}$ and $\xi_i^{\wedge}$, Equations (21) and (22) can be rewritten as

$$\min \frac{1}{2} \parallel \omega \parallel^2 + C \sum_{m-1}^{N} (\xi_m^{v} + \xi_m^{\wedge}) \tag{23}$$

$$s.t. -\varepsilon - \xi_i^{\vee} \leq y_i - \omega \cdot \phi(x_i) - b \leq \varepsilon + \xi_i^{\wedge} \tag{24}$$

$$\text{s.t.} \quad \begin{array}{c} f(x_i) - y_i \leq \varepsilon + \xi_i^{\vee}, \\ y_i - f(x_i) \leq \varepsilon + \xi_i^{\wedge} \\ \xi_i^{\vee} \geq 0, \xi_i^{\wedge} \geq 0, i = 1, 2, \ldots, m \end{array} \tag{25}$$

SVM aims to solve two-classification problems. The actual problem is often a complex nonlinear problem, and the "dimension increase" is used to deal with the nonlinearity between data [35]. The dimensional data are converted and mapped to a high-dimensional space, and then converted into a low-dimensional space after the high-dimensional space is classified. The purpose of introducing the kernel function can solve this kind of conversion operation. Using the SVR machine to make predictions, different kernel functions are selected for modeling, and the differences between the kernel functions are compared. The kernel function categories and characteristics are shown in Figure 7.

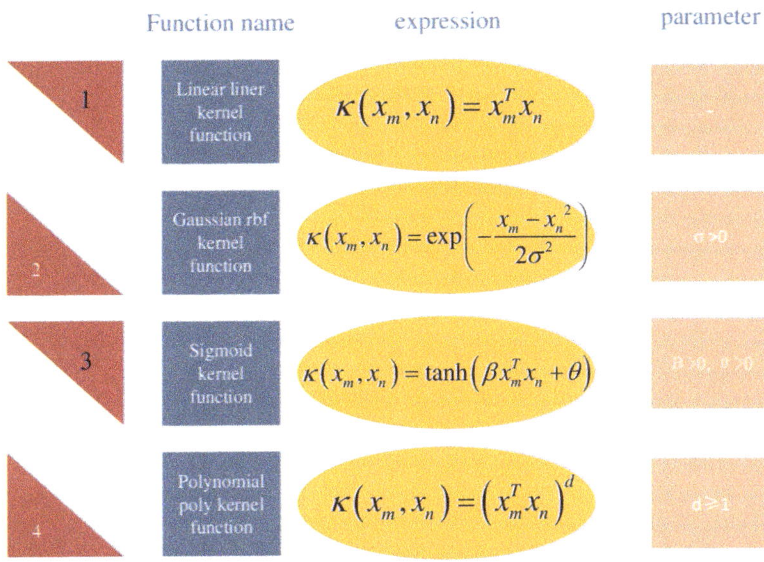

Figure 7. Schematic diagram of kernel function names and characteristic parameters.

Using the duality principle [36] and introducing the kernel function, the SVR model is obtained, as shown in Equation (26):

$$f(x) = \sum_{i=1}^{m} (\hat{\alpha}_i - \alpha_i) k(x_i, x) + b \tag{26}$$

3.5. LSTM-SVR Hybrid Model Construction

Carbon emission (CO_2) concentration data have the characteristics of time series and non-linearity. The LSTM-SVR hybrid model is proposed to improve the prediction accuracy of CO_2 concentration. Using the LSTM-SVR hybrid model to predict the CO_2 concentration, the specific steps are as follows:

(1) Acquisition of CO_2 concentration data and meteorological factor data;

(2) Preprocessing the acquired data to eliminate errors or abnormal factors in the data;

(3) Use the LSTM model to train and predict the processed data to generate a set of corresponding prediction values $\widetilde{D_t}$;

(4) By making the difference between the processed data *Dt* and the predicted value $\widetilde{D_t}$, the error value e_t at time t can be obtained;

(5) The SVR model is used to perform regression prediction on the error value e_t at time t, and cross-validation and grid search algorithms are used to find the optimal kernel function parameter g and penalty factor c of the SVR model; the predicted value is obtained, that is, the error value e_t is corrected, and the corrected error value is $\hat{e}_t$;

(6) The corrected error value $\hat{e}_t$ is combined with the predicted value $\widetilde{D_t}$ of LSTM, and finally the predicted value D_t^*, $D_t^* = \hat{e}_t + \widetilde{D_t}$ of the mixed model is obtained. The model framework is shown in Figure 8.

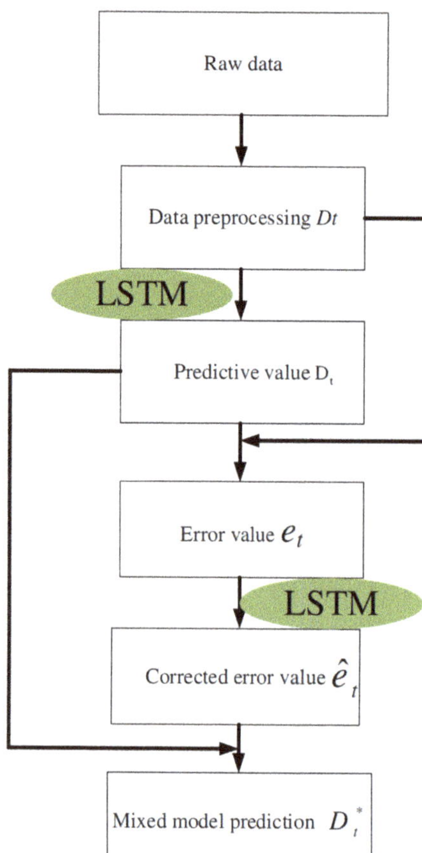

Figure 8. LSTM-SVR hybrid model framework diagram.

4. Experimental Results and Analysis

4.1. SVR Model Fitting Results

The SVR model is used to perform non-linear regression classification and fitting on the carbon emission-related information of all provinces and cities in the "Yangtze River Economic Belt" basin. Under different kernel functions, the fitting results of the model are shown in Figures 9–13.

The experiment finally obtained the RMSE of the test set to be 0.715. This evaluation index is the result of calculating the normalized data. The fitting result of the Gaussian kernel function is shown in Figure 10.

Figure 11 indicates that the phenomenon of over-fitting is more serious, and the function parameters need to be adjusted. After adjustment, the training set score of the final model is 0.663, the test set score is 0.634, and the RMSE of the test set is 0.682.

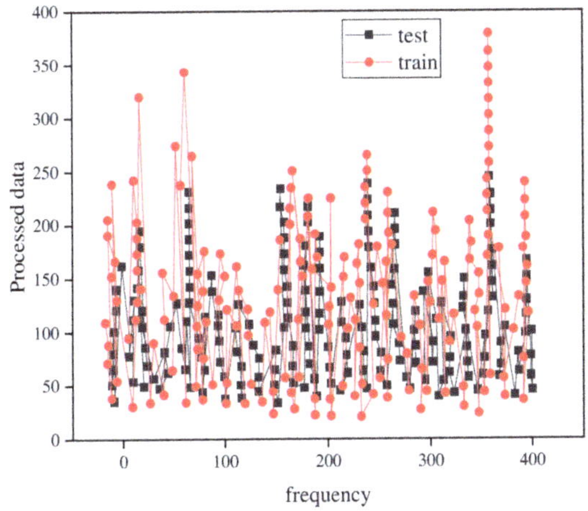

Figure 9. Schematic diagram of linear kernel function fitting results.

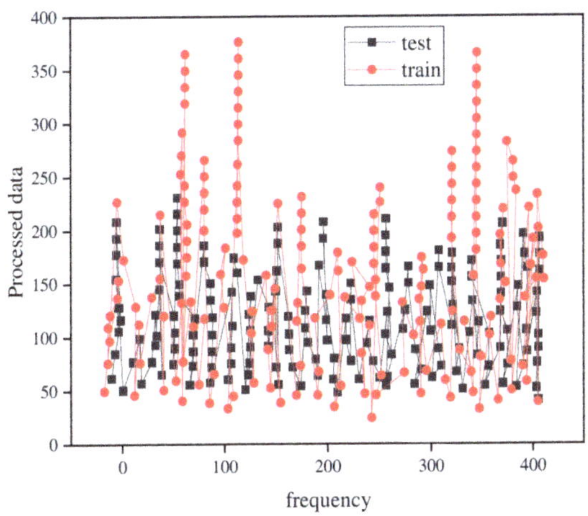

Figure 10. Schematic diagram of the Gaussian kernel function fitting results.

Figure 11 reveals that the fitting effect of the sigmoid kernel function is relatively poor, and the scores of the test set and the training set are both negative and relatively low. Meanwhile, the experiment needs to adjust the parameters to achieve a reasonable fitting result.

Figure 12 proves that the performances of the test set and the training set are different, that the training set score is reasonable, and that the test set is low. In all the above figures, the fitting results of the different kernel functions are different, and the specific function fitting score and the RMSE comparison are shown in Figure 13.

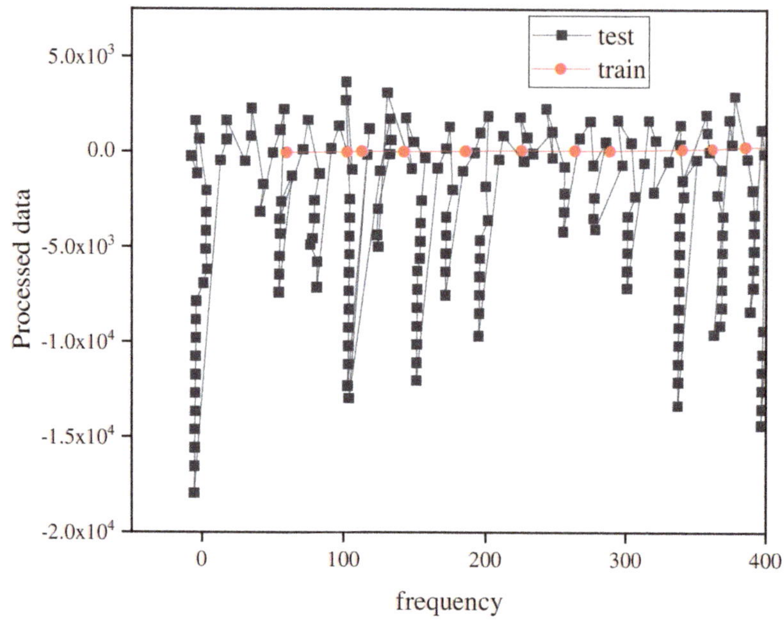

Figure 11. Schematic diagram of the sigmoid kernel function fitting results.

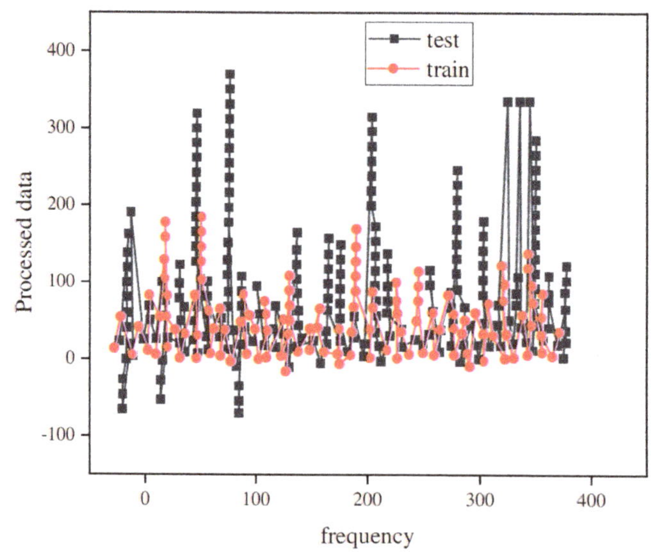

Figure 12. Schematic diagram of the polynomial kernel function fitting results.

Figure 13 visually shows that the training performance scores of the four different kernel functions of the model are relatively small. This is due to the constant use of network search in the process of model training to adjust the parameters of the model.

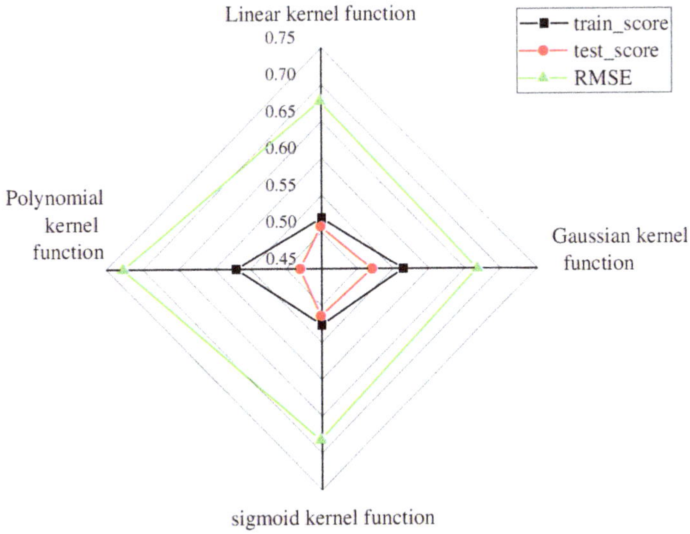

Figure 13. Comparison of the model performance of four types of kernel functions.

4.2. Deep Learning Network Model Training Results

This experiment used the LSTM neural network to train the model. After 60 iterations, the loss of the model and its performance on the training and test sets were obtained—these are shown in Figure 14.

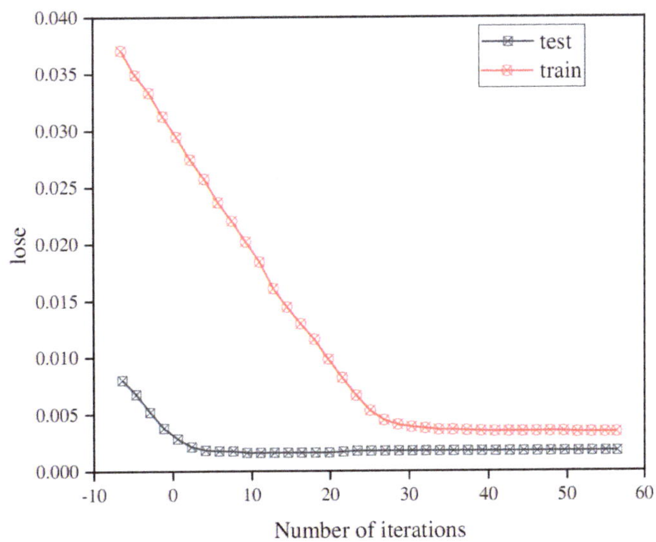

Figure 14. Comparison of the training and testing losses for multiple iterations of LSTM.

Figure 14 shows that the loss gradually stabilizes after multiple iterations. Figure 15 shows the fitting trend graph of the model trained by the LSTM neural network.

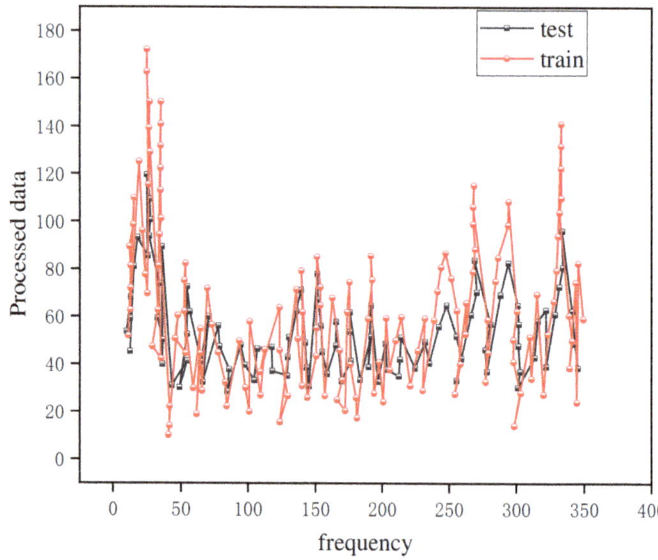

Figure 15. Schematic diagram of the fitting trend results after LSTM training.

4.3. Carbon Emission Forecast Results

According to the SVR model and the LSTM training model, the experimental training on the experimental data set, the comparison between the prediction curve of the carbon emission information-related indicators in the Yangtze River Economic Belt and the real curve is shown in Figures 16–18.

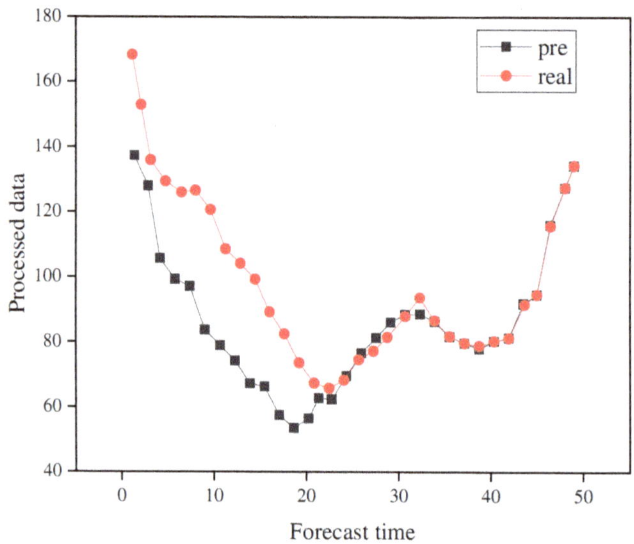

Figure 16. Schematic diagram of the comparison between the industry investment scale information and "Yangtze River Economic Belt" carbon emission information prediction and the actual comparison.

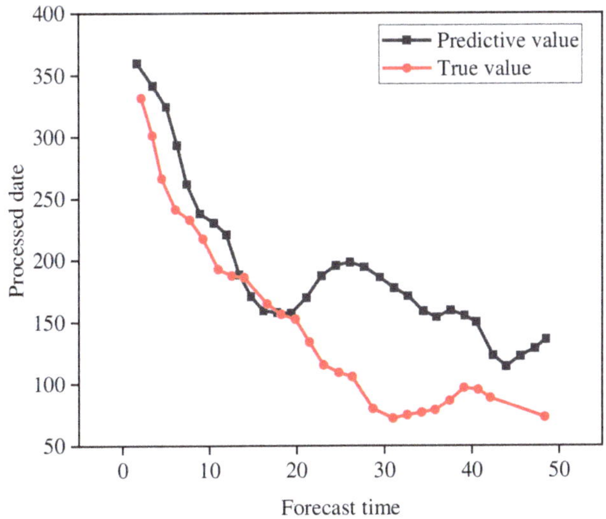

Figure 17. Schematic diagram of the comparison between the labor efficiency output information and carbon emission information prediction and the actual comparison of the "Yangtze River Economic Belt".

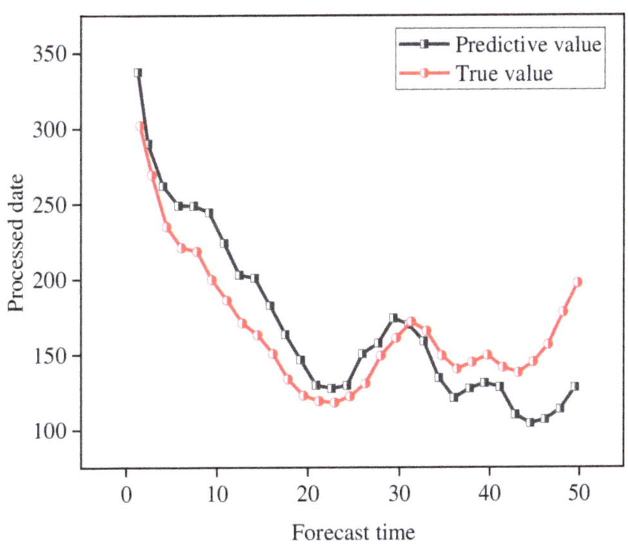

Figure 18. A schematic diagram of the comparison between the scale of carbon emissions information and the carbon emissions information prediction and the actual comparison of the "Yangtze River Economic Belt".

In Figure 16, the scale of the industry investment is used to predict carbon emissions. There are consistent overall trends. This also reflects the fact that the economic development and growth of the Yangtze River Economic Belt will inevitably increase carbon emissions.

Figure 17 suggests that the labor efficiency output information is not very accurate for the prediction of information on carbon emissions, and that there are certain errors. The reason for this is that the labor efficiency output has spatial lag.

In Figure 18, the information on the scale of carbon emissions is more accurate in predicting the trend of carbon emissions. In summary, in the combination of the SVR model and the LSTM model proposed herein, the relevant index information for predicting carbon emissions in the Yangtze River Economic Zone is relatively accurate. Meanwhile, the ZB1 and ZB2 indicator information play a key role in predicting carbon emissions.

4.4. Policy Recommendations

In response to the proposed factors influencing carbon emissions and the prediction results of the carbon emissions model, the following policy recommendations are proposed as follows, taking account of China's specific national conditions.

1. Strengthen carbon market capacity building:

Chinese central state-owned enterprises have the responsibility and obligation to implement and fulfill the state's requirements in terms of corporate, political, and social responsibility. Chinese central state-owned enterprises should actively implement the national climate change policy requirements. They also should take the lead in promoting research into carbon asset management, low-carbon development strategies and paths, and complete the country's binding targets for greenhouse gas emissions.

This requires Chinese central state-owned enterprises to interpret carbon emission policies. Meanwhile, they also need to study the extent of their influence on the company's corporate development strategy selection, production decision making, technological progress, energy conservation, and environmental protection. Furthermore, companies should explore the impact of carbon emission policy structure on the behavior of industrial enterprises, and study the relationship between carbon emission policy and corporate competitiveness. Additionally, companies should also make full use of China's carbon emission policies to reduce the cost of reducing carbon emissions while creating a green and clean energy company, effectively increasing the value of carbon assets, and preparing for the cultivation of new industries in the future.

2. Strengthen corporate carbon asset management:

Only three (sub-)sectors are included in the first batch of the unified national carbon market. The pilot carbon emission control companies that were previously included in the pilot areas will also be included (including the Tianjin branch of CNOOC (China) Co., Ltd.).

With the improvement in the carbon market and the increasing pressures of the conditions for implementing a carbon tax, no company will be beyond reducing its carbon emissions. Therefore, it is particularly important to do a good job in terms of corporate carbon asset management in advance. A company's carbon emissions should be known as soon as possible by utilizing inventory and verification. Companies should make emission reduction or response measures as early as possible following relevant policies and development trends to minimize the impact of carbon emissions on the company. Companies with surplus carbon emission allowances can also strive for additional benefits for these companies.

3. Properly assess the impact of new projects on carbon emissions:

Carbon emission assessment projects are a new assessment method that has been gradually developed in recent years. Regardless of whether it is the carbon market or the implementation of a carbon tax, for companies with high carbon emissions, it will affect the production and operation of the company to a certain extent. For new or renovated and expanded projects, it is recommended that carbon emission assessment is carried out in the preliminary research stage of the project. The main purposes of the assessment are: 1—predict the carbon emission cost of investment projects and assess the degree of impact of carbon emissions on the economics of the project; and 2—enable a better-informed choice of measures and paths to meet carbon emission requirements through the assessment.

4. Deploy large-scale carbon emission reduction technologies as early as possible:

CO_2 emission reduction is a long-term and arduous task. Soon, work can be carried out on energy optimization, production energy conservation, and the development and utilization of clean energy by tapping the company's internal emission reduction potential. However, more CO_2 emission reduction and utilization methods are needed to achieve long-term and effective emission reduction, i.e., "post-processing" technology. Therefore, it is necessary to deploy carbon emission reduction utilization technology research and development as soon as possible, and make full preparations for emission reduction technologies and programs. While fulfilling the goals required by the state, this will also enhance the core competitiveness of the enterprise and promote its low-carbon transformation as well as its sustainable development.

In summary, in the process of carbon emission management, due to the complexity of the factors influencing carbon emission, it is a great challenge to carry out forecasting, and it will also consume a lot of human effort and material resources. The use of deep learning to build a carbon emission prediction model can predict the carbon emissions of the Yangtze River Economic Zone with high accuracy, reduce the human and material investment in carbon emission management, and provide a reference for carbon emission management.

5. Conclusions

According to the model test results, the "Yangtze River Economic Belt" basin and the industry investment scale, the labor efficiency output, carbon emission intensity, and other indicators that affect carbon emissions are relatively accurate in the carbon emission information forecast [37]. Therefore, the proposed method concludes that the SVR model for solving complex nonlinear problems can achieve a relatively excellent prediction effect under the training of LSTM. Due to the complexity of the "carbon sources" of the carbon emissions of the research object, the information affecting carbon emissions has the characteristics of diversity, dynamics, and big data. On the other hand, deep learning algorithms have strong fusion and changeable algorithms. The use of a deep learning network to process the information of the prediction model is complicated, which is main shortcoming of this study [38]. Therefore, the study hereby puts forward expectations that the prediction of carbon emission information is crucial to the country's mid-to-long-term "carbon peak" strategy. The deep learning network must be used to accurately predict carbon emissions within a specific economic region, and then be promoted nationwide. This process requires the concerted efforts of researchers from related fields to work together.

Author Contributions: Conceptualization, H.H. and X.W.; methodology, X.W.; software, X.W.; formal analysis, H.H.; investigation, X.C.; resources, X.C.; data curation, H.H.; writing—original draft preparation, H.H.; writing—review and editing, X.C.; visualization, X.C.; supervision, X.C.; project administration, X.C. All authors have read and agreed to the published version of the manuscript.

Funding: This research was supported by the National Natural Science Foundation of China (Grant No. 41271516).

Institutional Review Board Statement: The study was conducted according to the guidelines of the Declaration of Helsinki, and approved by the Institutional Review Board of Anhui Normal University.

Informed Consent Statement: Informed consent was obtained from all individual participants included in the study.

Data Availability Statement: The raw data supporting the conclusions of this article will be made available by the authors, without undue reservation, to any qualified researcher.

Conflicts of Interest: The authors declare no conflict of interest.

References

1. Xie, H.; Zheng, X.T.; Long, S.M. The characteristics and mechanism of the rapid and slow changes of the equatorial Pacific thermocline under the background of global warming. *J. Ocean Univ. China (Nat. Sci. Ed.)* **2021**, *51*, 12–19.
2. Chen, Y.M.; Wang, Z.T.; Luo, H.L.; Qin, Y. Research progress in biofuel production from microalgae. *Chem. Eng. Technol.* **2021**, *11*, 10.
3. Hong, J.K.; Li, Y.C.; Cai, W.G. Simulation of China's carbon peak path from a multi-scenario perspective—By the RICE-LEAP model. *Resour. Sci.* **2021**, *43*, 639–651.
4. Jalaee, S.A.; Shakibaei, A.; Akbarifard, H.; Horry, H.R.; GhasemiNejad, A.; Robati, F.N.; Derakhshani, R. A novel hybrid method by cuckoo optimization algorithm and artificial neural network to Forecast world's CO_2 emission. *MethodsX* **2021**, *8*, 101310. [CrossRef]
5. Shivam, K.; Tzou, J.C.; Wu, S.C. A multi-objective predictive energy management strategy for residential grid-connected PV-battery hybrid systems by machine learning technique. *Energy Convers. Manag.* **2021**, *237*, 114103. [CrossRef]
6. Kee, K.K.; Lim, Y.S.; Wong, J.; Chua, K.H. Impact of nonintrusive load monitoring on CO_2 emissions in Malaysia. *Bull. Electr. Eng. Inform.* **2021**, *10*, 1803–1810. [CrossRef]
7. Ning, L.; Pei, L.; Li, F. Forecast of China's Carbon Emissions by ARIMA Method. *Discret. Dyn. Nat. Soc.* **2021**, *21*, 1–12. [CrossRef]
8. Xie, B.C.; Tan, X.Y.; Zhang, S.; Wang, H. Decomposing CO_2 emission changes in thermal power sector: A modified production-theoretical approach. *J. Environ. Manag.* **2021**, *1*, 111887. [CrossRef] [PubMed]
9. Liu, X.; Meng, X.; Wang, X. Carbon Emissions Prediction of Jiangsu Province by Lasso-BP Neural Network Combined Model. *IOP Conf. Ser. Earth Environ. Sci.* **2021**, *769*, 17–22. [CrossRef]
10. Acheampong, A.O.; Boateng, E.B. Modelling carbon emission intensity: Application of artificial neural network. *J. Clean. Prod.* **2019**, *225*, 833–856. [CrossRef]
11. Wang, Z.; Zhao, Z.; Wang, C. Random forest analysis of factors affecting urban carbon emissions in cities within the Yangtze River Economic Belt. *PLoS ONE* **2021**, *16*, e0252337. [CrossRef]
12. Yan, P.; Lu, H.; Chen, Y.; Li, Z.; Li, H. A stack-based set inversion model for smart water, carbon and ecological assessment in urban agglomerations. *J. Clean. Prod.* **2021**, *319*, 128665. [CrossRef]
13. Huang, L.; Liu, S.; Yang, Z.; Xing, J.; Zhang, J.; Bian, J.; Li, S.; Sahu, S.K.; Wang, S.; Liu, T.-Y. Exploring Deep Learning for Air Pollutant Emission Estimation. *Geosci. Model Dev. Discuss.* **2021**, *14*, 4641–4654. [CrossRef]
14. Rai, P.; Shukla, G.; Manohar, K.; Bhat, J.A.; Kumar, A.; Kumar, M.; Cabral-Pinto, M.; Chakravarty, S. Carbon storage of single tree and mixed tree dominant species stands in a reserve forest—Case study of the Eastern Sub-Himalayan Region of India. *Land* **2021**, *10*, 435. [CrossRef]
15. Mohammed, S.; Gill, A.R.; Alsafadi, K.; Hijazi, O.; Yadav, K.K.; Hasanf, M.A.; Khan, A.H.; Islamf, S.; Cabral-Pintoh, M.S.M.; Harsanyi, E. An overview of greenhouse gases emissions in Hungary. *J. Clean. Prod.* **2021**, *314*, 127865. [CrossRef]
16. Zhang, P.; Zhu, X.; He, Q.Y. Analysis on the temporal and spatial differentiation and balance pattern of ecosystem service supply and demand in the Yangtze River Economic Belt. *Ecol. Sci.* **2020**, *39*, 155–166.
17. Huang, Z.X.; Wang, F.F.; Cao, W.Z. Analysis and prediction of factors affecting the level of ecological civilization in the Yangtze River Economic Zone—By VAR, GWR-BP neural network combined model. *Econ. Geogr.* **2020**, *40*, 199–209.
18. Xia, L.L.; Yan, X.Y.; Cai, Z.C. Research progress and prospects of greenhouse gas emission reduction and organic carbon fixation in farmland soils in my country. *J. Agric. Environ. Sci.* **2020**, *39*, 178–185.
19. Rubin, E.S.; Rao, A.B. A technical, economic, and environmental assessment of amine-based CO_2 capture technology for power plant greenhouse gas control. *Environ. Sci. Technol.* **2019**, *36*, 4467–5542.
20. Guo, Y.H.; Wang, F.Y.; You, H.Q. Effects of different carbon sources on the growth and active ingredient accumulation of Salvia miltiorrhiza and Tibetan Salvia miltiorrhiza hairy roots. *Chin. J. Chin. Mater. Med.* **2020**, *45*, 43–48.
21. Lathika, N.; Rahaman, W.; Tarique, M.; Gandhi, N.; Kumar, A.; Thamban, M. Deep water circulation in the Arabian Sea during the last glacial cycle: Implications for paleo-redox condition, carbon sink and atmospheric CO_2 variability. *Quat. Sci. Rev.* **2021**, *257*, 106853. [CrossRef]
22. Mayes, R.T.; VanCleve, S.M.; Kehn, J.S.; Delashmitt, J.; Langley, J.T.; Lester, B.P.; Du, M.; Felker, L.K.; Delmau, L.H. Combination of DGA and LN Columns: A Versatile Option for Isotope Production and Purification at Oak Ridge National Laboratory. *Solvent Extr. Ion. Exch.* **2021**, *39*, 166–183. [CrossRef]
23. Erdogan, S. Dynamic nexus between technological innovation and buildings Sector's carbon emission in BRICS countries. *J. Environ. Manag.* **2021**, *293*, 112780. [CrossRef] [PubMed]
24. Shan, A.; Fan, X.; Wu, C.; Zhang, X.; Fan, S. Quantitative Study on the Impact of Energy Consumption Based Dynamic Selfishness in MANETs. *Sensors* **2021**, *21*, 716. [CrossRef] [PubMed]
25. Gu, S.Y.; Wu, Y.W. Input-output method to calculate and analyze China's tourism carbon emissions. *North Econ. Trade* **2020**, *423*, 156–161.
26. Cai, B.F.; Zhu, S.L.; Yu, S.M. Interpretation of "IPCC 2006 National Greenhouse Gas Inventory Guidelines 2019 Revised Edition". *Environ. Eng.* **2019**, *37*, 4–14.
27. Hu, Z.; Gong, X.; Liu, H. Analysis on the influencing factors and changing trend of household consumption carbon emission—Taking Shaanxi Province as an example. *Ecol. Econ.* **2020**, *36*, 28–34.

28. Ehrlich, L.; Ledbetter, D.; Aczon, M.; Laksana, E.; Wetzel, R. 966: Continuous Risk of Desaturation Within the Next Hour Prediction Using a Recurrent Neural Network. *Crit. Care Med.* **2021**, *49*, 480. [CrossRef]
29. Lin, K.; Zhao, Y.; Tian, L.; Zhao, C.; Zhang, M.; Zhou, T. Estimation of municipal solid waste amount by one-dimension convolutional neural network and long short-term memory with attention mechanism model: A case study of Shanghai. *Sci. Total Environ.* **2021**, *791*, 148088. [CrossRef]
30. Xing, H. Research on the Multivariable Driving Factors of Carbon Emissions from Energy Consumption in the Yangtze River Economic Zone—By the Extended STIRPAT Model. *Resour. Dev. Mark.* **2020**, *36*, 4–10.
31. Liu, Z.Q.; Xu, H.F.; Wang, C.Y. Research on the normalization method of scientific research scores by Sigmoid function. *J. Xinxiang Univ. (Nat. Sci. Ed.)* **2019**, *36*, 19–22.
32. Zou, Y.B.; Lei, B.J.; Zang, Z.X. Automatic threshold selection method by the maximization of normalized mutual information. *Acta Autom. Sin.* **2019**, *45*, 1373–1385.
33. Gonzaga, A.D.S.; Cordeiro, R. The similarity-aware relational division database operator with case studies in agriculture and genetics. *Inf. Syst.* **2019**, *82*, 71–87. [CrossRef]
34. Chen, J. Several applications of Eviews software in unary linear regression model prediction. *J. Foshan Univ. Sci. Technol. (Nat. Sci. Ed.)* **2019**, *37*, 6–10.
35. Tang, L.; Tian, Y.; Li, W.; Pardalos, P.M. Valley-loss regular simplex support vector machine for robust multiclass classification. *Knowl.-Based Syst.* **2021**, *216*, 106801. [CrossRef]
36. Ding, S.; Sun, Y.; An, Y.; Jia, W. Multiple birth support vector machine by recurrent neural networks. *Appl. Intell.* **2020**, *50*, 2280–2292. [CrossRef]
37. Liu, L.; Chu, M.; Gong, R.; Peng, Y. Nonparallel support vector machine with large margin distribution for pattern classification. *Pattern Recognit.* **2020**, *106*, 107374. [CrossRef]
38. Chung, E.; Leung, W.T.; Pun, S.-M.; Zhang, Z. A multi-stage deep learning based algorithm for multiscale model reduction. *J. Comput. Appl. Math.* **2021**, *394*, 113506. [CrossRef]

land

Review

Social Vulnerability Assessment for Landslide Hazards in Malaysia: A Systematic Review Study

Mohd Idris Nor Diana [1,*], Nurfashareena Muhamad [1], Mohd Raihan Taha [2], Ashraf Osman [3] and Md. Mahmudul Alam [4]

1 Institute for Environment and Development, Universiti Kebangsaan Malaysia, Bangi 43600, Selangor, Malaysia; fasha@ukm.edu.my
2 Department of Civil Engineering, Universiti Kebangsaan Malaysia, Bangi 43600, Selangor, Malaysia; profraihan@ukm.edu.my
3 Department of Engineering, The Palatine Centre, Durham University, Stockton Road, Durham DH1 3LE, UK; ashraf.osman@durham.ac.uk
4 School of Economics, Finance and Banking, Universiti Utara Malaysia, Sintok 06010, Kedah, Malaysia; mahmudul@uum.edu.my
* Correspondence: nordiana@ukm.edu.my; Tel.: +60-3-89214862

Abstract: Landslides represent one of the world's most dangerous and widespread risks, annually causing thousands of deaths and billions of dollars worth of damage. Building on and around hilly areas in many regions has increased, and it poses a severe threat to the physical infrastructure and people living within such zones. Quantitative assessment of social vulnerability in Malaysia is worrying because it has been given less attention than hazard-related studies. Therefore, this study's objective is to find out the indicators used for social vulnerability assessment in the context of a landslide in Malaysia. The analysis is critical for understanding the measures of social vulnerability, given that the incorporation of climate change and disaster risk mitigation issues in urban planning and management are considered priorities in ensuring a stable population growth and avoiding economic disruption. A systematic study on the Scopus and Web of Science repositories was conducted based on the PRISMA Report analysis method. This article concluded that there are six important indicators of social vulnerability in the context of landslide in Malaysia.

Keywords: social vulnerability assessment; landslide; social indicator; disaster risk reduction; Malaysia

Citation: Nor Diana, M.I.; Muhamad, N.; Taha, M.R.; Osman, A.; Alam, M..M. Social Vulnerability Assessment for Landslide Hazards in Malaysia: A Systematic Review Study. *Land* **2021**, *10*, 315. https://doi.org/10.3390/land10030315

Academic Editor: Marina Cabral Pinto

Received: 11 February 2021
Accepted: 28 February 2021
Published: 19 March 2021

1. Introduction

In recent years, extreme events have increased in intensity and frequency globally, leading to rising economic losses and casualties. It is believed that these events will continue to accelerate in future climate scenarios. An accurate understanding of the physical and socioeconomic drivers of these extreme events is crucial and can ultimately enhance adaptive strategies. The frequency and intensity of geophysical events is increasing. This is the result of the interaction between humans and the environment. Climate change and increasingly aggressive human activities contribute to the vulnerability of catastrophic hazards to humans, their infrastructure, and the environment [1]. Faced with ever-increasing societal impacts arising from such events, a wealth of research and analysis has focused on understanding causal processes and outcomes [2]. Landslides are a type of geophysical event that plays a significant role in the evolution of a landscape [3]. However, landslides do pose a serious threat to local populations given that these events are being triggered increasingly by a changing climate and more unpredictable weather patterns. In recent years, it has become clear from previous research that the location, abundance, activity, frequency of landslides as well as the social and economic consequences are increasing over time and more people are exposed to the risks [4–10]. It was reported in [11] that geophysical disasters such as landslides are the deadliest. The presence of humans, infrastructure, and other forms of vulnerabilities in one location will make things worse.

Historically, efforts to reduce landslides are physically oriented resulting in a proliferation of technocratic approaches in the literature, while financial losses and social vulnerability from the geophysical events continue to increase. Over time, this gave rise to an alternative explanation that mounting losses are related less to the dynamics of the events but more to the vulnerability of exposed human populations [2]. Although assessing the magnitude and intensity of disasters is critical, the nature of population demographics and various socioeconomic contexts may also lead to a greater risk of disasters. Understanding the complexities of vulnerability to disasters, including those caused by geophysical events, is at the heart of disaster risk reduction. Efforts to reduce disaster risk involve various disciplines and should be viewed from numerous perspectives to provide long-term benefits. A comprehensive disaster risk reduction strategy that incorporates physical and socio-economic aspects is the key determinant of vulnerability.

In spite of very high importance of socioeconomic data to assess landslide vulnerability, there are lack of social data documented for analysis and mapping in Malaysia. Therefore, the objective of this study is to find out the indicators that are used for social vulnerability assessment in the context of landslides in Malaysia. The analysis is critical for understanding the measures of social vulnerability, since the incorporation of climate change and disaster risk mitigation issues in urban planning and management are a priority for ensuring stable population growth and evading economic disruption.

2. Literature Review

The definition of vulnerability is "the quality of being vulnerable (able to be easily hurt, influenced, or attacked), or something that is vulnerable" [12]. Vulnerability means the risk of being vulnerable or easily hurt by something or someone. Vulnerability is a concept that has being used over a long period of time, and it has been recognised in much research covering various fields of endeavour [13], for instance, the social sciences, economics, psychology, and engineering. It should be noted that there is no consensus regarding how vulnerability is defined [14]. It has, in fact, been interpreted in many ways according to the subject area being investigated.

According to [15], vulnerability refers to situations where individuals and societies are exposed to social, economic, and cultural risks and in essence the dangers posed by harm to them. All people and all communities at some point cannot avoid risk or harm, so at best each individual needs to prepare for every situation. Moreover, stress that social vulnerability is partly the result of social difference or social inequality, which affects or forms the susceptibility of different groups to harm or at risk and regulates their capacity to react to a certain situation [16]. There is inequality in every society and the unequal distribution of wealth and resources is something that has permeated all of human history. For instance, in a farmer's perspective, inequality can take many forms such as unequal distribution of wealth, water allocation, rights to land and water, taxation inequity, economic poverty, land tenure issues, and much more. The definition of climate vulnerability according to the Intergovernmental Panel on Climate Change (IPCC) is "... the degree to which geophysical, biological and socio-economic systems are susceptible to, and unable to cope with, adverse impacts of climate change" [17]. The concept of vulnerability has been refined over the decades so that people understand the disasters and hazards that occur in communities susceptible to this kind of situation. Vulnerability is something that can help people achieve a level of sustainable development realistically. Economic development or progress should be engaged with as long as the natural environment in which they occur can be sustained.

For this reason, vulnerability can be defined as individuals, households, or communities that are dealing with external shock from the outside and are unexpected [18]. Vulnerability is present in both internal and external factors that influence the lives of individuals and communities. Furthermore, vulnerability can be understood as the capacity of individuals, groups or communities to reciprocate, cooperate, survive, and recover from the impact of environmental events that have happened around them [19]. Landslides are very indicative of how the characteristics of a social group can overcome this kind of disaster but also reflect the harsh realities of social vulnerability to natural events.

2.1. Social Vulnerability to Disaster

Vulnerability is broadly defined as the potential to suffer loss or harm. The theory includes structural vulnerability of buildings, physical exposure of people, and places to natural events, while social vulnerability describes different kinds of susceptibility based on social, economic, and political factors [20,21]. Vulnerability and exposure are dynamic, varying in temporal and spatial scales, and depend on economic, social, geographical, demographic, cultural, institutional, governance, and environmental variables [22]. Analyses of vulnerability in the engineering context of landslide or slope (or any disaster) are quite common [23,24]. The study by [25] has described vulnerability as the characteristic of a person or group and their situation that influences their capacity to anticipate, cope with, resist, and recover from the impact of a natural hazard (an extreme natural event or process). Despite its importance in disaster risk reduction, there is still a lack of approaches that contribute to a better understanding of social vulnerability hidden in dynamic contextual conditions [26].

The definition of social vulnerability within the disaster framework was introduced in the 1970s when researchers realised that exposure included socioeconomic factors affecting group resilience [27,28]. Social vulnerability is useful as an indicator in determining the differential recovery potential from disasters. Social vulnerability normally employed individual characteristics of people such as age, race, health, income, type of dwelling unit, and employment [29]. Social vulnerability is a concept that can explain social imbalances that are happening in society in some parts of the world. It is one of the results of social inequalities that occur in many communities. Factors affecting social disparities evident in a society include: lack of resources such as information, knowledge, and technology; limited access to political power or representation; social capital; social networks and connections; beliefs and customs; building stock and age of infrastructure; and type and density of infrastructure and lifelines [30]. Next, the 18 social vulnerability indicators was introduced as follows: socioeconomic status (income, political power, and prestige), gender, race and ethnicity, age, commercial and industrial development, employment loss, rural/urban, residential property, infrastructure and lifelines, renters, occupation, family structure, education, population growth, medical services, social dependence, and social needs population [16].

The design of these indicators depends on their expected use, and it must be relevant to the hazard context, methodologies, and data availability [31]. However, social vulnerability exists based on the underlying characteristic of a population, and it does not rely on the hazard or susceptibility of an area. Apart from indicators, numerous indices have been developed in order to measure social vulnerability. Many pioneer researchers have devoted much effort to formulating the concept of social vulnerability. Social Vulnerability Index (SoVI) was introduced [32] to quantify social vulnerability through an empirical basis to compare social differences within a community regarding social variables selected to mitigate the disadvantageous effects of certain events. It was asserted that socially vulnerable communities are more likely to be adversely affected in disaster events because they are much less likely to recover from them and more likely to die [33]. Even though the SoVI was devised with the United States in mind, many studies have adapted SoVI for a variety of contexts, no matter the nature of the population or places being investigated.

2.2. Landslides: Malaysia's Experience

Malaysia is located in the south-east of Asia. It is divided into two archipelagos, Peninsular Malaysia and Borneo Island. Malaysia is a tropical country with a warm and humid climate throughout the year. Over a recent 20-year period (1998–August 2018),

Malaysia has witnessed 51 disaster events [34–43]. During that time, 281 people died, more than 3 million people were affected, and disasters caused nearly US$2 billion in damage [44]. Flood, landslides, drought, and forest fires are common in Malaysia, while the annual rainfall is the main contributor due to two monsoon periods, i.e., South West (SW) and North East (NE) occurring between April and October and from November to March, respectively. These monsoons contribute to high annual rainfalls amounting to 2000–4000 mm with a maximum of about 200 rainy days [45]. The amount of rainfall varies from one rainy day to the next [46]. The rain and consistently high temperatures throughout the year lead to intensive and extensive weathering of features on the ground. These combinations of climate and geological conditions together with other causative factors such as slope angle, drainage conditions, geological boundaries, etc. [47] have led to landslides becoming one of Malaysia's most common natural disasters.

The most common trigger for landslides is heavy or prolonged rainfall, but seismicity, river undercutting, freeze-thawing cycles, and human activity may also cause substantial and destructive landslides. As reported [48], Malaysia recorded 171 landslides between 2007 and March 2016, according to data from the US National Aeronautics Space Administration (NASA), making the country the world's 10th highest in terms of landslide frequencies. In recent years, Malaysia has experienced several landslides resulting from extreme tropical rainfall. Landslides have occurred in several parts of the country, such as Paya Terubong (Penang), Highland Towers (Kuala Lumpur), Hulu Langat, and Pos Dipang (Perak). These landslides incur significant property loss and hundreds of lives. In 2017, 6000, people were severely affected by a flash flood and landslide in the Kundang, Selangor area, which left many stretches of roads, infrastructure, and assets badly damaged [49]. When the population density of towns increases, highland or hilly terrain development also increases and this puts more stress on the natural environment. Urban areas are then exposed to a high risk of landslides [50]. Significant landslides in Malaysia were recorded from 1993 to 2020 (see Table 1).

Table 1. Series of significant landslide occurrences in Malaysia.

No.	Year	Location	Consequences
1	1993	Highland Tower, Ulu Klang, Selangor	48 deaths and 2 injuries. One building collapsed
2	1993	Pinggiran Bukit Segar, Kuala Lumpur	One family evacuated their house
2	1993	Pantai Remis, Perak	No record
4	1994	Taman Puchong Perdana, Puchong, Selangor	10 families evacuated
5	1995	Taman Keramat Permai, Ampang, Selangor	No damage recorded
6	1995	Kuala Lumpur—Karak Highway	20 deaths, 22 injuries, and ten cars damaged
7	1996	North-South Expressway (NSE) near Gua Tempurung, Perak	No record
8	1996	Pos Dipang, Kampar, Perak	44 people were killed
9	1996	Ampang Jaya, Selangor	No record
10	1999	Puncak Athenaeum Condominium, Ampang, Selangor	Minor landslide, road access to the hilly residential area affected
11	1999	Mutiara Condominium, Ampang, Selangor	No record
12	1999	North-South Expressway, Kuang, Selangor	Thousands of vehicles stranded. Road closure lasting one day
13	2000	Jalan Bukit Antarabangsa, Ampang, Selangor	No record
14	2001	Kampung Sungai Chinchin, Gombak, Selangor	A house partly destroyed
15	2002	Jalan Ipoh, Kuala Lumpur	Covering three-lane road leading from Selayang to Rawang
16	2002	Taman Hillview, Ampang, Selangor	Eight deaths and five injuries
17	2003	Taman Bukit Jaya, Ampang, Selangor	No fatalities
18	2004	Taman Melati, Gombak, Selangor	1 death
19	2004	Jalan Seri Penchala 1, Kuala Lumpur	24 houses evacuated
20	2006	Taman Zooview, Ampang, Selangor	Four deaths
21	2006	Taman Bukit Serdang, Seri Kembangan, Selangor	Damaged section of the road measured 50 m × 25 m
22	2006	Bukit Tunku, Kuala Lumpur	No record
23	2006	Taman Esplanad, Kuala Lumpur	Two houses damaged
24	2008	Taman Bukit Mewah, Ampang, Selangor	4 deaths

Table 1. *Cont.*

No.	Year	Location	Consequences
25	2008	Ulu Kelang, Selangor	Four deaths and 15 injuries
26	2008	Kuala Kubu Bharu, Batang Kali, Selangor	Two sisters were buried alive when a landslide hit a bungalow
27	2008	Kemensah Heights, Ampang, Selangor	No fatalities
28	2008	Bukit Ceylon, Kuala Lumpur	One worker killed
29	2008	Pantai Dalam, Kuala Lumpur	One killed, 4 injured, and 19 families evacuated
30	2009	Taman Cheras Awana, Cheras, Selangor	Destroyed 3 cars and a motorcycle, 10 families evacuated
31	2010	Ukay Perdana, Ampang, Selangor	No fatalities
32	2010	Taman Bukit Mulia, Ampang, Selangor	No fatalities
33	2011	Puncak Setiawangsa, Kuala Lumpur	88 residents of bungalows, shop houses, and double-storey terrace houses ordered to move out
34	2011	Jalan Semantan, Kuala Lumpur	Six cars were buried and five vehicles were damaged
35	2011	Pekan Batu 14 Hulu Langat, Selangor	16 deaths
36	2011	Kampung Tengah, Puchong, Selangor	5 houses affected
37	2012	Taman Desa Sentosa, Hulu Langat, Selangor	Endangered four occupants of the Perkid Welfare Home for girls
38	2012	Taman Mulia Jaya, Ampang, Selangor	Water seeped through the sewerage system.
39	2013	Putra Heights, Subang Jaya, Selangor	Several vehicles submerged in mud
40	2015	KM 52.4 of the Kuala Lumpur-Karak Expressway between Lentang and Bukit Tinggi, Pahang and Gombak-Bentong old roads	Lentang-Bukit Tinggi stretch of the expressway was closed to traffic
41	2016	Karak Highway	Blocked all lanes in both directions on the highway and four vehicles were trapped in the landslide
42	2016	Bau-Puncak Borneo, Sarawak	Comprising mainly Bidayuh settlements and Padawan Ring Road critically affected
43	2017	Tanjung Bungah, Penang Island	Killed 11 construction workers
44	2018	Jalan Bukit Kukus, Georgetown, Penang Island	Killed nine construction workers
45	2019	Taman Batu Permai	No record
46	2019	Jalan Lee Woon, Ampang, Selangor	A house evacuated
47	2019	Genting Highland, Pahang	Affected a portion of the Jalan Genting-Amber Court slip road and no access to the resort
48	2020	Taman Kelab Ukay, Bukit Antarabangsa	40 residents were ordered to leave their homes
49	2020	Taman Silibin Indah, Ipoh	Killed one construction worker
50	2020	Sungai Penchala, Kuala Lumpur	3 families ordered to leave their homes
51	2020	Jalan Gombak to Genting Highland	The main road was closed for repairs
52	2020	Ulu Beram, Jalan Lapok	Residents cut off due to damaged roads
53	2020	Tapah to Ringlet	Fallen trees blocked the main road
54	2020	Tapah to Cameron Highland	Fallen trees blocked the main road
55	2020	Jalan Simpang Pulai to Cameron Highland	The retaining wall suffered damage and part of the structure collapsed
56	2020	The Banjaran Hotspring Retreat, Tambun	Killed 2 guest house
57	2020	Jalan Lojing-Gua Musang	Closed half of main road
58	2020 2021	Jalan Keningau-Kimanis, Sabah Jalan Raub-Bukit Fraser	Two houses were damaged and no casualties Road closed and 13 vehicles trapped
59	2021	Taman Bukit Kempas, Johor	Tank water pipe broke, and 42 people vacated the residence
60	2021	Kemaman, Terengganu	The restaurant was hit by a rock, no casualties
61	2021	Jalan raya Timur Barat, Ipoh	One hallway closed
62	2021	Kampung Garong, Padawan, Sarawak	2 houses were damaged, and a house half buried in the ground
63			
64	2021	Kota Kinabalu	10 landslides were reported in seven villages, involving four districts, namely, Kota Kinabalu, Kota Marudu, Pitas, and Kudat. No casualties

Source: [51–60].

In Malaysia, there have been numerous landslide events in the mountains, along the valleys, rivers, and coastal regions [61,62] but the most massive have generally been associated with rivers. Findings from the literature have shown that landslides occur frequently along hilly areas in the rainy season. There is a strong correlation between the density of drainage and distance to the river due to landslides in the mountainous region being triggered by erosion-related phenomena [63]. Development on hilly areas in Malaysia has increased the risk and likelihood of landslides [64]. Hilly areas are attractive for building residential areas, hotels, or resorts. This poses a severe threat to the physical infrastructure and population living within that area. This situation will lead to many casualties and significant financial losses if these hilly regions are struck by landslides [65].

Global landslides cause billions of dollars' worth of infrastructure damage and thousands of deaths annually. The estimated number of deaths is 1000 per year and destruction of property amounting to approximately US$4 billion [66]. Meanwhile, losses due to landslides in Malaysia have cost more than US$1 billion since 1973 [67]. Emergency preparedness plays a part in reducing the effects of disasters. The most effective preparedness at the initial stage was to make the right decision to reduce the number of deaths and damage to property in communities. The rescue team provided some emergency response and preparedness training for each member of the community so that their reactions were practical. In Malaysia, there are several agencies involved in dealing with landslides such as Malaysia Civil Defence Force (MCDF), Fire and Rescue Department of Malaysia, National Disaster Management Agency (NADMA), and others. Furthermore, the Ministry of Housing and Local Government has issued a guideline for any physical development on the hilly terrain area in Malaysia. Table 2 summarises the criteria of the biological effect based on the slope gradient, slope classification for engineering work, and the description of development activities.

Table 2. Malaysian Guideline on physical development in hilly terrains.

Slope Gradient (α)	Slope Classification For Engineering Work	Description
Below 15°	Class 1	Compliance with: i. Development Guidelines in Hill Areas 1997 (issued by the local government) ii. Erosion and Dirt Control Guidelines, 1996 (issued by the Department of Environment) iii. Environmentally Friendly Drainage Manual 2000 (issued by the Department of Irrigation and Drainage)
15–25°	Class 2	EIA report prepared by EIA consultants registered with the Department of Environment for development exceeding 50 ha. For class 1 and II development projects only subject to section 34A, the Environmental Quality Act 1974 must be provided EIA.
25–35°	Class 3	Requires an additional environmental impact assessment study. Proposes the conduct of landslide vulnerability assessment, which may serve as an alternative tool to establish a sustainable development environment.
Above 35°	Class 4	Development projects within this area are not permitted at all, except for road construction, which is inevitable. However, an environmental impact assessment is required. Proposes the conduct of landslide vulnerability assessment, which may serve as an alternative tool to establish a sustainable development environment data

Source: [68].

Malaysia has its share of landslides and most of the landslide studies conducted focus on the engineering perspective. Socioeconomic aspects should be taken into account to evaluate the vulnerability of the community, especially one at high risk of experiencing such catastrophic effects, but previous research concentrated more on describing the disaster types [61,69], susceptibility, and risk assessment [70,71]. The level of quantitative evaluation of social vulnerability in Malaysia is worrying due to the lack of social data documented for analysis and mapping. Therefore, the objective of this study is to find out indicators that are used for social vulnerability assessment in the context of a landslide in Malaysia. The analysis is critical for understanding the measures of social vulnerability, since the incorporation of climate change and disaster risk mitigation issues in urban planning and management are a priority for ensuring stable population growth and evading economic disruption.

The representativeness of Malaysia as an important case for research, though can be critical in other cases, is not an issue for his study. What we are trying to demonstrate is that in analysing landslide risk, the human part is an integral part and should be incorporated as detailed in this study. The methodology used in this study is a pioneer for landslide risk assessment. Assessing the landslide risk with the proposed methodology can be a crucial tool for engineers and policy-makers in developing a site, particularly in hilly areas, for population development. Thus, it must be done at its locality, per se, in order to assess the real risk of landslide. More importantly this methodology can serve to highlight the importance of public education to increase the level of knowledge of the population on the hazard and mitigation of possible landslide events in their area. Limited literature found on social vulnerability mapping to climate-driven disasters in the country. The socio-economic aspect is the most apparent after disasters as different patterns of damages, losses, and suffering maybe experience differently by certain groups of the population.

3. Materials and Methods

This section incorporates five significant sub-sections that explain the following: PRISMA, resources, inclusion and exclusion criteria, systematic review procedure, and data extraction and interpretation. The methodology technique to retrieved articles is the one suggested by [72].

3.1. PRISMA

The systematic review in this article was guided by the PRISMA method, and this abbreviation stands for "Preferred Reporting Items for Systematic Reviews and Meta-Analyses." PRISMA has mainly been utilised by healthcare personnel create systematic reviews and meta-analyses. As well as the medical field, PRISMA has been employed by environmental management experts to undertake systematic reviews.

3.2. Resources

This study used two primary journal databases, specifically Scopus and Web of Science (WoS). Scopus is a bibliographic database for journal articles and consists of abstract and citation sources. This database covers journals from scientific, technical, medical and social sciences and currently has more than 5000 publishers worldwide and more than 22,000 titles. Web of Science (WoS) is a database producing Clarivate Analytics, which includes articles from 256 disciplines such as science, social science, arts, humanities, etc. WoS offers full-text articles, reviews, editorials, abstracts, proceedings and book chapters. WoS includes more than 33,000 journals published from the year 1900 to the present day. Other databases like JSTOR and Google Scholar were considered for this research.

3.3. Systematic Review Process

The systematic review process includes four main stages to acquire relevant: identification, screening, eligibility, and data extraction.

3.3.1. Identification

The first process of undertaking systematic reviews is identification. Identification means finding the most relevant studies, using keywords, dictionary terms, thesaurus, encyclopaedias, etc. The keywords used help to build the "search string" for the research (Table 3). Subsequently, 13 articles were found in JSTOR using the term "social vulnerability index." From the Scopus database, in total, 147 articles related to the search string were discovered while a total of 69 items emerged from Web of Science (WoS). Meanwhile, 29 studies were found in Google Scholar search engine, where the data covers a huge range of subjects and is essentially a superset of WoS and Scopus [73].

Table 3. Search string.

Databases	Keyword Used
Scopus	TITLE-ABS-KEY (("socia* vulnerabilit* inde*" OR "soci* economi* vulnerabilit* inde*" OR "soci* vulnerabilit* inde*" OR "socia* vulnerabilit*" OR "SoVI" OR "SeVI" OR "SVI")) AND (landslid* OR rockslid* OR earthfal*)
Web of Science (WoS)	TS = (("socia* vulnerabilit* inde*" OR "soci* economi* vulnerabilit* inde*" OR "soci* vulnerabilit* inde*" OR "socia* vulnerabilit*" OR "SoVI" OR "SeVI" OR "SVI") AND (landslid* OR rockslid* OR earthfal*))
JSTOR	((((((("socia* vulnerabilit* inde*") OR ("soci* economi* vulnerabilit* inde*")) OR ("soci* vulnerabilit* inde*")) OR ("socia* vulnerabilit*")) OR ("SoVI")) OR ("SeVI")) AND ("landslide"))
Google Scholar	("social vulnerability") ("social vulnerability index") ("socio economic vulnerability index") (landslide)

3.3.2. Screening

The second part of the systematic review process is screening. Here, it is necessary to gather all the articles related to the study topic and exclude all irrelevant items. Table 4 shows the inclusion and exclusion criteria that need to be followed in finding related articles. The total of 258 articles was screened using the inclusion and exclusion criteria including literature type, language, timeline, countries and territories, and the subject area. For the first criterion of the literature type, this study decided to focus on journal research articles and excluded papers resembling review articles, book chapters, and conference proceedings. Meanwhile, for language, the chosen one was English, and all other non-English articles were excluded. The criterion for publication was the period from 2010 to 2020 only, and the geographical criterion was Southeast Asia, Southwest Asia and Europe. Lastly, for the subject area, this study only chooses articles from social sciences, environmental science, science, and agriculture. From the inclusion and exclusion criteria, the number of articles that have been excluded is 199, in total (Figure 1).

Table 4. Inclusion and Exclusion criteria.

Criterion	Eligibility	Exclusion
Literature type	Journal (Research articles)	Journals (review article), book chapter, conference proceeding
Language	English	Non- English
Timeline	2010 to 2020	<2010
Countries and territories	Southeast Asia, Southwest Asia and Europe countries	Non-Southeast Asia, non-Southwest Asia and Non-Europe country
Subject Area	Social Science, Environmental Science, Agricultural	Other than Social Science, Environmental Science, Agricultural

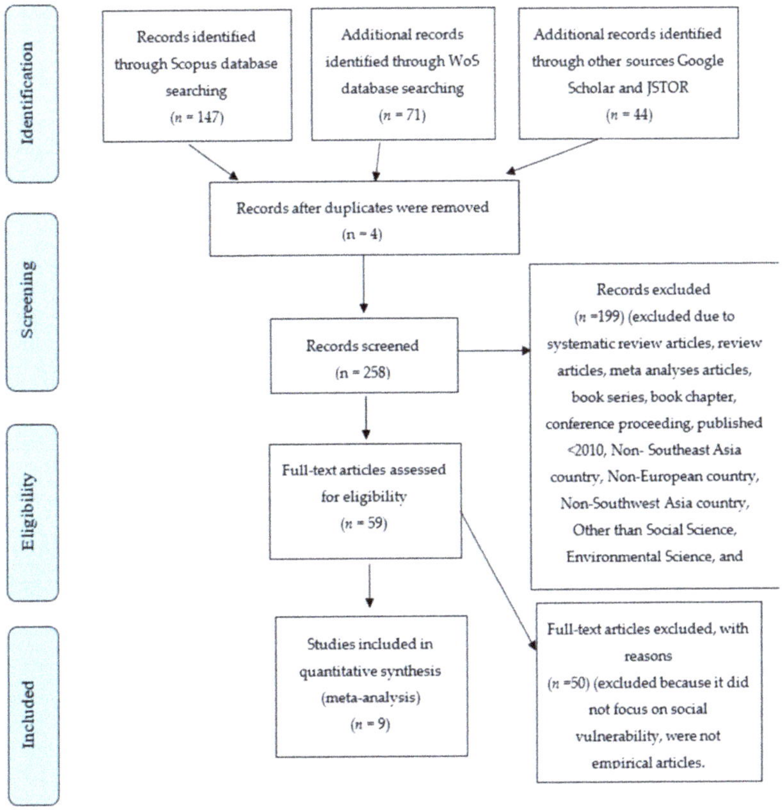

Figure 1. Literature searches based on Preferred Reporting Items for Systematic Reviews and Meta-Analyses (PRISMA) is guidelines (adapted from [74]).

3.3.3. Eligibility

For the third stage eligibility, a total of 59 articles were used. Title, abstract, and the content of each paper are important and need to be examined thoroughly to make sure it fulfils the inclusion criteria and review objective. In total 50 articles have been excluded because they did not fit this criterion. Therefore, the criteria of selected articles to be analysed is focus on the social vulnerability study and the empirical articles. It is because the purpose of this study is to define the indicators used to assess social vulnerability in the context of landslides in Malaysia. The research is important for understanding social vulnerability interventions, as the inclusion of climate change and disaster risk mitigation problems in urban/rural planning and management. More specifically, this approach will help to illustrate the value of public education in growing the population's level of awareness about the risk and mitigation of potential landslide events in their area. Even though the occurrence of landslides is different due to the climatic conditions among the countries for article analysis, however, due to the lack of research on the formation of social vulnerability indicators in Southeast Asian countries, alternatively, this study has expanded its study to Southwest Asian and European countries.

3.3.4. Data Extraction

After the remaining articles were assessed and analysed, the researcher started to extract the data. First, this was done by reading the abstract of the article, and then the researcher read the full text to start identifying themes and sub-themes related to the objective. After that, themes and sub-themes were organised to establish a typology for the article.

4. Results

According to the results shown in Table 5, in total, 9 articles were chosen for this study. The nine authors of the articles include [75–83] in this study. Besides, the selected articles were published in the years ranging from 2011 to 2020. It aims to identify research trends on social indicators that are constantly being studied and considered for the purpose of forming a social vulnerability index for certain area and community. Next, with reference to countries covered, two studies are from Nepal, and the rest are one study each from Portugal, England, Italy, Pakistan, India, China, and Indonesia. It comprises the name of authors, the country of studies, title of articles, and the objective of the studies by scholars.

Table 5. List of articles analysed for systematic review.

Author	Country	Year	Title	Objective
[75]	Nepal	2020	A geospatial analysis of multi-hazard risk in Dharan, Nepal	Produce individual hazard assessment for the rapidly growing city of Dharam city and calculate its level of social vulnerability
[76]	Nepal	2019	An analysis of social vulnerability to natural hazards in Nepal using a modified social vulnerability index	To quantify social exposure at the local level using indicators relevant to Nepal's distinct social and physical landscape
[77]	Portugal	2015	Application of social vulnerability (SoVI) and delineation of natural risk	To go further into the biological risk analysis in the Greater Lisbon area using a multi-hazard approach
[78]	England	2019	Evaluation of social vulnerability to natural hazards: A case of Barton on Sea, England	The current study examines the social vulnerability of Barton-on-Sea by conducting a survey-based analysis
[79]	Italy	2016	Mapping social vulnerability to natural hazards in Italy: A suitable tool for risk mitigation strategies	The study aims to define a social vulnerability index (SVI) for Italy by applying an inductive approach
[80]	Pakistan	2018	Socioeconomic determinants of landslide risk perception in Murree hills of Pakistan	The aim is to assess the determinants of landslide risk perceptions in the Murree Hills of Pakistan
[81]	India	2020	Study of integrated social vulnerability index SoVLint of the hilly region of Uttarakhand, India	This study focuses on producing a map for the hilly district of Uttarakhand showing the vulnerabilities measured by natural, social, and economic indicators
[82]	China	2011	Social vulnerability assessment of natural hazards on county-scale using high spatial resolutions satellite imagery: A case study in the Luogang district of Guangzhou, South China	This study examines the social vulnerability assessment of natural hazards on a county-scale using high spatial resolutions satellite imagery
[83]	Indonesia	2018	Quantitative assessment of social vulnerability for landslide disaster risk reduction using GIS approach (Case study: Cilacap Regency, Province of Central Java Indonesia)	To examine social exposure for landslide disaster risk reduction using a GIS approach

Sources: Author analysis, 2020.

4.1. Indicators Used to Measure Social Vulnerability in a Landslide

There are 14 indicators serving to measure social vulnerability when a landslide occurs. Included here are age, gender, ethnicity, built environment, income, family structure, education, employment, occupation, urban or rural, disability, migration, medical, and population (Table 6).

Table 6. List of indicators use as social vulnerability index.

Indicators	Variables	Reference
Age	The elderly population (>65 years), children under 5 years old, dependency ratio, elderly index, resident population aged 5–14, resident population aged 15–19, mean age (years) of the resident population	[77,79,80,82,83]
Gender	Females, a household that is run by a woman, a household with land owned by females	[75,77]
Ethnicity	Population by ethnic, minority population	[75–77]
Built Environment	A household without piped water connection, electricity, reinforced cement concrete (RCC) foundation, sewage water, and the population lived in a home with quality external walls.	[76]
Income	The income per capita, the ratio of high income to low income	[80]
Family structure	Female-headed household, the average number of people per household	[75,78]
Education	People cannot read and write, high level of education, low level of education	[76–81,83]
Employment	Female labour force employed, labour force employed, unemployment rate	[78,81,83]
Occupation	Employed in agriculture, forestry, fishing, mining and quarrying; employed in manufacturing and construction; employed in transportation, communication and other public utilities; employed in accommodation and food services	[75]
Urban/Rural	Population density	[77]
Special Needs Population	Population with disabled person (auditory, visual, motor, or mental disability), the person who is disabled and/or unemployed or without any economic activity, a person with disability that is more than 60%, a person who is disabled, and under 4 or above 65 years of age, permanently disabled and unable to work	[75–77]
Migration	Foreign population, absentee population	[77,81]
Health	Medical services, health problems, distance from the hospital	[77,81,82]
Population	Population growth	[79,83]

Sources: Author analysis, 2020.

In this study, there are five main indicators that are focused on, these being age, ethnicity, education, disability, and health. These are the variables that most scholars measure when investigating landslides. They are explained in more detail below.

4.1.1. Age

The first component that has been discussed in [77] is "urban, age (elderly), and gender." Variable for age includes the proportion of resident population aged 65 and over, proportion of resident population aged 4 and younger, proportion of residents aged 5–14, and proportion of resident population aged 15–19. The study shows a negative result for elderly people, which means they are more susceptible to vulnerability. There was reported [79] that focuses more on four component indicators—age, employment, population growth, and education. He also stated that aging index is one component that represents the age indicator.

The variables include population of people aged 65 and above and those aged 15 and younger. The aging phenomenon that is very evident in Italy has resulted from the depopulation of people in mountain areas, people leaving the land, migration, and the lure of promising jobs in the industrial and service sectors. Italy's people are generally living longer and the average birth rate has declined. According to the study by [80] there are five main indicators affecting the landslide risk perception: age, income, education level, location, and experience. In addition, the study shows that age of respondent wields an effect on the perception of landslides.

4.1.2. Ethnicity

According to [76], the ethnicity indicators focus on the Dalit population and minority population such as Muslims and Sikhs/Punjabis. They found that this group was less than 5% of the total population in Nepal, and it is considering as disadvantage groups. In [77], "nationality and ethnicity" is one of the five main indicators in that particular study. The variables for ethnicity indicator include person of African origin living in the country, foreign nationality, and resident who was born outside the country as a marginal group. Like age, ethnicity can be an indicator in the social vulnerability index and help assess what is happening in a given society.

4.1.3. Education

Education has always been regarded as one of the key vulnerabilities all communities have to deal with. Educated people are more likely to have advantages in everything they do compared to people without or with little education. There are three main variables relating to education as follows [75]: percentage of the population who can read and write, percentage who completed school certificate (SLC), and percentage who completed a college or university degree. In the study by [77], one of the indicators "development and education" included variables such as the proportion of illiterate people. The community can be very vulnerable when the proportions of literate and illiterate are dangerously disproportionate.

Furthermore, the level of education and qualification can affect vulnerability in one community. The higher the qualification in education that someone has, the more unlikely it is that they will experience vulnerability from any hazards. According to [78], an individual who has enough education and knowledge regarding about a certain issue will generally better understand the nature of a hazard and its likely effect on them. Not only can education affect individuals' knowledge of certain issues but it also helps to reduce poverty, improve health, get more and better job opportunities, higher salaries, etc.

4.1.4. Special Need Population

The population with special needs is usually much more vulnerable than people without a disability. Disability can be a huge factor for assessing vulnerabilities, especially when disasters or hazards occur. As mentioned by [76], this factor is closely linked to socioeconomic status, education and built environment, and ethnicity—all components of vulnerability assessment. It is shown by the variance for socioeconomic status (45.12%), education and built environment (19.74%), ethnicity (10.98%), and disability (10.78%).

4.1.5. Health

Health is one of the major indicators of this study. Variables such as medical services, health problems, and distance from the hospital are important factors of measuring social vulnerability as mentioned by [77,81,82]. Being healthy and having a good public healthcare system is important for communities that are more vulnerable to a disaster or hazard. Poor public health systems can simply make problems worse, and lead to more accidents and disruptions.

5. Discussion

There are not many studies concerning the Social Vulnerability Index (SoVI) with reference to landslides. Based on the research undertaken, articles regarding landslide in the context of social vulnerability index usually consider other types of hazard or where landslides are bracketed with other natural disasters. Articles based only on social vulnera bility and landslides are difficult to locate. Social vulnerability or the social vulnerability index has many types—not only SoVI but also referred to as SEVI or SVI. Even though the focus is only on SoVI in this paper, the researcher has taken note of other types of social vulnerability index.

There are 14 indicators that have been employed to measure social vulnerability in the context of landslide including (see Table 5): age, gender, ethnicity, built environment, income, family structure, education, employment, occupation, urban or rural, disability, migration, medical, and population. Based on the analysis, the researcher only focuses on five main indicators that have been used by many scholars: age, ethnicity, education, special needs population, and health. These were chosen because they are very relevant to the more vulnerable in society, especially where inequalities and imperiled areas are very evident.

5.1. Education

There are a few factors that affect social vulnerability including lack of access to: resources such as information, knowledge, and technology; social networks and connections with other individuals; social capital; and infrastructure [83]. In this study, education emerges as a major indicator employed in other studies regarding social vulnerability and landslides. Education is a bridge to success for many people, and it can refer to both formal and informal education. Education can also mean information, knowledge, and technology regarding the scope of discussion. The importance of education is to help people achieve more success and status in society, get a better job and understand the issues involved in a hazard or disaster. Furthermore, it helps individuals to be prepared for any circumstances. According to [84], people who have better response mechanisms, always prepare and constantly recover from a disaster, and this is certainly the case for those individuals, households, and societies with better and more widespread higher education outcomes compared to others.

5.2. Age

Indicators such as age can also mean susceptibility to social vulnerability. Older and very young people are more vulnerable to hazards and disasters than people in the middle. A higher proportion of senior citizens means that a society is at greater risk of disaster and the strategies needed to repair any given situation, simply because older people are more vulnerable to hazards than other age groups. Older people normally need a lot of physical and emotional care and support services. They can also be more disadvantaged compared to other age groups. The indicators that have been collected from previous studies do not represent the population or the place.

5.3. Ethnicity

Racism or ethnic discord is one of the factors of disaster risk, and especially for minority groups such as migrants and/or non-residents in a given location [85]. They are also known as marginalised groups, considered to be inferior in terms of their economic status, health, social relationships, and environment. If this situation continues, it will result in lasting social, political, and economic losses [86]. Although a mixture of socio-spatial and biophysical influences forms people's susceptibility to environmental hazards, race/ethnicity, and class have been central to understanding social dynamics during hazard events [87].

5.4. Special Needs Population

Special needs populations such as people with a disability are the most-at-risk persons when a disaster occurs. Disability means that the person with a physical or mental condition has limited movements, senses, or ability to participate in activities. Characteristics that are considered to be a disability are deafness, blindness, diabetes, autism, epilepsy, depression, and HIV. According to [88], disability emerges from the connection between people with health problems, such as cerebral palsy, Down syndrome, depression, as well as personal and environmental influences, including negative attitudes, limited transport facilities, public service facilities, and insufficient social support systems. They are generally the first victims of natural disasters. Indeed, early warning systems that alert the public may not actually reach the disabled individuals in time. The death toll from a disaster is two to four times larger than for those who are not disabled [89].

5.5. Healthcare Accessibility

Those with health problems are particularly vulnerable to landslides. They require constant attention and healthcare services to ensure their safety and good health. Therefore, access to health services such as hospitals, healthcare clinics, and pharmacies is an important need for this community. One of the principal components of emergency management is healthcare management to cope with disasters [90]. In disaster prevention activities, well targeted healthcare supply chain management can function effectively and efficiently. A substantial number of disaster casualties or even fatalities could be absorbed as long as healthcare services are up and running when a disaster occurs [91].

All the variables are listed above give an essential role in determining the security of a community based on social inductors. However, the results of the author's study found that income indicators and social capital are less emphasised. Income indicators referring to those with low incomes and belonging to the group below 40% of Malaysia's income are very vulnerable to disasters. For example, the floods that occur every year have caused suffering because they cannot work, and the worst consequences, they will lose their jobs. The study [92] found that the income sub-domain is the largest contributor and gives high value to the index of endangered livelihoods of rural communities in Pahang in 2014. Low-income conditions will also affect the period for them to recover after a catastrophic event. The results of the author's research found that there are no studies that explore social asset indicators. Social assets carry meaning as resources available to individuals and groups through membership in social networks. If the household has a higher position in a group or social institution, he or she will produce higher social strengths and resources [93]. Longer membership history as well as more participation in other social groups make it easier for access to information, business opportunities, social strength, and influence. The ability to access other assets is also simpler [94]. The evolution of social capital through the interaction of relationships between people and groups in community social networks [95,96]. Social networking means the interaction of an individual with other individuals, organisations, and groups to obtain information and assistance on something related to their livelihood [96,97]. The lack and absence of these elements within the social life environment of an individual will contribute to their vulnerability factors, as emphasised by [98,99]. Social capital influences, the sustainable life they possess significantly to strengthen the ability to develop a network of cooperation between groups both internally and externally and through enhancing the institutional capacity of community groups to improve the well-being of society.

State government agencies, local governments, and community leaders are the most familiar with the people in their communities. The social vulnerability index's importance is design to assist them in ensuring the security and well-being of their population. The SVI components can help the state and local people involved in all phases of the disaster sequence, in particular, landslides. Knowledge of locations and community information that is vulnerable to landslides can help planners in identifying target groups and accelerating assistance in efforts to reduce and impact property damage and loss of life, as well as prepare for disaster events. The stakeholders and management planners can setting the evacuation centre to places in secure condition to those are needs emergency assistances such as elderly people, single mothers with kids and infants, no transportations people and migrants whose are not influent in local language. In the recovery process, local governments may recognise communities that may require additional funding for human services or as a mitigation gauge to avoid a need for more costs due to the post support [100]. The slower to recover are those with the socioeconomically low-income community with hazardous areas of landslide occurrence. Therefore, the analysis results show that there are seven indicators as outlined that should be used as a social vulnerability index in measuring the level of susceptibility of landslides events. It consisted of education, age, ethnicity, special need population, healthcare accessibility, income, and social asset indicators. Future research will examine how SoVI can be used in the planning and mitigation processes to help target disaster management interventions as part of the system. Besides, the SoVI outcome can lead to geological mapping of disaster risk management in Malaysia's decision-making systems based on specific zones.

6. Conclusions

In this study, we have reviewed a selection of socioeconomic vulnerability components. At the searching stage, 258 articles were found in key databases, and after inclusion and exclusion criteria using the PRISMA guideline, only nine articles were chosen as being valid to this research. Fourteen variables were listed, and five variables of social vulnerability, which were typically used by scholars, proved to be relevant to Malaysia. Not all places or locations have the same experiences of landslides, and so the level of social vulnerability will differ and how these are measured. Although people may experience the same hazard or disaster, it does not mean that all individuals go through the same processes of destruction, recovery, evaluation, etc., as others. There are individuals who experience much higher social vulnerability than others, and it depends on which indicators are employed. As a climatic condition and the landslides occurrences in Malaysian context, there are seven indicators underlined which are education, age, ethnicity, special needs population, health accessibility, income, and social capital. These are the important indicator to measure the social vulnerability index to the high-risk communities towards landslide hazard. The result of these indicator measurement should be useful to authorities to include it as a complementary data to their geological mapping of disaster risk management based on the location of the landslide events. Furthermore, that is why, this study is important for understanding the social vulnerability index in the context of landslides in Malaysia.

Author Contributions: Conceptualization, M.I.N.D.; methodology, M.I.N.D.; software, M.I.N.D.; validation, M.I.N.D., M.R.T., and N.M.; formal analysis, M.I.N.D.; investigation, M.M.A.; resources, M.I.N.D.; data curation, M.M.A.; writing—original draft preparation, M.I.N.D.; writing—review and editing, M.M.A.; visualization, N.M.; supervision, M.R.T.; project administration, M.R.T. and A.O.; funding acquisition, M.R.T. and A.O. All authors have read and agreed to the published version of the manuscript.

Funding: This research was funded by Dana Padanan Antarabangsa—(MyPAIR)—Natural Environment Research Council (NERC), grant number NEWTON/1/2018/TK01/UKM/2.

Institutional Review Board Statement: Not applicable.

Informed Consent Statement: Not applicable.

Data Availability Statement: Not applicable.

Acknowledgments: The authors acknowledged the technical support by Nurul Atikah Zulkepli and Siti Nursakinah Selamat for the searching material of this study. We are also thanking to four anonymous reviewers whose give the comments/suggestion helped improve and clarify this manuscript.

Conflicts of Interest: The authors declare no conflict of interest.

References

1. Peduzzi, P. The Disaster Risk, Global Change, and Sustainability Nexus. *Sustainability* **2019**, *11*, 957. [CrossRef]
2. Burton, C.; Betts, R.A.; Jones, C.D.; Williams, K. Will fire danger be reduced by using Solar Radiation Management to limit global warming to 1.5 °C compared to 2.0 °C? *Geophys. Res. Lett.* **2018**, *45*, 3644–3652. [CrossRef]
3. Parker, R.N.; Densmore, A.L.; Rosser, N.J.; de Michele, M.; Li, Y.; Huang, R.; Whadcoat, S.; Petley, D.N. Mass wasting triggered by the 2008 Wenchuan earthquake is greater than orogenic growth. *Nat. Geosci.* **2011**, *4*, 449–452. [CrossRef]
4. Sultana, N. Analysis of Landslide-induced Fatalities and Injuries in Bangladesh: 2000–2018. *Cogent Soc. Sci.* **2020**, *6*, 1737402. [CrossRef]
5. Dai, F.C.; Lee, C.F.; Ngai, Y.Y. Landslide risk assessment and management: An overview. *Eng. Geol.* **2002**, *64*, 65–87. [CrossRef]
6. Guzzetti, F. Landslide fatalities and the evaluation of landslide risk in Italy. *Eng. Geol.* **2000**, *58*, 89–107. [CrossRef]

7. Haque, U.; da Silva, P.F.; Devoli, G.; Pilz, J.; Zhao, B.; Khaloua, A.; Wilopo, W.; Andersen, P.; Lu, P.; Lee, J.; et al. The human cost of global warming: Deadly landslides and their triggers (1995–2014). *Sci. Total Environ.* **2019**, *682*, 673–684. [CrossRef]
8. Keefer, D.K.; Larsen, M.C. Assessing Landslide Hazards. *Science* **2007**, *316*, 1136–1138. [CrossRef] [PubMed]
9. Petley, D.N.; Dunning, S.; Rosser, N.; Hungr, O. The analysis of global landslide risk through the creation of a database of worldwide landslide fatalities. In *Landslide Risk Management*; Hungr, O., Fell, R., Couture, R., Eberhardt, E., Eds.; Balkema Publishers: Rotterdam, The Netherlands, 2005; pp. 367–374.
10. Schuster, R.L.; Highland, L.M. *Socioeconomic Impacts of Landslides in the Western Hemisphere*; U.S. Geological Survey: Reston, VA, USA, 2001; p. 47.
11. CRED. Disaster Risk Profile: South East Asia-EM-DAT. The Emergency Events Database—Université Catholique de Louvain (UCL). 2018. Available online: http://emdat.be/emdat_db/ (accessed on 3 September 2020).
12. Vulnerability. Cambridge University Press. 2020. Available online: https://dictionary.cambridge.org/dictionary/english/vulnerability (accessed on 22 February 2020).
13. Smit, B.; Wandel, J. Adaptation, adaptive capacity and vulnerability. *Glob. Environ. Chang.* **2006**, *16*, 282–292. [CrossRef]
14. Hufschmidt, G. A comparative analysis of several vulnerability concepts. *Nat. Hazards* **2011**, *58*, 621–643. [CrossRef]
15. Bankoff, G.; Frerks, G.; Hilhorst, D. *Disasters, Development & People*; Routledge: London, UK, 2004; pp. 1–233.
16. Cutter, S.L.; Boruff, B.J.; Shirley, W.L. Social vulnerability to environmental hazards. *Soc. Sci. Q.* **2003**, *84*, 242–261. [CrossRef]
17. IPCC. *Climate Change 2007: Impacts, Adaptation, and Vulnerability*; Contribution of Working Group II to the Fourth Assessment Report of the Intergovernmental Panel on Climate Change; Parry, M.L., Canziani, O.F., Palutikof, J.P., van der Linden, P.J., Hanson, C.E., Eds.; Cambridge University Press: Cambridge, UK, 2007.
18. Henninger, N. *Mapping and Geographic Analysis of Human Welfare and Poverty Review and Assessment*; World Resources Institute: Washington, DC, USA, 1998.
19. Idrus, S.; Lim, C.-S.; Hadi, A.-S. Kemudahterancaman (Vulnerability) Penduduk Terhadap Perubahan Guna Tanah di Selangor. *Malays. J. Environ. Manag.* **2005**, *5*, 79–98.
20. O'Keefe, P.; Westgate, K.; Wisner, B. Taking the naturalness out of natural disasters. *Nature* **1976**, *260*, 566–567. [CrossRef]
21. Cutter, S.L. The secret history of natural disaster. *Glob. Environ. Change Part B Environ. Hazards* **2001**, *3*, 29. [CrossRef]
22. IPCC. Summary for Policymakers. In *Managing the Risks of Extreme Events and Disasters to Advance Climate Change Adaptation*; Field, C.B., Barros, V., Stocker, T.F., Qin, D., Dokken, D.J., Ebi, K.L., Mastrandrea, M.D., Mach, K.J., Plattner, G.-K., Allen, S.K., et al., Eds.; A Special Report of Working Groups I and II of the Intergovernmental Panel on Climate Change; Cambridge University Press: Cambridge, UK; New York, NY, USA, 2012; pp. 3–21.
23. Roslee, R.; Jamaludin, T.A.; Simon, N. Landslide vulnerability assessment (LVAs): A case study from Kota Kinabalu, Sabah, Malaysia. *Indones. J. Geosci.* **2017**, *4*, 49–59. [CrossRef]
24. CIDB. *Guidelines for Landslide Vulnerability Assessment and Risk Analysis for Critical Infrastructure in Malaysia*; CIDB: Kuala Lumpur, Malaysia, 2020; 54p.
25. Wisner, B.; Blaikie, P.; Cannon, T.; Davis, I. *At Risk: Natural Hazards, People Vulnerability and Disasters*, 2nd ed.; Routledge: London, UK, 2004.
26. Mavhura, E.; Manyena, B.; Collins, A.E. An approach for measuring social vulnerability in context: The case of flood hazards in Muzarabani district, Zimbabwe. *Geoforum* **2017**, *86*, 103–117. [CrossRef]
27. Juntunen, L. Addressing social vulnerability to hazards. *Disaster Saf. Rev.* **2006**, *4*, 3–10.
28. Emrich, C.T.; Cutter, S.L. Social vulnerability to climate-sensitive hazards in the Southern United States. *Am. Meteorol. Soc.* **2011**, *3*, 193–208. [CrossRef]
29. Cutter, S.L.; Emrich, C.T.; Morath, D.P.; Dunning, C.M. Integrating social vulnerability into federal flood risk management planning. *J. Flood Risk Manag.* **2013**, *6*, 332–344. [CrossRef]
30. Putnam, R.D. *Bowling Alone: Collapse and Revival of the American Community*; Simon & Schuster: New York, NY, USA, 2000.
31. Fekete, A. Social Vulnerability (Re-)Assessment in Context to Natural Hazards: Review of the Usefulness of the Spatial Indicator Approach and Investigations of Validation Demands. *Int. J. Disaster Risk Sci.* **2019**, *10*, 220–232. [CrossRef]
32. Roder, G.; Sofia, G.; Wu, Z.; Tarolli, P. Assessment of Social Vulnerability to Floods in the Floodplain of Northern Italy. *Weather Clim. Soc.* **2017**, *9*, 717–737. [CrossRef]
33. Johnson, K. SAMHSA Disaster Technical Assistance Center Supplemental Research Bulletin Greater Impact: How Disasters Affect People of Low Socioeconomic Status. *Phys. Health Health Probl.* **2017**. Available online: https://www.samhsa.gov/sites/default/files/programs_campaigns/dtac/srb-low-ses.pdf (accessed on 3 March 2021).
34. Lee, S.; Pradhan, B. Probabilistic landslide hazards and risk mapping on Penang Island, Malaysia. *J. Earth Syst. Sci.* **2006**, *115*, 661–672. [CrossRef]
35. Nhu, V.H.; Mohammadi, A.; Shahabi, H.; Ahmad, B.; Al-Ansari, N.; Shirzadi, A.; Geertsema, M.; Kress, V.R.; Karimzadeh, S.; Kamran, K.V.; et al. Landslide detection and susceptibility modeling on cameron highlands (Malaysia): A comparison between random forest, logistic regression and logistic model tree algorithms. *Forests* **2020**, *11*, 830. [CrossRef]
36. Ohkubo, T.; Tani, M.; Akojima, I.; Takahashi, S.; Yamakura, T.; Itoh, A.; Kanzaki, M.; Lee, H.S.; Ashton, P.S.; Ogino, K. Spatial pattern of landslides due to heavy rains in a mixed Dipterocarp forest, North-Western Borneo. *Tropics* **2007**, *16*, 59–70. [CrossRef]
37. Sharir, K.; Roslee, R.; Ern, L.K.; Simon, N. Landslide Factors and susceptibility mapping on natural and artificial slopes in Kundasang, Sabah. *Sains Malays.* **2017**, *46*, 1531–1540. [CrossRef]

38. Yusof, N.M.; Pradhan, B. Landslide susceptibility mapping along PLUS expressways in Malaysia using probabilistic based model in GIS. *IOP Conf. Ser. Earth Environ. Sci.* **2014**, *20*. [CrossRef]
39. Ya'Acob, N.; Tajudin, N.; Yusof, A.L.; Ali, D.M.; Sarnin, S.S. Rainfall thresholds for possible landslide occurrence in Ulu Kelang, Selangor, Malaysia using TRMM satellite precipitation estimates. *IOP Conf. Ser. Earth Environ. Sci.* **2018**, *169*. [CrossRef]
40. Alnaimat, A.; Choy, L.K.; Jaafar, M. An Assessment of Current Practices on Landslides Risk Management: A Case of Kuala Lumpur Territory. *Geogr. Malays. J. Soc. Space* **2017**, *13*, 1–12.
41. Mollaee, S.; Pirasteh, S.; Ramli, M.F.; Rizvi, S.M.A. Identifying effecting factors and landslide mapping of cameron highland Malaysia. In Proceedings of the 2013 5th International Conference on Geo-Information Technologies for Natural Disaster Management, GiT4NDM 2013, Mississauga, ON, Canada, 9–11 October 2013; pp. 73–83. [CrossRef]
42. Qasim, S.; Osman, S.B.S. Causal factors of Malaysian landslides: A narrative study. *Res. J. Appl. Sci. Eng. Technol.* **2013**, *5*, 2303–2308. [CrossRef]
43. Sulaiman, M.S.; Nazaruddin, A.; Salleh, N.M.; Abidin, R.Z.; Miniandi, N.D.; Yusoff, A.H. Landslide occurrences in Malaysia based on soil series and lithology factors. *Int. J. Adv. Sci. Technol.* **2019**, *28*, 1–26.
44. CFE-DHMA. *Malaysia: Disaster Management Reference Handbook, Center for Excellence in Disaster Management and Humanitarian Assistance*; Pearl Harbour-Hickam: Hawaii, HI, USA, 2019; 95p.
45. Jamaluddin, S.; Abd Aziz, J. Fitting Daily Rainfall Amount in Malaysia Using the Normal Transform Distribution. *J. Appl. Sci.* **2007**, *7*, 1880–1886.
46. Muhammad, N.S.; Julien, P.Y.; Salas, J.D. Probability structure and return period of multiday monsoon rainfall. *J. Hydrol. Eng.* **2016**, *21*, 04015048. [CrossRef]
47. Dou, J.; Tien Bui, D.; Yunus, A.P.; Jia, K.; Song, X.; Revhaug, I. Optimization of causative factors for landslide susceptibility evaluation using remote sensing and GIS data in parts of Niigata, Japan. *PLoS ONE* **2015**, *10*, e0133262. [CrossRef] [PubMed]
48. Sim, L.L.; Adrian, C.; Trisha, N. Malaysia among Countries Especially Prone to Landslides. 4 December 2018. Available online: https://www.thestar.com.my/news/nation/2018/12/04/msia-ranks-highly-for-landslides-country-experienced-185-occurrences-annually-in-past-10-years#:~{}:text=According%20to%20data%20from%20the,)%20and%20China%20(426) (accessed on 27 February 2021).
49. Prevention Web Malaysia. Disaster Risk Reduction is Everyone's Business. Dr Khamarrul Azahari Razak, UTM Geohazards Leader. 2 May 2019. Available online: https://www.preventionweb.net/news/view/65076 (accessed on 7 September 2020).
50. Rahman, H.; Mapjabil, J. Landslides disaster in Malaysia: An overview. *Health* **2017**, *8*, 58–71.
51. Majid, N.; Taha, M.; Selamat, S. Historical landslide events in Malaysia 1993–2019. *Indian J. Sci. Technol.* **2020**, *13*, 3387–3399. [CrossRef]
52. Saadatkhah, N.; Kassim, A.; Lee, M. Spatial patterns of precipitation, altitude and monsoon directions in Hulu Kelang area, Malaysia. *EJGE C* **2014**, *19*, 521–534.
53. Syah Hairizal Kamalul Arifin. Dua Rumah Terlibat Tanah Runtuh di Lereng Bukit di Papar. 10 November 2020. Available online: https://www.astroawani.com/berita-malaysia/dua-rumah-terlibat-tanah-runtuh-di-lereng-bukit-di-papar-267692 (accessed on 18 February 2021).
54. Bernama. Police: 13 Vehicles Trapped by Fraser's Hill landslides, No Injuries. 3 January 2021. Available online: https://www.thestar.com.my/news/nation/2021/01/03/police-13-vehicles-trapped-by-fraser039s-hill-landslides-no-injuries (accessed on 19 February 2021).
55. Nik Nur Izzatul Hazawani Nik Adnan. Tanah Runtuh, Kedai di Medan Selera Geliga Dihempap Bongkah Batu. 4 January 2021. Available online: https://www.utusan.com.my/nasional/2021/01/tanah-runtuh-kedai-di-medan-selera-geliga-dihempap-bongkah-batu/ (accessed on 18 February 2021).
56. Bernama. Hujan Lebat Punca Tanah Runtuh di JRTB. 7 January 2021. Available online: https://www.astroawani.com/berita-malaysia/hujan-lebat-punca-tanah-runtuh-di-jrtb-276444 (accessed on 18 February 2021).
57. Khairul Mohd Ali. 42 Penduduk Kosongkan Rumah Ekoran Tanah Runtuh. Utusan Malaysia Online. 11 January 2021. Available online: https://www.utusan.com.my/berita/2021/01/42-penduduk-kosongkan-rumah-ekoran-tanah-runtuh/ (accessed on 18 February 2021).
58. Bernama. Jalan Bukit Fraser-Raub Ditutup. 16 January 2021. Available online: https://www.hmetro.com.my/mutakhir/2021/01/664819/jalan-bukit-fraser-raub-ditutup (accessed on 18 January 2021).
59. Bernama. Banjir Sabah: 25 Jalan Terjejas, 10 Kejadian Tanah Runtuh. 9 February 2021. Available online: https://www.sinarharian.com.my/article/119525/BERITA/Semasa/Banjir-Sabah-25-jalan-terjejas-10-kejadian-tanah-runtuh (accessed on 18 February 2021).
60. Bernama. Women Saves Four Grandchildren as House Collapes during Landslides in Padawan, Sawarak. 13 January 2021. Available online: https://www.thestar.com.my/news/nation/2021/01/13/woman-saves-four-grandchildren-as-house-collapses-during-landslide-in-padawan-sarawak (accessed on 19 February 2021).
61. Abdul Rahman, H. An overview of environmental disaster in Malaysia and preparedness strategies. *Iran. J. Public Health* **2014**, *43*, 17–24.
62. Ahmad, J.; Lateh, H.; Saleh, S. Landslide hazards: Household vulnerability, resilience and coping in Malaysia. *J. Educ. Hum. Dev.* **2014**, *3*, 149–155. [CrossRef]

63. Sun, X.; Chen, J.; Han, X.; Bao, Y.; Zhan, J.; Peng, W. Application of a GIS-based slope unit method for landslide susceptibility mapping along the rapidly uplifting section of the upper Jinsha River, South-Western China. *Bull. Eng. Geol. Environ.* **2020**, *79*, 533–549. [CrossRef]
64. Soma, A.S.; Kubota, T. The Performance of Land Use Change Causative Factor on Landslide Susceptibility Map in Upper Ujung-Loe Watersheds South Sulawesi, Indonesia. *Geoplanning J. Geomat. Plan.* **2017**, *4*, 157. [CrossRef]
65. Mukhlisin, M.; Semarang, P.N.; Nizam, K.; Maulud, A.; Taha, M.R.; Water, S.; Using, R.; Pivot, C. GIS Based Landslide Hazard Mapping Prediction in Ulu Klang, Malaysia. *ITB Sci. J. A* **2010**, *42*, 163–178. [CrossRef]
66. Pradhan, B.; Youssef, A.M. Manifestation of remote sensing data and GIS on landslide hazard analysis using spatial-based statistical models. *Arab. J. Geosci.* **2010**, *3*, 319–326. [CrossRef]
67. Abdullah, C.H. Landslide risk management in Malaysia. *Disaster Manag. Hum. Health Risk III* **2013**, *133*, 255–265. [CrossRef]
68. Ministry of Housing and Local Government. Corporate Communications Unit, Kuala Lumpur. 2009. Available online: https://www.kpkt.gov.my/resources/index/user_1/siaran_media/ogos/SIARAN_AKHBAR_Garis_Panduan_Tanah_Tinggi09.pdf (accessed on 30 December 2020).
69. Shaluf, I.M.; Ahmadun, F.R. Disaster types in Malaysia: An overview. *Disaster Prev. Manag. Int. J.* **2006**, *15*, 286–298. [CrossRef]
70. Elmahdy, S.I.; Mostafa, M.M. Natural hazards susceptibility mapping in Kuala Lumpur, Malaysia: An assessment using remote sensing and geographic information system (GIS). *Geomat. Nat. Hazards Risk.* **2013**, *4*, 71–91. [CrossRef]
71. Jamaludin, S.; Hussein, A.N. Landslide hazard and risk assessment: The Malaysian experience. *IAEG* **2006**, *455*, 1–10.
72. Mohamed Shaffril, H.A.; Samah, A.A.; Samsuddin, S.F.; Ali, Z. Mirror-mirror on the wall, what climate change adaptation strategies are practiced by the Asian's fishermen of all? *J. Clean. Prod.* **2019**, *232*, 104–117. [CrossRef]
73. Martín-Martín, A.; Orduna-Malea, E.; Thelwall, M.; López-Cózar, E.D. Google scholar, web of science, and scopus: A systematic comparison of citations in 252 subject categories. *J. Informetr.* **2018**, *12*, 1160–1177. [CrossRef]
74. Moher, D.; Liberati, A.; Tetzlaff, J.; Altman, D.G. The Prisma Group. Preferred reporting items for systematic reviews and meta-analyses: The PRISMA statement. *PLoS Med.* **2009**, *6*, e1000097. [CrossRef] [PubMed]
75. Aksha, S.K.; Juran, L.; Resler, L.M.; Zhang, Y. An analysis of social vulnerability to natural hazards in Nepal using a modified social vulnerability index. *Int. J. Disaster Risk Sci.* **2019**, *10*, 103–116. [CrossRef]
76. Aksha, S.K.; Resler, L.M.; Juran, L.; Carstensen, L.W., Jr. A geospatial analysis of multi-hazard risk in Dharan, Nepal. *Geomat. Nat. Hazards Risk* **2020**, *11*, 88–111. [CrossRef]
77. Guillard-Gonçalves, C.; Cutter, S.L.; Emrich, C.T.; Zêzere, J.L. Application of Social Vulnerability Index (SoVI) and delineation of natural risk zones in Greater Lisbon, Portugal. *J. Risk Res.* **2015**, *18*, 651–674. [CrossRef]
78. Kantamaneni, K. Evaluation of social vulnerability to natural hazards: A case of Barton on Sea, England. *Arab. J. Geosci.* **2019**, *12*, 1–11. [CrossRef]
79. Frigerio, I.; Amicis, M.D. Mapping social vulnerability to natural hazards in Italy: A suitable tool for risk mitigation strategies. *Environ. Sci. Policy* **2016**, *6*, 187–196. [CrossRef]
80. Qasim, S.; Qasim, M.; Shrestha, R.P.; Khan, N. Socioeconomic determinants of landslide risk perception in Murree hills of Pakistan. *Aims Environ. Sci.* **2018**, *5*, 305–314. [CrossRef]
81. Kumar, D.; Bhattacharjya, R.K. Study of integrated social vulnerability index SoVIint of hilly region of Uttarakhand, India. *Environ. Clim. Technol.* **2020**, *24*, 105–122. [CrossRef]
82. Zeng, J.; Zhu, Z.Y.; Zhang, J.L.; Ouyang, T.P.; Qiu, S.F.; Zou, Y.; Zeng, T. Social vulnerability assessment of natural hazards on county-scale using high spatial resolution satellite imagery: A case study in the Lougang district of Guangzhou, South China. *Environ. Earth Sci.* **2012**, *65*, 173–182. [CrossRef]
83. Wijaya, A.P.; Hong, J.H. Quantitative assessment of social vulnerability for landslide disaster risk reduction using GIS approach. *Int. Arch. Photogramm. Remote Sens. Spat. Inf. Sci.* **2018**, *XLII*, 703–709. [CrossRef]
84. Blaikie, P.; Canon, T.; Davis, I.; Wisner, B. *At Risk: Natural Hazards People's Vulnerability and Disasters*; Routledge: London, UK, 1994.
85. Elstad, J.L. Inequalities in health related to women's marital, parental, and employment status: A comparison between the early 70s and the late 80s, Norway. *Soc. Sci. Med.* **1996**, *42*, 75–89. [CrossRef]
86. Bolin, B. Race, Class, Ethnicity, and Disaster Vulnerability. In *Handbook of Disaster Research. Handbooks of Sociology and Social Research*; Springer: New York, NY, USA, 2007. [CrossRef]
87. Winant, H. *The World Is a Ghetto. Basic Books;* Latin American Faculty of Social Sciences Ecuador Headquarters–FLACSO: New York, NY, USA, 2001.
88. World Health Organization. Disability. 2021. Available online: https://www.who.int/health-topics/disability#tab=tab_1 (accessed on 29 January 2021).
89. Castro, C.-P.; Sarmiento, J.-P.; Edwards, R.; Hoberman, G.; Wyndham, K. Disaster risk perception in urban contexts and for people with disabilities: Case study on the city of Iquique (Chile). *Nat. Hazards* **2017**, *86*, 411–436. [CrossRef]
90. Pourhosseini, S.S.; Ardalan, A.; Mehrolhassani, M.H. Key Aspects of Providing Healthcare Services in Disaster Response Stage. *Iran. J. Public Health* **2015**, *44*, 111–118. [PubMed]
91. Syahrir, I.; Vanany, I. Healthcare and Disaster Supply Chain: Literature Review and Future Research. *Procedia Manuf.* **2015**, *4*, 2–9. [CrossRef]

92. Nurul Ashikin, A.; Nor Diana, M.I.; Siwar, C.; Alam, M.M.; Yasar, M. Community Preparation and Vulnerability Indices for Floods in Pahang State of Malaysia. *Land* **2021**, *10*, 198. [CrossRef]
93. Dehghani Pour, M.; Barati, A.A.; Azadi, H.; Scheffran, J. Revealing the role of livelihood assets in livelihood strategies: Towards enhancing conservation and livelihood development in the Hara Biosphere Reserve, Iran. *Ecol. Indic.* **2018**, *94*, 336–347. [CrossRef]
94. Singgalen, Y.A.; Sasongko, G.; Wiloso, P.G. Community participation in regional tourism development: A case study in North Halmahera Regency—Indonesia. *Insights Reg. Dev.* **2019**, *1*, 318–333. [CrossRef]
95. Po, J.Y.T.; Hickey, G.M. Cross-scale relationships between social capital and women's participation in decision-making on the farm: A multilevel study in semi-arid Kenya. *J. Rural Stud.* **2020**, *78*, 333–349. [CrossRef]
96. Kumar, A.; Cabral-Pinto, M.; Kumar, A.; Kumar, M.; Dinis, P.A. Estimation of Risk to the Eco-Environment and Human Health of Using Heavy Metals in the Uttarakhand Himalaya, India. *Appl. Sci.* **2020**, *10*, 7078. [CrossRef]
97. Kumar, A.; Mishra, S.; Kumar, A.; Singhal, S. Environmental quantification of soil elements in the catchment of hydroelectric reservoirs in India. *Human Ecol. Risk Assess. An Int. J.* **2017**, *23*, 1202–1218. [CrossRef]
98. Majid, N.A.; Nazi, N.M.; Idris ND, M.; Taha, M.R. GIS-based livelihood vulnerability index mapping of the socioeconomy of the Pekan community. *Sustainability* **2019**, *11*, 6935. [CrossRef]
99. Nor Diana, M.I.; Chamburi, S.; Mohd Raihan, T.; Nurul Ashikin, A. Assessing local vulnerability to climate change by using Livelihood Vulnerability Index: Case study in Pahang region, Malaysia. *IOP Conf. Ser. Mater. Sci. Eng.* **2019**, *506*. [CrossRef]
100. Kumar, A.; Cabral-Pinto, M.M.S.; Chaturvedi, A.K.; Shabnam, A.A.; Subrahmanyam, G.; Mondal, R.; Gupta, D.K.; Malyan, S.K.; Kumar, S.S.; A Khan, S.; et al. Lead Toxicity: Health Hazards, Influence on Food Chain, and Sustainable Remediation Approaches. *Int. J. Environ. Res. Public Health* **2020**, *17*, 2179. [CrossRef] [PubMed]

land

MDPI

Article

Land Financialization, Uncoordinated Development of Population Urbanization and Land Urbanization, and Economic Growth: Evidence from China

Yunyang Ji [1], Xiaoxin Guo [2,*], Shihu Zhong [2] and Lina Wu [2]

1 School of Public Finance & Taxiation, Guangdong University of Finance & Economics, Guangzhou 510320, China; 20196031@gdufe.edu.cn
2 School of Public Economics and Administration, Shanghai University of Finance and Economics, Shanghai 200433, China; zhongshihu@163.sufe.edu.cn (S.Z.); 2015310055@live.sufe.edu.cn (L.W.)
* Correspondence: guoxiaoxin@163.sufe.edu.cn

Received: 10 November 2020; Accepted: 26 November 2020; Published: 29 November 2020

check for updates

Abstract: In recent years, it has become common practice for Chinese local governments to inject land assets into financing platform companies and use them as mortgage or credit guarantees to obtain bank loans and issue urban investment bonds, which is known as "land financialization". This study investigates the impact and mechanism of land financialization on the uncoordinated development of population urbanization and land urbanization in China. Theoretical analysis and empirical analysis results based on the data of prefecture-level cities in China from 2006 to 2015 demonstrate that land financialization by local governments is a significant cause of the uncoordinated development of population urbanization and land urbanization, and the pressure of urban economic development will strengthen this negative impact. Extended analysis further reveals that in areas where population urbanization and land urbanization are uncoordinated, land financialization, while promoting urban spatial expansion, will lower land use efficiency and have an inverted U-shaped influence on economic growth due to a weak agglomeration effect. The above conclusion shows that urbanization driven by debt-based investment is unsustainable. Efforts should be made to establish a financialization system that propels sound urbanization and to build a stable input linkage between land financialization and the supply of urban public service.

Keywords: land financialization; uncoordinated development of population urbanization and land urbanization; pressure of urban economic development; land use efficiency; urban economic growth

1. Introduction

Urbanization is a vital symbol showcasing economic development in a country and is the universal choice used by developing countries to promote economic development. Existing research shows that urbanization can boost economic growth through promoting the accumulation of elements and the spillover of knowledge, expanding consumption and investment and delivering an agglomeration effect and economies of scale [1–3]. Further, it has become a critical issue facing countries across the world to facilitate economic growth based on faster and better-quality urbanization.

As the largest developing country, China's urbanization level has soared over the past 40 years, with its urbanization rate growing from 17.92% in 1978, the fledgling years of China's reform and opening up, to 59.58% in 2018. The number of permanent residents in urban areas has reached 831.4 million (China Statistical Yearbook, 2018). Urbanization has become a gigantic engine, after industrialization, for China's economic growth [4,5]. However, there is no denying that some contradictions and problems appear along with China's urbanization. The major problem is that population urbanization lags behind land urbanization [6,7] (China's New Urbanization Plan (2014–2020), 2014). According to the author's estimation based on data from the China Urban Construction Statistical Yearbook, the pace of urban land expansion in China measured by built-up area is 1.44 times that of the increase in permanent residents from 1990 to 2000. The difference between the two from 2000 to 2015 further expands to 1.94 times, far beyond the reasonable level of 1.12 [8]. Additionally, in terms of the annual average growth rate, the difference between the annual average growth rate of urban built-up area in China and that of urban population is 3.21% from 1990 to 2006. This figure climbs to 3.47% from 2006 to 2015. Meanwhile, among three regions in China, the difference between the increased rate of land urbanization and population urbanization in the western area is the largest, followed by the central and eastern area (see Figure A1).

Existing literature shows that the uncoordinated development of population urbanization and land urbanization will result in a series of long-term negative effects on socioeconomic development, such as an increasingly distorted economic structure [6,9] and a widening income gap among urban and rural residents [10]. Ultimately, this uncoordinated development will obstruct urban-rural economic sustainable development and the healthy development of the city [11,12]. The cases in point are mushrooming "ghost cities", "empty cities", and the low occupancy rate in some development zones in China [13]. Indeed, promoting the coordinated development of land urbanization and population urbanization has become a critical issue in encouraging future urbanization in China. Resolving this issue will require the examination of its internal logic.

What are the reasons for the uncoordinated development of population urbanization and land urbanization in China? Existing literature has focused on the effect of some institutional factors, including the hukou system's obstruction of population urbanization in China [14,15], urban spatial expansion facilitated by the land-dependent fiscal system [12,16], and land urbanization propelled by regional competition under the fiscal decentralization system [17,18]. Moreover, the urbanization model in China is different from those in Western countries, such as governance structures for farmland conversion [19], the hukou system, internal migration, the taxation system, and the key role of local government in development [20]. Therefore, the behavior and intention of local governments should be considered in order to understand urbanization in China [21]. The above literature laid the foundation for our analysis of the uncoordinated development of population urbanization and land urbanization, but the impact of the land financialization mode, which local governments rely on, on the development of urbanization has been, to some extent, ignored.

In actual fact, local governments in China have established a huge number of financing platform companies in recent years, which directly or indirectly assume the functions of land reserve, development, and transfer in various cities [22]. It has become common practice for Chinese local governments to inject land assets into financing platform companies and use them as mortgage or credit guarantees to obtain bank loans and issue urban investment bonds, which is known as "land financialization". This land financialization model greatly enhances the financing effect of land transfer [23]. Although they take the same form as land capitalization, land financialization differs from land transfer greatly, because the former is a borrowing behavior, which creates implicit debts, while the latter is a one-time deal. According to the statistics for 84 major cities in China, the area and revenue from land mortgages have continued to rise and gradually exceed those of land transfer since the market-orientated reform of industrial land in 2006 (see Figure A2). Large-scale land financialization provides funding support for urban construction and land development, forming a Chinese-style urban construction investment and financialization model characterized by positive feedback between land financialization and urban infrastructure investment [24].

Thus, the question must be asked, does land financialization by Chinese local governments affect land urbanization and population urbanization? If the impact exists, does this kind of financialization lead to the uncoordinated development of population urbanization and land urbanization? Moreover, in what way does urbanization propelled through land financialization by local governments affect urban economic growth? These answers are the mainly logic connection of this study.

In effect, urbanization propelled via land financialization by Chinese local governments is, in essence, designed to promote urban construction investment through borrowing. Land financialization, while accelerating urban construction, generates huge debts for local governments [25]. As of the end of 2017, the local debts released by the authority amounted to 16.47 trillion yuan. In addition, urban development in China follows a "supply-driven" model led by local governments, where urban construction predates the agglomeration of industry and people. This means that economic agglomeration and economies of scale will determine whether a city can realize sustainable development. If local governments fail to attract enough people and companies after the construction is completed, construction investment which relies on borrowing will not be translated into an effective tax base and land tax revenue, which will probably trigger fiscal risks. In particular, in cities where unbalanced development is noted, population urbanization which is lagging behind reflects the insufficient spatial conglomeration of the population. The large-scale "enclosure for cities" movement gives rise to a low efficiency in land use [26]. Economies of scale are stunted, thus obstructing sustainable economic development in China.

This study focuses on the impact of land financialization by local governments on the uncoordinated development of population urbanization and land urbanization in China and its underlying mechanism. The contributions of this study mainly include: first, from the research perspective, previous research proceeds from land financialization and examines its stimulating effect on urban spatial expansion, while our study explores the uncoordinated development of population urbanization and land urbanization during urban construction and land development funded by local governments' debt-based financialization. This study also puts under the microscope the moderating effect of urban development pressure on the uncoordinated development of population urbanization and land urbanization caused by land financialization. Second, by combining the realistic background of "debt-oriented" urbanization financing and "supply-oriented" urban construction, this study analyzes the potential negative effect of land financialization on urban economic development and its working mechanism. Lastly, as financing platform companies are the main carrier allowing local governments to conduct land financialization, this study calculates the land financialization scale through the interest-bearing debt of financing platform companies for empirical research. This constitutes a beneficial attempt in the absence of relevant data.

2. Institutional Background and Hypotheses

2.1. The Formation and Evolution of the Financing of Urban Development in China

Local governments are major players in urban construction investment and public service supply in China [27–29]. The fiscal revenue and spending caused by Chinese local governments have a huge impact on urban development [30]. Since China's tax-sharing reform in 1994, financial power has been transferred to the central government, while responsibilities and spending have been laid on the shoulders of local governments [31]. The budgetary fiscal deficit of local governments has grown with each passing year, pushing local governments to expand extra budgetary sources of revenue. In the meantime, the newly revised Land Management Law of 1998 made Chinese local governments the only monopoly in land supply. While the market-orientated reform of urban housing in 2000 and the "bidding, auction, and listing" land transfer system that started in 2002 have made market-orientated land transfer possible in China. These reform measures have bred the commercial value of land, which gradually rises as the important asset for local governments. In this stage, land-related fiscal revenue for local governments mainly includes three components: land transfer fees, taxes related to the construction industry and the real estate industry (including farmland occupancy tax, urban land use tax, land appreciation tax, deed tax, and property tax), and land mortgage loans. Therefore, the fiscal revenue of Chinese local governments and urban construction has become increasing dependent on land.

Originally, Chinese local governments mainly obtained land transfer revenue through selling land use rights in primary markets. According to the China Land and Resources Statistical Yearbook 2001 to 2014, land transfer fees, known as "the secondary fiscal revenue", soared to 4.37 trillion yuan in 2013 from 59.5 billion yuan in 2000. The share of land transfer revenue in local fiscal revenue has increased from 9.3% to 63%. In some cities, such as Nanjing, Hefei, and Hangzhou, etc., the dependence on land finance (the ratio of budget revenue to land transfer fees) even approaches 100%, and remains at that level. Along with the development of the real estate market, the relevant tax revenue accordingly increased rapidly. Land finance therefore came into place. While compensating for fund shortage in local governments, the revenue also exerted far-reaching impacts on socioeconomic development in China. Said revenue was utilized in urban infrastructure construction and provided a funding guarantee for urbanization [32].

However, in China, along with deepening regulations in the real estate market and the growing pressure on economic downturn, land transfer revenue and relevant tax fees could hardly keep up with the substantial spending in urban construction, which ushered in the tipping point of land finance. According to Research on Real Estate's Contribution to Fiscal Revenue 2015[1], the tax revenue in real estate in 2015, including property tax, urban land use tax, deed tax, farmland occupancy tax, and land appreciation tax, amounted to around 1.3 trillion yuan, a decrease of 3.6% compared to 2014, the first-ever decline in the past 15 years. Land transfer fees in 2015 registered negative growth for the fourth time in the past 15 years. The land revenue of local governments fell by nearly one half. Against this backdrop, establishing a local financialization platform through mortgaging land to financial institutions or using land transfer revenue as a guarantee to issue urban investment bonds has become an alternative for local governments to obtain land financing. Since then, the financing of urban development has transformed from land finance to land financialization in China. Particularly, after the financial crisis in 2008, China encouraged the establishment of local financing platform companies in order to raise supporting capital for the "funds for the "four-trillion stimulus plan". The number of local financing platform companies surged to 8221 at the end of 2009 from 3000 in the second half of 2008. In 2009 alone, more than 2000 new financing platform companies were added. These platform companies raised 75% of the supporting capital [33]. Land, as the major asset, played a key role in fund collecting.

Essentially, land and land-related revenue are deeply involved, from the establishment of local financialization platforms and asset injection, guarantees, and mortgages in borrowing and issuing bonds, to the repayment of principle and interest in the future. Extensive empirical research has verified the guarantee function of land in the issuance of urban investment bonds and its mortgage function in obtaining loans from banks. For example, the research of Shu, Xie, Jiang, and Chen (2018) [16] and Zhong, Chen, and Huang (2016) [34] indicates that the greater the price of urban land transfer or revenue, the higher the possibility of issuing urban investment bonds. The scale of the bonds issued will also accordingly be larger. This is because land use rights are often used as the guarantee, and land transfer revenues are promised to pay back debt during the issuance of urban investment bonds. Zhang, Nian, and Liu (2018) [23] found that there was a significant positive relationship between land transfer revenue and urban investment bonds. According to the debt auditing report by the National Audit Office from 2011 to 2013, as of the end of 2010 the outstanding obligation that local governments promised to pay through land transfer revenue stood at 2.547 trillion yuan, amounting to 38% of the total liable debts. This figure grew to 3.49 trillion yuan at the end of 2012—an increase of nearly 37% within two years. The outstanding obligation promised to be paid through land transfer revenue is shown in Figure 1. We can find that, by the end of 2012, the proportion

---

1 Shanghai E-House Real Estate Research Institute, Local government's dependence on Land Finance, 29 February 2016. https://finance.sina.com.cn/roll/2016-02-29/doc-ifxpvzah8335807.shtml.

of outstanding obligation promised to be paid through land transfer revenue in Zhejiang Province, Tianjin Municipality, and Beijing Municipality exceeds 60%.

Figure 1. Debts promised to be paid by land transfer revenue in Chinese provinces. Note: data were retrieved from the Local Debts Audit Report of provinces in 2014.

The above evidence shows that large-scale urban land expansion and land transfer revenue provide a credit basis for Chinese local governments to obtain debt financialization and thus play a catalytic role in the expansion of local government debt [35,36]. Against the backdrop of severe dependence on debt-based investment, local governments' passion for "operating city" and "operating land" will be galvanized through constructing various development zones, new areas, and new cities, turning agricultural land into construction land and ultimately triggering precocious urban construction. Following the ideas of Chang and Lu (2017) [37], as of early 2014, 272 out of 280 prefecture-level cities in China had new built-up cities. The number of cities with new cities under construction accounts for 90% of all prefecture cities in China—the area of which amounts to 66,300 km^2.

On the other hand, the employment of debt-based capital through land financialization by local governments showcases an obvious orientation to land urbanization. The majority of the capital was invested in areas related to land urbanization, such as municipal construction, land purchasing and storage, and transportation, with a proportion of approximately 70.38%. Only 10.5% of the capital was, however, invested in areas relevant to population urbanization, such as education, science, culture, healthcare, and affordable housing (National Audit Report on Debt). This means that land financialization does not promote population urbanization in the way that it does in facilitating urban spatial expansion and "land urbanization". On the contrary, the rapid rises in urban land price and housing price caused by land financialization may even suppress population urbanization. Research Group on China's Economic Growth (2011) [38] points out that the land urbanization and land-based urbanization financialization model led by local governments will accelerate land discount, heightening issues such as soaring land prices and housing prices while also having a crowd-out effect on the population and industries.

Based on the above analysis, we propose the first hypothesis.

Hypothesis 1. *The stimulating effect of land financialization activities by local governments on land urbanization is stronger than that on population urbanization, resulting in the uncoordinated development of population urbanization and land urbanization.*

China's special system of the political centralization of state power and the separation of economic power effectively resolves the incentives for local governments [39,40]. However, it also brings about "yardstick competition" among different regions [41,42]. An official government promotion mechanism focusing on economic performance, which has been gradually formed since the 1980s, leads to "competition for growth" among government officials at a local level. The competition for promotion has been turned into competition in economic growth among various regions [43,44], putting pressure on economic development. Li and Zhou (2005) [45] found, based on empirical research, that local government officials with better economic performance to their credit during their tenure do receive more promotion opportunities. Considering this, local governments tend to realize more public spending related to economic development within a short period of time, such as urban infrastructure construction and land development, to achieve excellent economic development performance. This move has also significantly distorted the fiscal expenditure structure of local governments, showcasing an emphasis on infrastructure and a neglect of human capital investment and public service [46]. Similarly, in the supply of public goods, local governments are passionate about providing public goods of an economic nature while overlooking those of a social nature [47] The research of Caldeira (2012) [48] demonstrates that there is competition in investment among prefecture-level cities in China. Large-scale infrastructure construction and government investment require ample funding support. Extrabudgetary debt funds from land financialization simply loosen budgetary constraints. Therefore, under the decentralized system and official government promotional mechanism, "yardstick competition" emerges in land financialization among local governments, which is indicated by greater passion about, and dependence on, land financialization in regions with larger pressure on economic development.

In comparison, land urbanization relies on government investment and infrastructure construction, while population urbanization is more dependent on the supply of public goods which are social in nature. The biased public spending structure originating from promotional incentives for local governments' officials will indirectly strengthen land urbanization orientation in spending the capital raised. A vicious cycle in the investment and financialization process centring on land has been formed, augmenting the uncoordinated development of population urbanization and land urbanization. Meanwhile, in cities with greater pressure on economic development, local governments would have preferred to follow a spending structure orientated towards land urbanization to obtain greater achievements.

Therefore, it can be inferred that, under the competition mechanism for the promotion of officials, local governments under great development pressure are more dependent on land financialization model to raise urban construction funds, and their expenditure is more inclined towards land urbanization, thus heightening the uncoordinated development of population urbanization and land urbanization. We therefore propose the second hypothesis.

Hypothesis 2. *The stimulating effect of land financialization on the uncoordinated development of population urbanization and land urbanization is more prominent in cities with greater pressure on economic development.*

2.2. *Land Financialization, the Uncoordinated Development of Population Urbanization and Land Urbanization, and Economic Growth*

Although each city has formed its own features in the course of development, the urbanization of these cities usually includes the continuous conglomeration of industries and populations, which promotes economic growth in cities, increased job opportunities, the inflow of population, and the expansion of urban land construction. We refer to this model as a market-led "demand-driven" model. Under this model, land expansion in cities is based on industrial conglomeration and population growth. Economies of scale can therefore be generated, realizing relatively high land use efficiency and ensuring sustainable economic development. In other words, sustainable economic growth is based on alternative and coordinated development between population urbanization and land urbanization.

However, in the government-led "supply-driven" urban development model in China, investment and construction rely on land financialization. Companies and populations are then attracted to the region through preferential policies on land and taxation [49]. This means that income from investment based on land financialization is decided by the conglomeration effect of urban construction on population and industry. However, when urban differentiation and regional difference become increasingly prominent, not all cities are able to attract enough companies and labour to the region. This is particularly true in areas where population urbanization and land urbanization are uncoordinated. Lagging-behind population urbanization itself reflects inadequate population conglomeration, ultimately resulting in a low land use efficiency [50,51]. In the meantime, land use efficiency is critical to sustainable economic development [52]. If precocious land urbanization could not deliver "increasing return to scale", sustainable economic growth in cities would face severe challenges.

Additionally, in terms of the effect of the transmission mechanism of urbanization on promoting economic development, population urbanization facilitates economic growth through population agglomeration and increasing consumption, while land urbanization does so via increments of fixed asset investments. In recent years, investments' contribution to economic growth has continued to decline during the adjustment of the economy structure. Increasing residential consumption is the key to realizing sustainable economic development. If population urbanization lags behind land urbanization over a long period of time, the effect of the demand structure, characterized by high investment and low consumption, on economic structure will have a threshold value according to the law of diminishing return on investment. The effect will shift from positive to negative once the demand structure passes the threshold.

To summarize, urbanization lacking industrial and population support could be temporarily sustained through government efforts and debt-based capital. Long-term economic growth relies on the performance of economic conglomeration and population urbanization. Land financialization leads to the uncoordinated development of population urbanization as well as land urbanization and is detrimental to sustainable economic development. Therefore, we propose the third hypothesis.

Hypothesis 3. *In areas with the uncoordinated development of population urbanization and land urbanization, due to weak conglomeration, land financialization will, while promoting urban spatial expansion, lower land use efficiency, generating an inverted U-shaped effect on economic performance.*

3. Model, Data and Variable

3.1. *Econometric Model*

Based on previous literature concerning land financialization, population urbanization, and land urbanization, this study constructs the following econometrics model to test the impact of land financialization on the uncoordinated development of population urbanization and land urbanization:

$$land\_urban_{it} = \alpha_1 + \beta_1 land\_financialization_{it} + \sum \gamma_1 X_{it} + \mu_i + \delta_t + \varepsilon_{it} \quad (1)$$

$$pop\_urban_{it} = \alpha_2 + \beta_2 land\_financialization_{it} + \sum \gamma_2 X_{it} + \mu_i + \delta_t + \varepsilon_{it} \quad (2)$$

$$unco\_development_{it} = \alpha_3 + \beta_3 land\_financialization_{it} + \sum \gamma_3 X_{it} + \mu_i + \delta_t + \varepsilon_{it} \quad (3)$$

$$\begin{aligned} unco\_development_{it} &= \alpha_4 + \beta_4 land\_financialization_{it} + \varphi pressure_{it} + \\ &\eta land\_financialization_{it} * pressure_{it} + \sum \gamma_4 X_{it} + \mu_i + \delta_t + \varepsilon_{it} \end{aligned} \quad (4)$$

where the dependent variable in model 1 is $land\_urban_{it}$, representing the land urbanization in city i at year t, which is measured by the logarithm of the urban built-up area. The dependent variable in model 2 is pop_urban_{it}, representing the population urbanization in city i at year t, which is measured by the logarithm of permanent residents in city districts. The dependent variable in models 3 and 4 is $unco\_development_{it}$, indicating the dummy variable of the uncoordinated development of population urbanization and land urbanization in city i at year t. The value 1 demonstrates the existence of the uncoordinated development of population urbanization and land urbanization, or the value will be 0. In models 1-4, the independent variable is $land\_financialization_{it}$, representing land financialization scale in city i at year t. The subscripts i and t represent the i-th city and the t-th year, respectively. X_{it} is the set of control variables, as discussed above. Further, μ_i represents the municipal fixed-effects and is used to control the unobservable but invariant characteristics of cities; δ_t represents the year fixed-effects, and is used to control the systematic differences in cities over time. ε_{it} represents other city-level natural endowments and socioeconomic factors that could potentially impact the uncoordinated development of population urbanization and land urbanization.

In models 1–3, we focus on the coefficients of β_1, β_2, and β_3, which represent the net effect of the land financialization on land urbanization, population urbanization, and the uncoordinated development of population urbanization and land urbanization, respectively. In model 4, the moderating variable is $pressure_{it}$, representing the pressure on urban economic development in city i at year t; we focus on the coefficient of η, which represents the moderating effect of urban economic development pressure on the uncoordinated development of population urbanization and land urbanization caused by land financialization.

The fixed-effect (FE) panel data method is first used to control factors that do not vary with time. However, there may be a reverse causality between land financialization and the uncoordinated development of population urbanization and land urbanization. Land financialization offers funding support for land development and spatial expansion and gives rise to the uncoordinated development of population urbanization and land urbanization. Meanwhile, increases in construction land further galvanize land-based mortgages and guarantees in cities with unbalanced development, adding fuel to land financialization. The Generalized Method of Moments (GMM) and two stage least square (2SLS) regression are therefore used to address this problem by introducing exogenous variables.

3.2. Data Source and Variable Selection

Our main data is drawn from the China Statistics Yearbook, the China Municipal Statistics Yearbook, the China Population and Employment Statistics Yearbook, the China Urban Construction Statistical Yearbook, and the China Land and Resources Statistical Yearbook between 2007 and 2016. Samples of Tibet and some other autonomous prefectures are excluded. Data on local financialization platforms comes from the Wind Economic Database (The Wind Economic Database pairs over 1.3 million macroeconomic and industry time series with powerful graphics and data analysis tools to give financial professionals the most comprehensive insights into China's economy). Missing values are filled using the interpolation method. It is important to note that, in the Chinese context, a city is often not a municipal unit (i.e., a large continuous urban area), but rather an administrative unit that with hierarchy ranking lower a province but higher a county in the Chinese administrative structure. A Chinese city usually comprises a main central urban area (with the same name as the "city") and a much larger surrounding rural area [53]. In this study, only the information regarding the city's central urban (municipal) area is used. Thus, our analysis is not very sensitive to changes in the city's administrative boundaries. The variable definitions and statistical descriptions of samples used in the analysis will be recorded as shown below.

3.2.1. Dependent Variables

The key dependent variables include: $\ln land\_urban_{it}$, the logarithm of city-level built-up area, which is the important index of land urbanization; $\ln pop\_urban_{it}$, the logarithm of city-level permanent residents, which is the important index of population urbanization; $unco\_development_{it}$, the dummy variable of the uncoordinated development of population urbanization and land urbanization. In reference to the practice of Xie (2016) [54], this study employs the elastic coefficient of urban built-up area growth (i.e., the growth rate of urban built-up area/the growth rate of permanent residents in a municipal district) to measure the lagged-behind level of population urbanization relative to land urbanization. Currently, the internationally recognized appropriate value for the elastic coefficient is 1.12. Further, following the research of Gail (2003) [8], this study defines the uncoordinated development of population urbanization and land urbanization as Equation (5) below. When $\frac{growth\ rate\ of\ urban\ built\text{-}up\ area}{growth\ rate\ of\ permanent\ residents\ in\ municipal\ district} \leq 1.12$, $unco\_development_{it}$ will equal 0, which means that the land urbanization growth rate relative to the population urbanization growth rate is within an appropriate range, and the uncoordinated development of population urbanization and land urbanization does not exist. When $\frac{growth\ rate\ of\ urban\ built\text{-}up\ area}{growth\ rate\ of\ permanent\ residents\ in\ municipal\ district} > 1.12$, $unco\_development_{it}$ will equal 1, which means that the population urbanization growth rate severely lags behind the land urbanization growth rate, and the uncoordinated development of population urbanization and land urbanization does exist. That is,

$$unco\_development_{it} = \begin{cases} 0, & \frac{Growth\ rate\ of\ urban\ built\text{-}up\ area}{Growth\ rate\ of\ permanent\ residents\ in\ municipal\ district} \leq 1.12 \\ 1, & \frac{Growth\ rate\ of\ urban\ built\text{-}up\ area}{Growth\ rate\ of\ permanent\ residents\ in\ municipal\ district} > 1.12 \end{cases}. \quad (5)$$

We have drawn up a map related to the development of population urbanization and land urbanization. This is shown as follows.

Figure 2 represent the lagged-behind level of population urbanization relative to the land urbanization of Chinese cities in 2015. Coordinated urbanization represents that the land urbanization growth rate relative to the population urbanization growth rate is within an appropriate range, and the uncoordinated development of population urbanization and land urbanization does not exist. Uncoordinated urbanization represents that the population urbanization growth rate severely lags behind the land urbanization growth rate, and the uncoordinated development of population urbanization and land urbanization does exist. We find that uncoordinated urbanization exists in most cities in China.

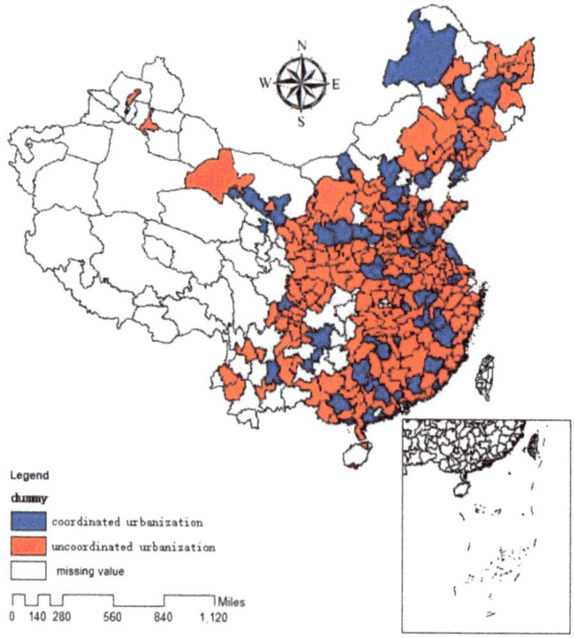

Figure 2. The uncoordinated development of population urbanization and land urbanization in Chinese cities.

3.2.2. Independent Variable

The land financialization scale in prefecture-level cities is the key independent variable in this study. In China, local governments have no right to issue government bonds to raise funds for infrastructure construction and other public welfare projects, so local governments have built some financing platform companies for financing. Further, local governments could rely on the financing platform companies to obtain bank loans through land mortgages and to issue urban investment bonds by taking land transfer revenue as a guarantee. Considering this, the present study selects the interest-bearing debt of local financing platform companies, including current liability and long-term liability, as the proxy variable for the land financialization scale. The specific calculation formula is shown below.

$$\begin{aligned} \textit{Land financialization level} = {} & \textit{short term liability (short term borrowing} + \textit{notes payable} + \\ & \textit{non current liabilities due within one year} + \textit{other current liabilities} + \\ & \textit{short term bond payable}) + \\ & \textit{long term liability (long term borrowing long term bond payable)} \end{aligned} \quad (6)$$

3.2.3. Moderating Variable

Referring to the existing literature, this study utilizes economic development catch-up pressure (including economic development level and fixed-assets investment etc.) to reflect the pressure of urban economic development. The calculation formula is shown below.

$$\textit{Economic development catch up pressure} = \frac{\textit{economic indicators of the prefecture level city which is one place ahead in ranking in the province}}{\textit{economic indicators in the respondent prefecture level city}} \quad (7)$$

where economic indicators include GDP per capita and the total investment in fixed assets as the proportion of GDP.

3.2.4. Control Variables

Control variables in the study mainly include factors affecting land urbanization and population urbanization, such as: (1) the economic development level ($\ln GDP_{it}$), which is measured by the logarithm of GDP per capita [55]; (2) industrial structure (ind_2_{it} *and* ind_3_{it}), which is represented by the ratio of the second and third industries' added value in GDP [14]; (3) population density ($\ln pop\_density_{it}$), which is measured by the logarithm of the city-level population density [53]; (4) fiscal expenditure ($\ln Fis\_expenditure_{it}$), which is represented by the logarithm of local public expenditure per capita [56]; (5) economic openness ($eco\_openness_{it}$), which is measured by the ratio of accumulated foreign direct investment (FDI) in capital stock [57]; (6) fixed-asset investment ($\ln fix\_investment_{it}$), which is measured by the logarithm of fixed-asset investment per capita [58].

Descriptive statistics of the key variables are shown in Table 1.

Table 1. Descriptive statistics of the data for 277 Chinese cities during the period 2006–2015.

Variable	Definition	Obs	Mean	Standard Error	Minimum	Maximum
$\ln land\_urban_{it}$	Logarithm of city-level built-up area	2770	9.544	0.405	8	11.27
$\ln pop\_urban_{it}$	Logarithm of city-level permanent residents	2770	4.549	0.708	2.703	6.750
$unco\_development_{it}$	Dummy variable of the uncoordinated development of population urbanization and land urbanization	2770	0.668	0.471	0	1
$land\_financialization_{it}$	Interest-bearing debt of local financing platform companies	2770	6.239	3.815	0	12.17
$\ln GDP_{it}$	Logarithm of GDP per capita	2770	10.15	0.729	7.306	13.11
ind_2_{it}	Ratio of the second and third industries' added value in GDP	2770	49.88	10.63	15.17	90.97
ind_3_{it}	Ratio of the second and third industries' added value in GDP	2770	36.33	8.461	8.580	75.84
$\ln pop\_density_{it}$	Logarithm of city-level population density	2770	422.2	309.8	4.700	2648
$\ln Fis\_expenditure_{it}$	Logarithm of local public expenditure per capita	2770	8.310	0.734	6.305	11.51
$eco\_openness_{it}$	Ratio of accumulated foreign direct investment (FDI) in capital stock	2770	7.556	2.140	0	13.89
$\ln fix\_investment_{it}$	Logarithm of fixed-assets investment per capita	2770	9.767	0.869	6.883	12.30
$\ln road_{it}$	Logarithm of road area per capita	2770	2.202	0.599	0.157	4.686
$\ln con\_land_{it}$	Amount of newly increased construction land	2770	5.664	1.713	0	12.41
$GDP\_pressure_{it}$	Pressure of urban economic development	2770	1.159	0.240	1	3.397
$inv\_pressure_{it}$	Pressure of urban investment	2770	1.077	0.118	1	2.262
$human\_rate_{it}$	Ratio of college students	2770	1.584	2.198	0	12.936

4. Measuring the Impact of Land Financialization on the Uncoordinated Development of Population Urbanization and Land Urbanization

4.1. Benchmark Results

Table 2 presents the regression results regarding the impact of land financialization on land urbanization and population urbanization. As can be seen, in models 1 and 2, $land\_financialization_{it}$ is significantly and positively correlated with $\ln land\_urban_{it}$. This indicates that land urbanization significantly facilitates land urbanization. Additionally, in models 3 and 4 $land\_financialization_{it}$ is positively correlated with $\ln pop\_urban_{it}$, but this is not significant. This reveals that the impact of land financialization on population urbanization is not significant. Therefore, land financialization by local governments may give rise to the uncoordinated development of population urbanization and land urbanization.

Table 2. Estimation results for the impact of land financialization on land urbanization and population urbanization.

	Model 1	Model 2	Model 3	Model 4
	FE	GMM	FE	GMM
Variable	$\ln land\_urban_{it}$		$\ln pop\_urban_{it}$	
$land\_financialization_{it}$	0.006 **	0.004 ***	0.001	−0.001
	(0.003)	(0.001)	(0.003)	(0.001)
$\ln GDP_{it}$	0.045	0.011	−0.105 **	0.056 ***
	(0.047)	(0.022)	(0.042)	(0.015)
$\ln pop\_density_{it}$	−0.000 ***	−0.000	0.000 *	0.000
	(0.000)	(0.000)	(0.000)	(0.000)
$\ln Fis\_expenditure_{it}$	0.022	0.111 ***	0.055 **	−0.091 **
	(0.046)	(0.037)	(0.025)	(0.038)
$\ln fix\_investment_{it}$	−0.026	0.005	−0.026	0.050
	(0.037)	(0.035)	(0.026)	(0.031)
ind_2_{it}	0.007 ***	−0.009 ***	0.002	−0.003
	(0.003)	(0.002)	(0.003)	(0.003)
ind_3_{it}	0.004	−0.009 ***	0.002	0.004
	(0.003)	(0.002)	(0.003)	(0.003)
$eco\_openness_{it}$	0.001	0.000	0.008 **	0.019 ***
	(0.006)	(0.002)	(0.003)	(0.007)
Lag		0.894 ***		0.914 ***
		(0.018)		(0.023)
_cons	8.681 ***	0.668 ***	4.892 ***	−0.072
	(0.406)	(0.212)	(0.433)	(0.231)
Time effect	Y	Y	Y	Y
Regional effect	Y	Y	Y	Y
Hansen p		0.907		0.304
Adjusted R^2	0.291		0.182	
N	2763	2491	2765	2493

Notes: figures in parentheses denote the standard errors of the respective coefficients, while ***/**/* indicate significance at the 1%/5%/10% levels, respectively.

We further investigate the impact of land financialization on the uncoordinated development of population urbanization and land urbanization. Table 3 presents the regression results using the linear probability model (LPM), the probit analysis method (PROBIT), logistic regression analysis (LOGIT), IV-PROBIT, and GMM. According to the results of models 1–3, $land\_financialization_{it}$ is significantly and positively correlated with $unco\_development_{it}$. This implies that land financialization significantly leads to the uncoordinated development of population urbanization and land urbanization in China.

Moreover, what can be seen from model 2 is that there exists a negative relation between population density and the uncoordinated development of population urbanization and land urbanization, thus demonstrating that increasing population density helps ease the uncoordinated development of population urbanization and land urbanization. The relation between fiscal spending and the uncoordinated development of population urbanization and land urbanization is positive and significant, which is attributable to the economic development orientation of fiscal spending and the urban construction model being dependent on government investment. Meanwhile, growth in the proportion of secondary industry also fuels the uncoordinated development of population urbanization and land urbanization, which is associated with the competition in attracting businesses and investments among local governments and the construction of large-scale industrial parks.

Table 3. Estimation results for the impact of land financialization on the uncoordinated development of population urbanization and land urbanization.

Variable	Model 1	Model 2	Model 3	Model 4	Model 5
	LPM	PROBIT	LOGIT	IV-PROBIT	GMM
$land\_financialization_{it}$	0.010 **	0.041 **	0.065 **	0.251 **	0.031 ***
	(0.004)	(0.017)	(0.029)	(0.113)	(0.008)
$\ln GDP_{it}$	0.010	−0.079	−0.101	0.038	0.073
	(0.048)	(0.290)	(0.498)	(0.299)	(0.239)
$\ln pop\_density_{it}$	−0.001	−0.003 *	−0.005 *	−0.003 *	0.000
	(0.000)	(0.002)	(0.003)	(0.002)	(0.000)
$\ln Fis\_expenditure_{it}$	0.141 **	0.655 **	1.121 **	0.508 *	0.174
	(0.061)	(0.285)	(0.487)	(0.306)	(0.286)
$\ln fix\_investment_{it}$	0.038	0.259	0.404	0.238	−0.351
	(0.042)	(0.185)	(0.309)	(0.192)	(0.236)
ind_2_{it}	0.011 **	0.050 **	0.088 **	0.034	0.007
	(0.004)	(0.023)	(0.038)	(0.025)	(0.014)
ind_3_{it}	0.009 *	0.041	0.074	0.024	0.023
	(0.005)	(0.028)	(0.046)	(0.030)	(0.019)
$eco\_openness_{it}$	−0.030 ***	−0.135 ***	−0.222 ***	−0.119 ***	0.010
	(0.009)	(0.043)	(0.073)	(0.045)	(0.036)
Lag					0.123 ***
					(0.030)
_cons	−1.603 **	−17.579	−13.567 *	−7.690 *	0.339
	(0.694)	(170.690)	(7.188)	(4.316)	(1.442)
Time effect	Y	Y	Y	Y	Y
Regional effect	Y	Y	Y	Y	Y
Hansen p	0.390	0.380	0.162		
Adjusted R^2					0.154
N	2770	2770	2770	2770	2493

Notes: figures in parentheses denote the standard errors of the respective coefficients, while ***/**/* indicate significance at the 1%/5%/10% levels, respectively.

Considering the estimation bias induced by the reciprocal causation that land financialization and the uncoordinated development of population urbanization and land urbanization always interact with each other, the spatial lagged term of land financialization as the instrumental variable is introduced, and the 2SLS method is employed to re-examine the relationships in our study. The result in model 4 shows that the coefficient of $land\_financialization_{it}$ remains significantly positive. In addition, considering institutional inertia and path dependence, this study introduces the one-phase lagged term of $unco\_development_{it}$ as the control variable. The *Hansen* test in model 5 demonstrates the efficacy of the instrumental variables. The regression coefficient remains positive in model 5, verifying the robustness of the result.

In summary, the above regression result corroborates hypothesis 1—that is, that the enabling effect of land financialization by local governments on land urbanization is stronger than that on population urbanization, which results in the uncoordinated development of population urbanization and land urbanization. The reason for this is, as explained previously, that one of the features of land financialization is that it binds government debts with land. Land transfer fees are not only the reference of credit standing for local governments' current borrowing, but a vital source of capital in repaying local governments' debt. Therefore, local governments are highly motivated to channel debt-based capital into land reserve and development activities. This forms a continuous cycle which propels the rapid expansion of urban space. However, local governments fail to establish a stable investment linkage between public services in cities such as education, science, culture, healthcare, and land financialization. Coupled with rising land and housing prices, the citizenization of rural-to-urban migrants moves ahead slowly, leading to the uncoordinated development of population urbanization and land urbanization.

4.2. Robustness Checks

This study conducts robustness checks on benchmark regression results, including replacing measurement indicators of land urbanization and population urbanization, changing sample range, and increasing the number of control variables. First, this paper refers to the practice of Xiong and Gao (2012) [59] and replaces the city district area, which represents land urbanization, with the urban road area to calculate the uncoordinated development of population urbanization and land urbanization. Detailed regression results are shown in model 1. Second, we replace the total population in urbanized city districts, which denotes population urbanization, with non-agricultural population calculated through household registration to obtain the uncoordinated development of population urbanization and land urbanization. Detailed regression results are given in model 2. Third, this study excludes 58 cities which have undergone administrative upgrading from county to district so as to expel the influence of administrative region adjustment. The regression results in Table 4 reveal that the impact of land financialization on the uncoordinated development of population urbanization and land urbanization is significantly positive, showcasing the robustness of the benchmark regression results. Fourth, we only employ samples after 2008 for regression to avoid potential influence by the surge of financialization platforms after the financial crisis in 2008. Results are shown in model 4. Finally, this paper utilizes new land supply for construction (*lnland*) as a control variable. The expansion of the urban built-up area is affected not only by factors from the demand side, such as population and economy, but also factors from the supply side, such as land for construction. Of particular note is that land for construction is determined by the administrative distribution of government in China and is therefore of greater significance. The results of model 5 indicate robustness.

Table 4. Robustness checks on the impact of land financialization on the uncoordinated development of population urbanization and land urbanization.

	Model 1	Model 2	Model 3	Model 4	Model 5
$land\_financialization_{it}$	0.046 **	0.041 **	0.038 **	0.046 *	0.040 **
	(0.018)	(0.020)	(0.018)	(0.025)	(0.017)
$\ln GDP_{it}$	−0.060	−0.112	0.036	0.089	−0.080
	(0.271)	(0.354)	(0.364)	(0.285)	(0.290)
$\ln pop\_density_{it}$	−0.003 *	−0.001	−0.002	−0.002	−0.003 *
	(0.002)	(0.001)	(0.002)	(0.002)	(0.002)
$\ln Fis\_expenditure_{it}$	−0.378	−0.019	0.628 **	1.063 ***	0.657 **
	(0.248)	(0.313)	(0.290)	(0.363)	(0.285)
$\ln fix\_investment_{it}$	0.082	0.182	0.176	0.309	0.252
	(0.191)	(0.204)	(0.189)	(0.233)	(0.185)
ind_2_{it}	0.012	−0.014	0.056 **	0.053	0.050 **
	(0.022)	(0.026)	(0.023)	(0.039)	(0.023)
ind_3_{it}	0.005	−0.014	0.047 *	0.068	0.041
	(0.028)	(0.032)	(0.028)	(0.047)	(0.028)
$eco\_openness_{it}$	−0.073 *	−0.058	−0.120 ***	−0.119 **	−0.135 ***
	(0.041)	(0.044)	(0.044)	(0.048)	(0.043)
$\ln con\_land_{it}$.					0.017
					(0.026)
_cons	−1.413	−8.447	−14.357	−15.958 ***	−17.907
	(606.945)	(203.141)	(701.848)	(5.341)	(269.780)
Time effect	Y	Y	Y	Y	Y
Regional effect	Y	Y	Y	Y	Y
N	2770	2770	2712	1939	2770

Notes: figures in parentheses denote the standard errors of the respective coefficients, while ***/**/* indicate significance at the 1%/5%/10% levels, respectively.

4.3. Moderating Effect of Urban Development Pressure on the Uncoordinated Development of Population Urbanization and Land Urbanization Caused by Land Financialization

China has implemented an official selection mechanism focusing on officials' achievements. The indicator that can best reflect the achievement of an official is GDP per capita. Besides this, investment currently remains the major "carriage" driving China's economic growth. Under the pressure of urban economic development, local governments have the impulse to boost investment and regard it as a vital means to promote economic development. Therefore, this study takes the backwardness of cities in terms of GDP per capita and fixed-asset investment relative to their counterparts to measure development pressure.

Hypothesis 2 states that the stimulating effect of land financialization on the uncoordinated development of population urbanization and land urbanization in cities with greater development pressure is more prominent—i.e., the pressure of urban economic development regulates land financialization's stimulating effect on the uncoordinated development of population urbanization and land urbanization. In order to test this hypothesis, the present study constructs Equation 3. If η, the coefficient of interaction term between land financialization and development pressure, is positive, the stimulating effect of land financialization on the uncoordinated development of population urbanization and land urbanization will be more obvious in regions with greater development pressure. Last, under the yardstick competition mechanism, there may be interaction on strategies for land financing among different regions. Therefore, we constructed a Spatial Dubin Model (SDM) to investigate the spillover effect of competition on land financing among different regions. Then, we added the spatial lagged term of land financing size into the regression model to examine whether competition on land financing among different regions affects urbanization unbalance. In terms of weight matrix, we constructed an economic weight matrix based on the GDP per capita of cities from the same province.

According to the result in Table 5, the coefficients of interaction terms are significantly positive in five models. From this, development pressure is constructed based on the GDP per capita in models 1 and 2, while it is constructed based on the proportion of fixed-asset investment in models 3 and 4. Development pressure caused by lagging-behind GDP or fixed-asset investment will intensify the motivation and incentive of local governments to expand urban construction and land development though land financialization. Hypothesis 2 is therefore proved. The regression results in Model 5 and 6 demonstrate emulation strategies for land financing among cities, which further aggravates the degree of uncoordinated development of population urbanization and land urbanization.

Table 5. Estimation results for development pressure, land financialization, and the uncoordinated development of population urbanization and land urbanization.

	Model 1 LPM	Model 2 PROBIT	Model 3 LPM	Model 4 PROBIT	Model 5 SDM	Model 6 LPM
	Catch-Up Pressure in GDP		Catch-Up Pressure in Fixed-Assets Investment		Yardstick Competition	
land_financialization	0.010 **	0.041 **	0.011 **	0.046 ***		
	(0.004)	(0.017)	(0.004)	(0.018)		
GDP_pressure$_{it}$	0.065	0.193				
	(0.151)	(0.647)				
land_financialization * *GDP_pressure*	0.069 ***	0.363 ***				
	(0.025)	(0.112)				
inv_pressure$_{it}$			0.199 **	1.061 **		
			(0.090)	(0.429)		
land_financialization * *inv_pressure*			0.033 *	0.140 *		
			(0.018)	(0.081)		
W * *land_financialization*					1.617 ***	0.251 **
					(0.225)	(0.113)
_cons	−1.574 **	−17.578	−2.007 ***	−19.402	10.805 **	−9.981 **
	(0.709)	(169.466)	(0.708)	(169.248)	(5.353)	(4.322)

Table 5. *Cont.*

	Model 1 LPM	Model 2 PROBIT	Model 3 LPM	Model 4 PROBIT	Model 5 SDM	Model 6 LPM
	Catch-Up Pressure in GDP		Catch-Up Pressure in Fixed-Assets Investment		Yardstick Competition	
Controls	Y	Y	Y	Y	Y	Y
Time effect	Y	Y	Y	Y	Y	Y
Regional effect	Y	Y	Y	Y	Y	Y
Adjusted R^2	0.319		0.320		0.768	0.164
N	2770	2770	2770	2770	2770	2770

Notes: figures in parentheses denote the standard errors of the respective coefficients, while ***/**/* indicate significance at the 1%/5%/10% levels, respectively.

5. Land Financialization, the Uncoordinated Development of Population Urbanization and Land Urbanization, and Economic Growth

This section analyzes the potential impact of land financialization, which stimulates the uncoordinated development of population urbanization and land urbanization, on urban economic sustainable development to examine the hidden risks behind the development model characterized by debt-based urbanization financialization and supply-driven urban construction.

The essence of urbanization driven by land financialization adopted by Chinese local governments is debt-based urban construction investment (Pan et al., 2017) [18]. During this process, the investment revenue for local governments comes mainly from economic growth due to population and industrial conglomeration. Currently, economic growth is indicated by the added value of secondary and tertiary industry. Therefore, this study, in reference to the practice of Lu (2011), divides the added value of secondary industry and tertiary industry by built-up area to obtain the land use efficiency and uses it to denote investment return on urban spatial expansion [50]. The basic logic here is that if land financialization decreases the land use efficiency, debt-based urban spatial expansion could not generate a scale effect and economic conglomeration. The return on investment is therefore insufficient. This study divides samples into the unbalanced group and the balance group. If land financialization causes a decline in land use efficiency in the unbalanced group compared to the balance group, hypothesis 3 is corroborated. Furthermore, we test whether there exists a threshold value for the effect of land financialization on economic growth—in other words, whether the effect of land financialization on economic growth turns to an inverted U shape.

To test hypothesis 3, we construct the following regression model.

$$land\_use_{it} = \alpha_4 + \beta_4 land\_financialization_{it} + \sum \gamma_4 X_{it} + \mu_i + \delta_t + \varepsilon_{it} \quad (8)$$

$$growth_{it} = \alpha_5 + \beta_5 land\_financialization_{it} + \lambda_5 land\_financialization_{it}{}^2 + \sum \gamma_5 X_{it} + \mu_i + \delta_t + \varepsilon_{it} \quad (9)$$

where $land_{useit}$ and $growth_{it}$. denote the land use efficiency and growth rate of GDP per capita, respectively. Control variables include city-level population density ($\ln pop\_density_{it}$), human capital ($human\_rate_{it}$), economic openness ($eco\_openness_{it}$), fiscal expenditure ($\ln Fis\_expenditure_{it}$), fixed-asset investment ($\ln fix\_investment_{it}$), and industrial structure (ind_2_{it} *and* ind_3_{it}).

Table 6 displays the impact of geographical factors on land use efficiency and urban economic growth. According to the regression result of model 1 in Table 6, land financialization lowers the land use efficiency in cities with the uncoordinated development of population urbanization and land urbanization. This might be because land financialization does not generate economic conglomeration or increasing returns to scale. The return on investment is low. The regression results of model 3 indicate that the first-order term of land financialization is significantly positive, while its second-order term is significantly negative. This signifies that the effect of land financialization on economic growth in cities with the uncoordinated development of population urbanization and land urbanization showcases an inverted U shape. That is to say, after surpassing the threshold value, the effect of land financialization on economic growth will transit from a positive one to a negative one. These characteristics are not found in cities with the coordinated development of population urbanization and land urbanization.

Table 6. Estimation results for land financialization, the uncoordinated development of population urbanization and land urbanization, and sustainable development.

	Model 1	**Model 2**	**Model 3**	**Model 4**
Variables	Land use efficiency		Economic growth rate	
	uncoordinated	coordinated	uncoordinated	coordinated
$land\_financialization_{it}$	−0.006 *	0.005	5.612 **	4.285
	(0.003)	(0.007)	(2.223)	(5.459)
$land\_financialization_{it}$^2			−2.733 **	−2.062
	0.020	0.160	(1.102)	(2.727)
	(0.056)	(0.175)		
$\ln pop\_density_{it}$	0.001 *	0.000		
	(0.000)	(0.000)		
$\ln Fis\_expenditure_{it}$	0.080 *	0.185 **	0.266	1.421
	(0.047)	(0.075)	(1.459)	(3.812)
$\ln fix\_investment_{it}$	0.129 ***	0.101	4.801 ***	2.117
	(0.034)	(0.063)	(1.213)	(1.850)
ind_2_{it}	0.021 ***	0.013		
	(0.004)	(0.008)		
ind_3_{it}	0.019 ***	0.019 **	−0.079	−0.262
	(0.005)	(0.008)	(0.078)	(0.202)
$eco\_openness_{it}$	−0.003	0.010	0.928 ***	0.847 *
	(0.007)	(0.013)	(0.268)	(0.443)
$\ln land\_urban_{it}$			3.581	12.835 *
			(4.519)	(6.517)
$human\_rate_{it}$			1.138 ***	3.095 ***
			(0.301)	(0.919)
$\_cons$	6.640 ***	4.995 ***	−51.324	−85.826
	(0.661)	(1.434)	(34.066)	(62.265)
Adjusted R^2	0.383	0.479	0.551	0.493
N	1850	920	1850	920

Notes: Figures in parentheses denote the standard errors of the respective coefficients, while ***/**/* indicate significance at the 1%/5%/10% levels, respectively.

6. Conclusions and Discussion

Nobel Prize-winning economist David Stiglitz has said that two events of the 21st century will have the greatest impact on the world: "America's high-tech industry and China's urbanization." The rapid development of urbanization in China cannot be separated from the promotion of local governments, and land financialization is an important means for the government to lead the urbanization process. For a long time, the gap between fiscal revenue and expenditure forces local governments to use land to obtain as much extrabudgetary income as possible, and this gradually forms the fiscal situation of relying on land. In recent years, the urbanization financing mode has been transformed from land transfer to land financialization, which has resulted in huge local government debts. By the end of 2019, the local government debt ratio has risen to 24.3%. What impact does this have on the pattern of urbanization?

In this paper, we investigate the effect and mechanisms of land financialization on the uncoordinated development of population urbanization and land urbanization in China for 277 Chinese cities during the period spanning 2006 to 2015. Moreover, the pressure of urban economic development is included as a mediation variable to measure how much it accounts for the effect of land financialization on the uncoordinated development of population urbanization and land urbanization. Finally, this paper examines the hidden risks behind the development model characterized by debt-based urbanization financialization and supply-driven urban construction.

The three main findings are that, first, land financialization significantly leads to the uncoordinated development of population urbanization and land urbanization. On the one hand, Chinese local governments' dependence on land-based urbanization financialization will intensify their motivation in land development and conveyance, leading to chaotic urban expansion. On the other hand, the expenditure structure of debt-based capital financed through land showcases an obvious orientation towards land urbanization and a weak "pushing" and "pulling" effect on the citizenization of rural-to-urban migrants. Population urbanization is relatively slow.

Second, the pressure of urban economic development positively regulates the stimulating effect of land financialization on the uncoordinated development of population urbanization and land urbanization. In cities with greater development pressure, the stimulating effect of land financialization on the uncoordinated development of population urbanization and land urbanization is more prominent.

Third, urbanization propelled by land financialization is unsustainable, because in regions with the uncoordinated development of population urbanization and land urbanization, land financialization, while promoting urban spatial expansion, lowers the land use efficiency due to weak conglomeration effect and exerts an inverted U-shaped influence on economic growth.

In the Chinese context, this finding is as expected. Since the financial crisis in 2008, local governments of China have established a huge number of financing platform companies, obtained bank loans and issued urban investment bonds through land mortgages or guarantees, and created large-scale invisible local government debts. This debt capital is mainly invested in urban public facilities and land development and becomes significant funding support for urbanization. A unique urbanization financialization model with Chinese characteristics has been gradually formed. This debt-based financialization model, which is highly reliant on land, further intensifies local governments' passion for "operating city" and "operating land", leading to a large-scale movement known as "enclosure for urbanization". Constructions for development zones, new cities, and new areas in various forms are fledgling. However, supply-driven urban construction without the due consideration of demand factors is destined to cause the uncoordinated development of population urbanization and land urbanization, which is unsustainable.

The conclusion of this study carries some policy implications. First, efforts should be made to explore financialization systems that facilitate sound urban development. A local tax revenue system with property tax being the main tax ought to be established during the early days to divert local governments from blind passion about land development. Besides this, a stable input linkage between land financialization and public service supply should be forged to deliver a pulling effect on the citizenization of rural-to-urban migrants. Furthermore, land, population, and industry are the most important elements in economic activities. Sustainable economic development can only be realized when land supply matches the trend of population and industrial conglomeration. The government should set up and improve cross-regional land resource regulating systems and land supply systems. Last, efforts should be made to tighten the budgetary constraints on local governments and to improve the utilization efficiency of debt capital and return on investment, while also establishing management systems that guard against and resolve local debt risks. On the whole, the key to new urbanization lies in population urbanization. The government should emphasize the quality of urbanization, promote integration between cities and industry, and support population conglomeration through enhanced public service to ensure the sustainable development of the urban economy.

Due to constraints in access to data, there are deficiencies in the research results which need further consideration. (1) Limitations in the measurement of variables: The existing studies on land financialization in China usually use urban investment bonds as a proxy variable, while we use the liabilities of local financing vehicles to estimate the scale of land financialization for the first time. There will still be a deviation from the real scale. In follow-up studies, additional micro-data of the land mortgage of local financing vehicles need to be compiled for further analysis. (2) Not considering the issue in an international context: It is a great challenge to define the boundaries of the urbanization process in the world, since we do not have the data on urbanization and land financialization in the world. To clearly identify the boundaries of the urbanization process in the world needs a great deal of theoretical analysis, logical deduction, and empirical evidence. We therefore chose to study the impact and mechanism of land financialization by local governments on the uncoordinated development of population urbanization and land urbanization in the context of China, following the practice in the literature [4]. We also have added the Spatial Dubin Model (SDM) to investigate the spillover effect of competition on land financing among different regions. Studying urbanization and financialization in an international context needs to be explored and new solutions developed in follow-up research.

Author Contributions: Conceptualization, Y.J. and X.G.; methodology, Y.J., L.W., and S.Z.; validation, Y.J., X.G., and S.Z.; formal analysis, Y.J. and X.G.; data curation, Y.J. and S.Z.; writing—original draft preparation, Y.J., X.G., and S.Z.; writing—review and editing, Y.J., L.W., X.G., and S.Z.; visualization, Y.J. and X.G.; supervision, Y.J.; project administration, Y.J.; funding acquisition, X.G. All authors have read and agreed to the published version of the manuscript.

Funding: This research was funded by Fundamental Research Funds for the Central Universities, grant number CXJJ-2019-363.

Conflicts of Interest: The authors declare no conflict of interest.

Appendix A

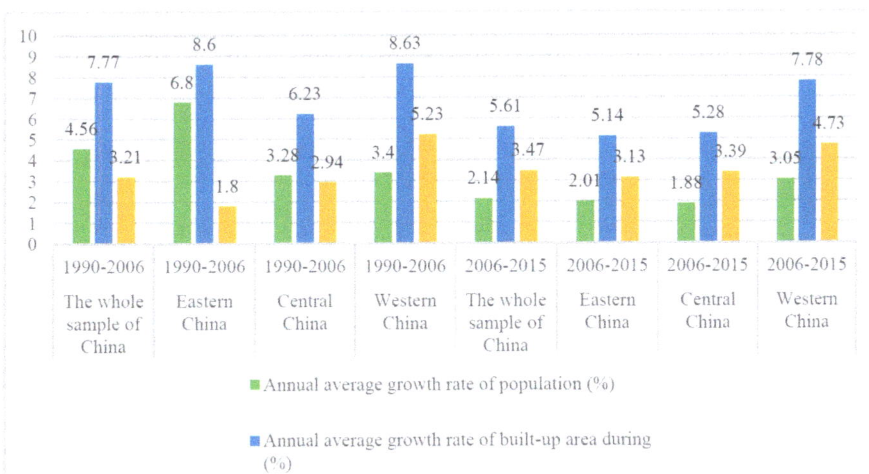

Figure A1. The trends of urban population and built-up area.

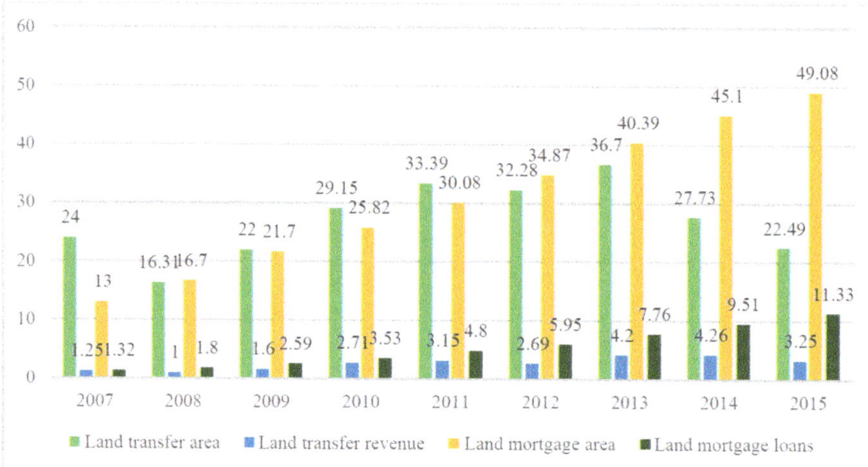

Figure A2. Land transfer and mortgage from 2007 to 2015. Note: data were retrieved from Report on China's Land Resources 2008–2016.

References

1. Duranton, G.; Puga, D. Micro-foundations of urban agglomeration economies. In *Handbook of Regional and Urban Economics*; Elsevier: Amsterdam, The Netherlands, 2004; Volume 4, pp. 2063–2117.
2. Bertinelli, L.; Black, D. Urbanization and growth. *J. Urban Econ.* **2004**, *56*, 80–96. [CrossRef]
3. Fujita, M.; Mori, T. The role of ports in the making of major cities: Self-agglomeration and hub-effect. *J. Dev. Econ.* **1996**, *49*, 93–120. [CrossRef]
4. Ding, C.; Lichtenberg, E. Land and urban economic growth in China. *J. Reg. Sci.* **2011**, *51*, 299–317. [CrossRef]
5. Du, D. The causal relationship between land urbanization quality and economic growth: Evidence from capital cities in China. *Qual. Quant.* **2017**, *51*, 2707–2723. [CrossRef]
6. Zhu, F.K.; Zhang, F.R.; Li, C.; Jiao, P.F.; Wang, J.X. Coordination and regional difference of urban land expansion and demographic urbanization in china during 1993–2008. *Prog. Geogr.* **2014**, *33*, 647–656.
7. Wang, Z.K. *The Imbalance in Regional Economic Development in China and Its Reasons*; Palgrave Macmillan: London, UK, 2015.
8. Gail, J. *Communication and Space*; China Building Industry Press: Beijing, China, 2003.
9. Liu, S. Risks and reform of Land-Based development model. *Int. Econ. Rev.* **2012**, *2*, 92–109.
10. Guan, X.; Wei, H.; Lu, S.; Dai, Q.; Su, H. Assessment on the urbanization strategy in China: Achievements, challenges and reflections. *Habitat Int.* **2018**, *71*, 97–109. [CrossRef]
11. Tan, M.; Li, X.; Xie, H.; Lu, C. Urban land expansion and arable land loss in China—A case study of Beijing–Tianjin–Hebei region. *Land Use Policy* **2005**, *22*, 187–196. [CrossRef]
12. Ye, L.; Wu, A.M. Urbanization, Land Development, and Land Financing: Evidence from Chinese Cities. *J. Urban Aff.* **2014**, *36*, 354–368. [CrossRef]
13. Jin, X.; Long, Y.; Sun, W.; Lu, Y.; Yang, X.; Tang, J. Evaluating cities' vitality and identifying ghost cities in China with emerging geographical data. *Cities* **2017**, *63*, 98–109. [CrossRef]
14. Au, C.-C.; Henderson, J.V. How migration restrictions limit agglomeration and productivity in China. *J. Dev. Econ.* **2006**, *80*, 350–388. [CrossRef]
15. Hertel, T.; Zhai, F. Labor market distortions, rural–urban inequality and the opening of China's economy. *Econ. Model.* **2006**, *23*, 76–109. [CrossRef]
16. Shu, C.; Xie, H.; Jiang, J.; Chen, Q. Is Urban Land Development Driven by Economic Development or Fiscal Revenue Stimuli in China? *Land Use Policy* **2018**, *77*, 107–115. [CrossRef]
17. Jin, H.; Qian, Y.; Weingast, B.R. Regional decentralization and fiscal incentives: Federalism, Chinese style. *J. Public Econ.* **2005**, *89*, 1719–1742. [CrossRef]

18. Pan, F.; Zhang, F.; Zhu, S.; Wójcik, D. Developing by borrowing? Inter-jurisdictional competition, land finance and local debt accumulation in China. *Urban Stud.* **2017**, *54*, 897–916. [CrossRef]
19. Tan, R.; Beckmann, V.; Berg, L.V.D.; Qu, F. Governing farmland conversion: Comparing China with the Netherlands and Germany. *Land Use Policy* **2009**, *26*, 961–974. [CrossRef]
20. Hamnett, C. Is Chinese urbanisation unique? *Urban Stud.* **2020**, *57*, 690–700. [CrossRef]
21. He, C.; Zhou, Y.; Huang, Z. Fiscal decentralization, political centralization, and land urbanization in China. *Urban Geogr.* **2016**, *37*, 436–457. [CrossRef]
22. Sun, J.; Chen, T.; Cheng, Z.; Wang, C.C.; Ning, X. A financing mode of Urban Rail transit based on land value capture: A case study in Wuhan City. *Transp. Policy* **2017**, *57*, 59–67. [CrossRef]
23. Zhang, L.; Nian, Y.; Liu, J. Land market fluctuations and local government debts: Evidence from the municipal investment bonds in China. *China Econ. Q.* **2018**, *17*, 1103–1126.
24. Zheng, S.; Sun, W.; Wu, J.; Wu, Y. Infrastructure investment, land leasing and real estate price: A unique financing and investment channel feo urban development in Chinese cities. *Econ. Res. J.* **2014**, *49*, 14–27.
25. Chen, J.; Wu, F. Housing and land financialization under the state ownership of land in China. *Land Use Policy* **2020**, 104844. [CrossRef]
26. Yuan, F.C.; Kang, H.J. The Imbalance Between Human and Land in the Process of New Urbanization and Its Breakthrough. *J. Chin. Acad. Goverance* **2016**, *4*, 47–52. (In Chinese)
27. Wu, Y.; Pagano, M.A. Municipal Communications Tax Revenue Reliance and Fiscal Impact of Possible Federal Preemption. *Public Budg. Finance* **2008**, *28*, 82–100. [CrossRef]
28. Brehm, S. Fiscal Incentives, Public Spending, and Productivity—County-Level Evidence from a Chinese Province. *World Dev.* **2013**, *46*, 92–103. [CrossRef]
29. Bo, L.; Mear, F.C.J.; Huang, J. New development: China's debt transparency and the case of urban construction investment bonds. *Public Money Manag.* **2017**, *37*, 225–230. [CrossRef]
30. Wu, F. Planning centrality, market instruments: Governing Chinese urban transformation under state entrepreneurialism. *Urban Stud.* **2018**, *55*, 1383–1399. [CrossRef]
31. Lin, G.C.; Li, X.; Yang, F.F.; Hu, F.Z.Y. Strategizing urbanism in the era of neoliberalization: State power reshuffling, land development and municipal finance in urbanizing China. *Urban Stud.* **2015**, *52*, 1962–1982. [CrossRef]
32. Cao, G.; Feng, C.; Tao, R. Local "Land Finance" in China's Urban Expansion: Challenges and Solutions. *China World Econ.* **2008**, *16*, 19–30. [CrossRef]
33. Bai, C.-E.; Hsieh, C.-T.; Song, Z. (Michael) The Long Shadow of a Fiscal Expansion. *SSRN Electron. J.* 2016. [CrossRef]
34. Zhong, T.; Chen, Y.; Huang, X. Impact of land revenue on the urban land growth toward decreasing population density in Jiangsu Province, China. *Habitat Int.* **2016**, *58*, 34–41. [CrossRef]
35. Weber, R. Selling City Futures: The Financialization of Urban Redevelopment Policy. *Econ. Geogr.* **2010**, *86*, 251–274. [CrossRef]
36. Wu, F. Land financialisation and the financing of urban development in China. *Land Use Policy* **2019**, 104412. [CrossRef]
37. Chang, C.; Lu, M. Misery of new town—Density, Distance and Debt. *China Econ. Q.* **2017**, *16*, 1621–1642.
38. Research Group on China's Economic Growth. Urbanization, Fiscal expansion and Economic growth. *Econ. Res. J.* **2011**, *46*, 4–20.
39. Qian, Y.; Roland, G. Federalism and the Soft Budget Constraint. *Am. Econ. Rev.* **1998**, *88*, 1143–1162. [CrossRef]
40. Maskin, E.; Qian, Y.; Xu, C. Incentives, information, and organizational form. *Rev. Econ. Stud.* **2000**, *67*, 359–378. [CrossRef]
41. Besley, T.; Case, A. Incumbent behavior: Vote-Seeking, Tax-Setting, and yardstick competition. *Am. Econ. Rev.* **1992**, *85*, 25–45.
42. Tsui, K.-Y. Local tax system, intergovernmental transfers and China's local fiscal disparities. *J. Comp. Econ.* **2005**, *33*, 173–196. [CrossRef]
43. Zhou, L. Governing China's Local Officials: An Analysis of Promotion Tournament Model. *Econ. Res. J.* **2007**, *7*, 36–50.
44. Zhang, J.; Wang, J.; Yang, X.; Ren, S.; Ranab, Q.; Haocdefg, Y. Does local government competition aggravate haze pollution? A new perspective of factor market distortion. *Socio Econ. Plan. Sci.* **2020**, 100959. [CrossRef]

45. Li, H.; Zhou, L.-A. Political turnover and economic performance: The incentive role of personnel control in China. *J. Public Econ.* **2005**, *89*, 1743–1762. [CrossRef]
46. Fu, Y.; Zhang, Y. Chinese Style Decentralization and Government Spending Bias: The Cost of Growth. *Manag. World* **2007**, *3*, 4–12.
47. Qiao, B.Y.; Fan, J.Y.; Feng, X.Y. Fiscal Decentralization and Compulsory Primary Education in China. *Soc. Sci. China* **2005**, *6*, 37–46.
48. Caldeira, E. Yardstick competition in a federation: Theory and evidence from China. *China Econ. Rev.* **2012**, *23*, 878–897. [CrossRef]
49. Démurger, S.; Sachs, J.D.; Woo, W.T.; Bao, S.; Chang, G.; Mellinger, A. Geography, economic policy, and regional development in China. *Asian Econ. Pap.* **2002**, *1*, 146–197. [CrossRef]
50. Lu, M. Cross-regional redistribution of the right to use Construction Land: A New Driving Force for China's Economic growth. *J. World Econ.* **2011**, *34*, 107–125.
51. Li, Y.; Jia, L.; Wu, W.; Yan, J.; Liu, Y. Urbanization for rural sustainability—Rethinking China's urbanization strategy. *J. Clean. Prod.* **2018**, *178*, 580–586. [CrossRef]
52. Halleux, J.-M.; Marcinczak, S.; Van Der Krabben, E. The adaptive efficiency of land use planning measured by the control of urban sprawl. The cases of the Netherlands, Belgium and Poland. *Land Use Policy* **2012**, *29*, 887–898. [CrossRef]
53. Chen, J.; Zhou, Q. City size and urban labor productivity in China: New evidence from spatial city-level panel data analysis. *Econ. Syst.* **2017**, *41*, 165–178. [CrossRef]
54. Xie, D. Local government competition, Monopolizes land supply and urbanization development imbalance. *J. Financ. Econ.* **2016**, *42*, 102–111.
55. Fred, Y.Y. A quantitative relationship between per capita GDP and scientometric criteria. *Scientometrics* **2007**, *71*, 407–413.
56. Duan, H.; Zhan, J.V. Fiscal transfer and local public expenditure in China: A case study of Shanxi Province. *China Rev.* **2011**, *11*, 57–88.
57. Au, C.-C.; Henderson, J.V. Are Chinese Cities Too Small? *Rev. Econ. Stud.* **2006**, *73*, 549–576. [CrossRef]
58. Qin, D.; Song, H. Sources of investment inefficiency: The case of fixed-asset investment in China. *J. Dev. Econ.* **2009**, *90*, 94–105. [CrossRef]
59. Xiong, C.; Gao, H. The incongruity between population urbanization and spatial urbanization. *Financ. Econ.* **2012**, *11*, 102–108.

land

Article

Research on the Impact of Land Circulation on the Income Gap of Rural Households: Evidence from CHIP

Congjia Huo [1] and Lingming Chen [1,2,*]

1 Department of Economics, School of Business, Hunan University of Science and Technology (HNUST), Xiantang 411201, China; huocongjia@mail.hnust.edu.cn
2 Department of Economics & Statistics, School of Economics and Management, Xinyu University (XYU), Xinyu 338004, China
* Correspondence: lingming1016@mail.hnust.edu.cn; Tel.: +86-158-7009-2285

Abstract: With the continued development of the economy, the income gap among Chinese rural households continues to widen. The land system plays a decisive role in developing "agriculture, rural areas and farmers" and land circulation is a factor in the increase in income inequality among farm households. Based on the 2013 China Household Income Project (CHIP), this article used the re-centered influence function (RIF) regression method to empirically test the impact of rural land circulation on the income gap of rural households in China in three regions: the central, eastern and western regions. The quantile regression tested the impact mechanism of income inequality of rural households from the perspective of labor mobility and land circulation. The empirical results showed that land circulation increases the income inequality of rural households. The theoretical mechanism test proved that the dynamic relationship between land circulation and labor mobility increases rural household income. However, this increase has a greater effect on rural households with a high income and a small effect on rural households with a low income, resulting in a further widening of the income gap. Therefore, while increasing the income of rural households through land circulation, the government should also consider income equity. Finally, this article puts forward the policies and opinions on land reform and provides a brief discussion on the future direction of development.

Keywords: land circulation; income gap; rural households income; re-centered influence function; quantile regression

check for updates

Citation: Huo, C.; Chen, L. Research on the Impact of Land Circulation on the Income Gap of Rural Households: Evidence from CHIP. *Land* **2021**, *10*, 781. https://doi.org/10.3390/land10080781

Academic Editors: Marina Cabral Pinto, Amit Kumar and Munesh Kumar

Received: 27 June 2021
Accepted: 22 July 2021
Published: 25 July 2021

1. Introduction

Since the founding of New China, Chinese farmers have experienced changes in farmland, agricultural management and farmland property rights systems. With the implementation of the rural revitalization strategy, the development of urban and rural areas has been coordinated, reforms have been comprehensively deepened, and farmers' incomes have continued to increase. However, this increase in income has also been accompanied by a continuous expansion of the income gap within rural areas. The "Report on the Development of Rural Households in China (2018)" posited that rural household income inequality is rising in China. The Gini coefficient increased from 0.45 in 2011 to 0.535 in 2017, significantly higher than the internationally recognized warning line of 0.4 [1]. Figure 1 shows the Gini coefficient and Theil index estimated from quintile data of the per capita disposable income of rural households. These data came from the "China Yearbook of Household Survey" from 2005–2019.

Figure 1. The Gini Coefficient and Theil Index of Per Capita Disposable Income of Rural Households in China from 2005 to 2019. Note: the data comes from the compilation and calculation of the Chinese Household Survey Yearbook.

It can be seen from Figure 1 that the Gini coefficient of rural household income has risen steadily, and the Gini coefficients calculated by quintile subgroups exceed 0.35, with a higher rise in the Theil index. In 2019, the Gini coefficient and Theil index declined slightly. China continues to explore a reform of the rural system, but the transformation of the rural economy has taken too long to exceed people's expectations. There have been twists, stagnations, and even deviations in the reform process. Although this slow economic transformation avoids the risk of economic and social disorder, it has also gained time for the formation and growth of interest groups. With a lag of political reforms, economic transformation has become the eternal theme of "always on the road". This unbalanced transformation path is fully reflected in the field of income distribution. Although the income of rural households continues to increase, the income gap within rural areas continues to widen, as does the income gap between urban and rural areas. There have been strange phenomena such as the intertwining of reasonable and unreasonable income gaps and the coexistence of open and transparent distribution models and hidden distribution mechanisms. In contrast, inequality in rural areas restricts economic development, reduces the welfare generated by increased incomes, and affects social stability.

There are many reasons for income inequality among rural households. Existing research mainly explores the physical capital, human capital and social capital owned by rural households. The land system plays a decisive role in the development of "agriculture, rural areas and farmers". The household contract responsibility system was implemented at the beginning of the founding of New China. Although it has greatly increased the enthusiasm of households' production, the government strictly prohibits the sale of land use rights. In recent years, the circulation system of land contract rights has become increasingly stable, and the government has liberalized the land circulation system. The process of urbanization and industrialization has accelerated, many rural laborers have flowed into cities and towns, and the relationship between people and land has been continuously adjusted. Figure 2 shows the change process of the land circulation scale in China from 1984 to 2014.

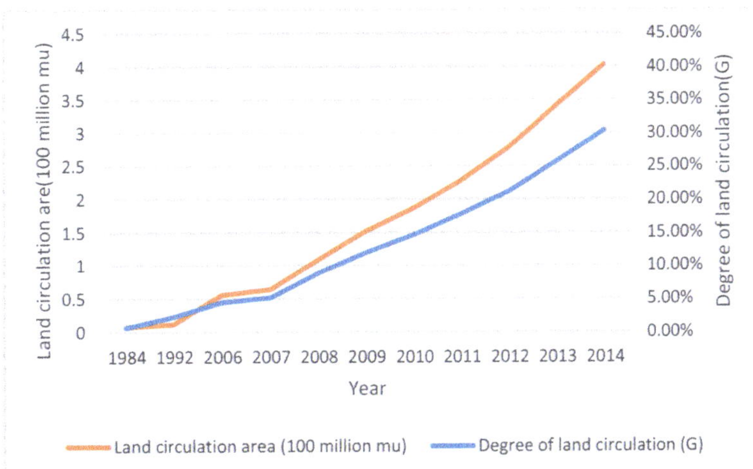

Figure 2. The scale of land circulation in China from 1984 to 2014. Note: calculated according to the annual statistical report of China's rural policies and reforms and the statistics of the Ministry of Agriculture, where the degree of land circulation is the proportion of the circulation land area to the total cultivable land area.

As shown in Figure 2, in 1984, the degree of land circulation was less than 1% in China. In 2003, the "Contracting Law" clarified, for the first time, the specific measures for the circulation of land management rights to protect said rights. This figure has grown rapidly, and by 2012 this figure exceeded 10%. In 2014, the land circulation area was 403 million mu. According to the published data in the "Statistical Annual Report on China's Rural Policies and Reforms (2019)," the area of rural land circulation accounted for 35.9% of the total area of contracted land in China in 2019 [2]. The substantial increase in the area of land circulation is reshaping the pattern of rural income. Households who have transferred into land carry out large-scale production and reduce land fragmentation. The young and middle-aged farmers who have transferred out their land choose non-agricultural operations and migrant workers, and their income is also increasing. The increasing scale of land circulation has reshaped the new pattern of rural man–land relationships, changed the original distribution pattern of land, reconstructed farmers' livelihood modes, and directly affected the income distribution of rural households.

China is currently in the process of a "hidden agricultural revolution". Population pressure has caused the per capita arable land in rural areas to steadily decline, so that it is difficult for farmers to maintain their livelihoods; as such, they must rely on auxiliary labor to meet their normal living needs. The increase in non-agricultural employment in rural areas is conducive to increasing the income of rural households. At the same time, with the advancement of new urbanization, the scale of land circulation has become larger and larger, which has promoted the large-scale development of agriculture. Therefore, discussing the effects of land circulation and labor mobility on the income inequality of rural households and comprehensively grasping the income distribution effects of land system reform can help to better promote the integration of urban and rural areas and realize the strategy of rural revitalization. This article attempts to answer two questions: first, does land circulation further widen the income inequality of rural households? Second, how does land circulation affect the income inequality of rural households?

The remainder of this paper is arranged as follows: Section 2 introduces the institutional background of land circulation and summarizes relevant literature; Section 3 discusses the impact mechanism of land circulation on the income inequality of rural households; Section 4 uses the data from the 2013 China Household Income Project (CHIP) to empirically analyze the impact of land circulation on the income gap of rural households, and to test its theoretical mechanism; Section 5 summarizes the conclusions and provides recommendations, and looks forward to future research directions.

2. Institutional Background and Literature Review

2.1. Institutional Background

The prerequisite for land circulation is that rural households have the legal right to transfer. Before China's reform and opening up, a collectivized rural land system was implemented, but this method greatly inhibited farmers' enthusiasm for production, resulting in extremely low agricultural productivity. After the reform and opening up in 1978, China began to implement the household contract responsibility system, which greatly increased the enthusiasm of farmers [3]. The property rights system of agricultural land in China has realized the "separation of the two rights," and the full collection of land property rights has come under collective ownership. In the "Minutes of the National Rural Work Conference" in 1982, the sale and lease of land was explicitly prohibited. In 1984, the national government proposed that "the right to land contracted management remains unchanged for 15 years." With the circulation of rural labor and the increase in non-agricultural business practices, the separation of the contracting and management rights of agricultural land became more serious. In 1993, the law began to recognize the separation of land contracting and management rights, allowing the paid circulation of land use rights. In 2003, the "Contracting Law" was introduced to clarify the specific measures for the circulation of land management rights. By 2005, land circulation rights accounted for 5%–6% of the contracted arable land area, and the figure even reached 8%–10% in developed coastal areas. With the rapid development of China's economy and the acceleration of urbanization, many young and middle-aged people from rural areas flocked to cities, further stimulating land circulation. According to statistics from the Ministry of Agriculture, the total area of contracted arable land was 228 million mu in China in 2011, accounting for 17.8% of the household contracted land area.

After preliminary exploration, the Ministry of Agriculture issued "Opinions" in 2011, and gradually began a nationwide pilot project on land right confirmation and registration. In 2014, China issued the document "Several Opinions of the State Council of the Central Committee of the Communist Party of China on Comprehensively Deepening Rural Reform and Accelerating the Advancement of Agricultural Modernization." "Separating ownership rights; contract rights; and the right to use contracted rural land" is the basic direction of the reform of the agricultural land property rights system in the new era. Therefore, this document has stabilized contract rights while allowing land management rights to be mortgaged to financial institutions for financing. In 2016, the government further proposed to equally protect the land management rights acquired by business entities following the circulation contract. This has ensured stable land management activities and has initiated a comprehensive reform of the "separating ownership rights; contract rights; and the right to use contracted rural land."

Land circulation can form a large-scale land operation, reduce the fragmented use of land [4], liberate rural productivity, and support urbanization. Therefore, the main purpose of the "separating ownership rights; contract rights; and the right to use contracted rural land" reform is to promote land circulation, improve agricultural productivity and competitiveness, and at the same time, to ensure the rights of operators. However, there are still significant limitations in the circulation of rural land. First of all, the definition of land property rights is unclear. The existing laws and regulations do not regulate the relationship between land circulation and land contracts, which has led to an increasing number of disputes about land contract rights, which virtually increases the cost of land circulation. Second is the structural limitation of urban and rural areas in China. Migrant workers and peasants who transfer out their land cannot obtain urban household registration and cannot enjoy complete education and medical benefits. Therefore, they tend to keep their land contract rights and entrust their relatives and friends to cultivate them instead. In addition, there are also other reasons, such as the reluctance of the elderly to leave their homeland and their attachment to the land, which also restricts the land circulation.

2.2. Literature Review

The separation of the right to use and ownership of land generates land rent [5]. Moreover, with diminishing returns to land, the price of land circulation determines the amount of land rent. The circulation period is short, so the price of land circulation depends only on the average annual income of the land. At the same time, the conditions for the generation of land rent are related to the price of agricultural products, the fertility of the land, the geographical location and the relationship between supply and demand [6]. When the transaction cost is zero, market transactions can achieve an effective allocation of land resources [7], so scholars cannot think that land rent is suppressed [8]. The generation of land rent is the basis of land circulation. China has changed the land structure of contracting production to households and liberalized land circulation, and land distribution has changed from absolute equality to relative equality. The land was redistributed to the peasants with higher productivity [9]. The land system has a profound effect on the income distribution of rural households [10]. Before further reforming the land system, it is essential for the government to carefully consider the impact of the reform on the income of rural households and income inequality. Therefore, research on the impact of land circulation on the income inequality of rural households has gradually become a popular topic. The research mainly focuses on two aspects, namely the income effect of land circulation and the distribution effect of land circulation.

2.2.1. Income Effect of the Land Circulation

The existing literature on land circulation and the income of rural households analyzes the mean effect of the impact of land circulation on the income of rural households under the framework of a linear model. There are two main points of view: first, the net income of peasants before and after land circulation has not changed significantly. There is a two-way causal relationship between the level of non-agricultural income of peasants and land circulation. The impact of early land circulation on the non-agricultural income of peasants is relatively weak [11]. Second, land circulation has generally increased the income of rural households, created income for rural households, and reduced poverty. Scholars have held this view in majority [12–14]. The land rent and income from migrant work brought about by land circulation accounts for most of the income growth of rural households, which is stable and sustainable [15,16]. The study by Xiao Han and Anlu Zhang [17] from the perspective of land circulation found that land circulation has a positive effect on the income of transfer-in rural households, but has no significant effect on the income of the transfer-out rural household. Fei Chen and Weijuan Zhai [18] pointed out that renting out land is beneficial for increasing the income of rural households and reducing the incidence of poverty. However, the welfare effect is significantly different between different family groups. Xiangyong Wang et al. [19] studied the changes in farmers' income before and after land circulation and found that land circulation increased farmers' property income.

In addition, the income effect of land circulation needs to be considered from two aspects. On the one hand, because of the characteristics and factor endowments of a rural household, the benefits they receive from the land circulation market are different [20]. Land circulation generally increased the income of rural households, but this may be the result of the pull of high-growth rural households, however, it exaggerates the role of land circulation. The results of testing that land circulation does not affect the income of rural households and may also be offset by increasing income and decreasing income. On the other hand, there is the problem of the "selection bias" of the sample. Whether a rural household participates in land circulation is a non-random "self-selection" behavior. However, this self-selection bias has not been corrected when examining the income effect of land circulation on the different types of rural households, which has also brought about estimation problems.

2.2.2. Income Distribution Effect of the Land Circulation

An increase in the absolute income level does not imply that land circulation has a positive effect on income distribution. There are two main views on the income distribution effect of land circulation. First, land circulation has widened the income gap of rural households. Early research considered that land circulation was the choice and behavior of rural households with a high income [21]. In other words, only the powerful class can receive high returns on land circulation [22], while the interests of small farmers are sacrificed, leading to further widening of the income gap [23]. The constraints of credit rationing make it difficult for poor rural households to obtain benefits from the land rental market [24]. Möllers and Meyer used the PSM method to analyze the impact of labor migration on income inequality and poverty in rural Kosovo [25]. In recent years, some scholars have used quantile regression to analyze the effect of land circulation on the income gap of rural households. For example, Junping Guo [26] used a quantile regression model to investigate regional income, finding that the circulation of farmland has widened the income gap between rural households in the eastern and central regions. From the perspective of different income classes, the transfer inflow of farmland has promoted an increase in the income of poor and low-income households, resulting in some high-income households. Although land transfer inflow and outflow can increase income, land circulation exacerbates income inequality [27–29]. The research for Changliang Shi [30] found that the income-increasing effect of land circulation is heterogeneous for rural households with different income levels. Rural households with middle and high-income levels obtain greater benefits from land circulation than rural households with a low income. Second, land circulation can narrow the income gap between rural households [31]. Guanghua Wan [32] constructed a regression decomposition framework using rural household data to study rural income inequality in China, and found that the land circulation between poor rural households reduces income inequality. Land circulation increases the income level of rural households with low income and can alleviate the inequality caused by non-agricultural employment [33]. However, some scholars believe that the income distribution effect of land circulation has a selection effect among heterogeneous rural households, and the mechanism of action for a rural household in different income ranges is different [34]. The disadvantage is that most studies only compare the income effects of rural households with different income levels who participate in land circulation, and fail to answer quantitatively how land circulation widens the income inequality of rural households.

In general, the existing literature agrees that land circulation can effectively increase the income of rural households, but the perception of the income distribution effect of land circulation is still ambiguous. Previous studies have made more use of quantile regression. However, the data are too rough to infer any effect of land circulation on the income gap of rural households by relying solely on the difference in the regression coefficients in different quantiles. Moreover, this approach would ignore the internal mechanism of the income distribution effect of land circulation and the mechanism of the effect of other economic and social factors on the income distribution of rural households in the process of land circulation. There is almost no literature involved in the research on its influence mechanism and degree of influence. Shi, C.L., et al. [35] and Yang. Z et al. [36] estimate the contribution of land circulation to the income gap of rural households using a Fields decomposition. However, the decomposition method is strictly restricted by the form of the income equation and the measurement index of the income gap, and the decomposition of the constant term has not been well handled and explained. Liang Y et al. [37] discussed the effects of labor mobility and land circulation on the income of rural households, respectively. They use the propensity score matching (PSM) model to explore the effect of land circulation on the income of rural households. This can effectively alleviate the bias problem caused by the "self-selection" of the income effect, but it does not deeply explore the mechanism of the dynamic relationship between labor mobility and land circulation on the income distribution of rural households.

Compared to previous literature, there are two possible contributions of this article: first of all, it provides a new perspective to deepen the understanding of the impact of land circulation. The existing literature studies focused on the effect of land circulation on agricultural productivity and rural household income. There were not many studies on income distribution and inequality. This article has supplemented this aspect. Second, we analyze the causes of rural household income inequality from the perspective of labor mobility, and provide new connotations for income distribution theory. Existing studies rarely consider the impact of the dynamic relationship between labor mobility and land circulation on the income inequality of rural households. However, the contradiction between more people and less land has always been one of the main contradictions in China's agricultural development. The relationship between the two may have a more profound impact on income distribution than traditional economic factors such as material capital and family characteristics. Therefore, this article uses the form of interaction terms to analyze the contribution rate of labor mobility and land circulation to different income levels, to deconstruct the mechanism of land circulation on the income inequality of rural households, to provide a theoretical basis and decision-making participation for broadening farmers' income increase channels.

3. Theoretical Mechanism

To analyze the theoretical mechanism of land circulation in the income gap of rural households, it is necessary first to clarify peasants' motivation to engage in land circulation. The decision of land circulation is based on the "cost-benefit" principle. When non-agricultural productivity is greater than agricultural productivity, a rural household will choose to transfer the outflow of the land. When the benefits of land operation from rural households exceeds the opportunity cost of farming the land, a rural household will choose to transfer the inflow of the land and expand the production of land on a large scale. To study the effect of land circulation on the income inequality of rural households, we can start from the income function of rural households. Assuming that the income of a rural household Y is completely determined by the scale of land circulation T, the size of the non-agricultural labor force population S, and the family characteristics Z, then the income of rural households can be expressed as:

$$Y_i = f(T, S, Z) \quad (1)$$

In order to obtain the income gap of rural households, now the variance of both sides of the equation is calculated simultaneously, and we can obtain:

$$Var(Y_i) = \delta^2 Var(T_i, S_i, Z_i) \quad (2)$$

Here, we fixed family characteristics, then δ in Equation (2) is the income effect of land circulation and the non-labor scale. In theory, land endowments in a perfectly competitive land market will not cause an income gap, and X is a constant at this time. However, in reality, the land market is incomplete. At this time, δ will vary with region, rural family, climate and time. δ_T is the income effect variable of the interaction term of land circulation and the non-labor scale. Assuming that the income effects of land scale, land circulation, and the non-labor scale are independent of one another, then the income gap can be expressed as:

$$\mu_\delta^2 Var(T_{iT}, S_{iT}, Z_{iT}) + \mu_T^2 Var(\delta_T) + Var(T_{iT}, S_{iT}, Z_{iT}) \cdot Var(\delta_T) \quad (3)$$

where μ_δ is the expected value of the income effect, and μ_T is the expected value of the interaction term between the scale of land circulation and the scale of non-agricultural labor. From Equation (3), it can be seen that the income gap of rural households is not only related to the expected values of both, but also depends on the scale of land circulation and the scale of non-agricultural labor. When the expectation of the income effect is 0, the income gap of rural households is also 0. When the expectation and the income effect are not 0, there will be an income gap in a rural household. Land circulation is the reallocation of land resources based on market-oriented means, focusing more on efficiency rather than fairness. Therefore, a rural household with different income levels faces unequal opportunities in the land circulation market and different benefits from land circulation. Analyzing the effect of land circulation from the perspective of non-agricultural labor mobility, the flow of labor is oriented: it only flows from low-income regions to high-income regions. For rural areas, the mobility of labor optimizes the allocation of resources and adjusts the structure of the agricultural industry. However, the flow of labor from agriculture to non-agricultural employment takes away the productivity of agricultural production, and most of the labor flow out is more competitive. In particular, many of the migrant workers who leave their hometowns are young and middle-aged. The elderly and children stay in their hometowns, and the remaining family members cannot afford excessive agricultural production. Therefore, rural households choose to transfer out their excess land to achieve optimal allocation of their household resources. This shows that there is a dynamic relationship between land circulation and labor mobility. When this dynamic relationship is higher, the wider the peasants' income channels, and the greater the impact on the income of rural households. The income gap of rural households is determined by the dynamic relationship between land circulation and labor mobility. From the characteristics of rural households with different incomes, for low-income families, they can only choose to cultivate the existing land, because they cannot afford to pay enough land rent and the cost of non-agricultural operations or migrant workers. In addition, compared to high-income households, low-income households are more dependent on land and are less willing to lease out the rights of using contracted land. Therefore, low-income rural households are easily excluded from the land circulation market. High-income rural households can avoid land fragmentation through land inflow, and large-scale agricultural production can further reduce costs and increase profits. As a result, the income of high-income rural households is becoming higher and higher, while the income of low-income rural households remains the same or rises slowly alongside economic development, meaning the income gap between rural households is further widening. Analyzing this mechanism specifically in terms of agricultural and non-agricultural operation productivity, it can be divided into two parts: on the one hand, in terms of agricultural productivity, high-income farmers are better able to afford the high cost of renting land and are more likely to acquire new technologies in the process of agricultural production. Combining the potential level of human capital and the ability to obtain market information, high-income rural households have higher returns on agricultural production. On the other hand, from the perspective of non-agricultural production activities, rural households with a low income have relatively weak anti-risk capabilities and experience more restrictions on employment in the non-agricultural market. For example, the education level of family members with a low income may be lower than the average. Meanwhile, rural households with a high income can integrate family resources through the land circulation to maximize their comparative advantages in non-agricultural fields. Since the 21st century, the process of urbanization has accelerated, the price of urban labor has become much higher than the income of agricultural production, and rural households with a high income have started moving to non-agricultural operations earlier. Therefore, although land circulation has increased the income level of most rural households, it has also widened the income inequality of rural households to a certain extent.

4. Empirical Research

4.1. Model Construction

The re-centered influence function (RIF) regression method proposed by Firpo [38] is different from other regressions methods in terms of the explained variables in said RIF regression. The explained variables in RIF regression can not only be the income level of residents or other statistics, but must also be the statistics of income inequality such as the quantile, variance and Gini coefficient obtained based on the influence function. Therefore, we can establish a direct relationship between the impact factors and the degree of income inequality. This article used RIF regression to discuss the effect of rural household land circulation on the rural household income gap. Considering that the Gini coefficient can better describe the degree of income inequality, this article used the Gini coefficient to measure the income gap in net income per capita of rural households. In the RIF regression framework, the benchmark regression model is constructed as follows:

$$Gini(inco) = \alpha_0 cland + \alpha_1 X + \varepsilon \quad (4)$$

where the explained variable *inco* is the per capita disposable income of rural households and *Gini* (*inco*) is the Gini coefficient of the disposable income of rural households; the explanatory variable *cland* is the land circulation area of rural households, which is the sum of the area of land transfer outflow and inflow; X is the control variable, which is used to mitigate the estimation error caused by the omitted variable(the selection of the control variable will be described in detail later); and ε is the random error term. The theoretical mechanism of land circulation affecting the income gap of rural households has been discussed above. Next, this article used quantile regression estimation to test this theoretical mechanism. Quantile regression was first proposed by Koenker and Bassett [39] in their systematic study, and it can accurately describe the effect of explanatory variables on the range of variation of the explained variables and the shape of the conditional distribution. Therefore, this article used quantile regression and established the following regression model:

$$Q_\tau[\ln inco|Y] = \beta_{0,\tau} + \beta_{1,\tau} blab * sland + \sum \beta_{i,\tau} CV + \omega_\tau \quad (5)$$

where the explained variable $Q_\tau[\ln inco|Y]$ is the logarithm of the per capita annual disposable income of rural households at the τ quantile. The explanatory variable is the product of the proportion of household non-agricultural labor force *bland* and the land area *sland* transfer out from rural households, which represents the dynamic relationship between labor mobility and land circulation. CV refers to the other control variables, which are the same as the control variables in Equation (4), and ω is a random disturbance term.

4.2. Data Source

The Chinese Household Income Project (CHIP) conducted five household surveys in 1989, 1996, 2003, 2008 and 2014, and collected data of the income and expenditure of urban and rural households from 1988, 1995, 2002, 2007 and 2013, respectively. This article selected the data from the 2013 CHIP, which covers 18,948 household samples and 64,777 individual samples, selected from 234 counties and districts of 126 cities in 14 provinces, including 7175 urban household samples, 11,013 rural household samples, and 760 outdoor migrant workers samples. According to the research content of this article, only the data of the rural households in the questionnaire were retained, and missing values and unreasonable data were excluded, thus, a total of 10,262 valid data were obtained. These valid data were distributed across 14 provinces, namely, Shanxi, Henan, Anhui, Hubei, Hunan, Gansu, Yunnan, Sichuan, Chongqing, Beijing, Liaoning, Jiangsu, Shandong and Guangzhou. After sorting, we divided the data into the central, western, and eastern regions, with 3963, 2705, and 3594 data, respectively.

4.3. Variables

Explained variable: per-capita annual disposable income of rural households (*inco*); the unit is yuan/person. The CHIP data selected "2013 household disposable income" divided by "total household population". Disposable income mainly included wage income, net operating income, property income and transfer income.

Explanatory variable: land circulation area (*cland*); the 1st unit is hm^2. We used the sum of the area of land transfer out (*sland*) and the area of land transfer in (*tland*) of rural households. From the CHIP data, we deleted the data of rural households who transferred the outflow and inflow of land at the same time. In addition, rural households were the only inflow and outflow of land.

The control variables mainly included two aspects, namely the characteristics of the head of the rural household and the characteristics of the family. The characteristics of the head of the household included:

① Gender (*gen*): with gender as a dummy variable, a value of 1 was assigned to males and 0 to females. Generally speaking, the income of male-headed rural households is greater than that of female-headed households, especially for low- and middle-income households.
② Age (*age*): age inevitably affects the income gap of rural households, which shows multiple effects.
③ Education level (*edu*): the highest educational level experienced by the head of the household was used to measure education level. As dummy variables, no elementary school was assigned a value of 1, elementary school a value of 2, junior high school a value of 3, senior high school a value of 4, vocational/technical school a value of 5, technical secondary school a value of 6, junior college a value of 7, undergraduate university a value of 8, and postgraduate university and above a value of 9.
④ Being a member of the Communist Party of China (*pol*): a value of 1 was assigned to being a member of the Communist Party of China (CPC), while a value of 0 was assigned to not being a member of the Communist Party of China (CPC).
⑤ Being a village cadre (*cad*): being a village cadre was assigned a value of 1, while not being a village cadre was assigned a value of 0.
⑥ Participation in a professional cooperative economic organization (*org*): participation was assigned a value of 1, while no participation was assigned a value of 0.
⑦ Health status (*hea*): this indicator was self-evaluated by the respondent, and was also used as a dummy variable and assigned the following values: 1 = very good; 2 = good; 3 = fair; 4 = bad; 5 = very bad.

Meanwhile, family characteristics included:

① Family size (*pop*): the total family population represents the family size.
② Non-agricultural labor force population (*lab*): the sum of non-agricultural business population and migrant workers was regarded as the family non-agricultural labor force population, and the proportion of the non-agricultural labor force (*blab*) was used as the non-agricultural labor force population over the total family population.
③ The net value of agricultural operating fixed assets (*agr*): this is an operating asset, presented in the CHIP data as the "estimated net value of agricultural operating fixed assets at current prices at the end of 2013".
④ Operating land area (*land*): refers to the area of land operated by a family. In addition, this article divided the provinces into three regions based on their location. The central region included the five provinces of Shanxi, Henan, Anhui, Hubei and Hunan; the western region included the four provinces of Gansu, Yunnan, Sichuan and Chongqing; the eastern region included the five provinces of Beijing, Liaoning, Jiangsu, Shandong and Guangdong. The descriptive statistics of the variables are shown in Table 1.

Table 1. Definitions of variables and descriptive statistics.

Variable	Variable Name	Variable Definition and Assignment	Mean	Standard Deviation
Explained variable	Household income per capita (*inco*)	Family annual total income/total population, in yuan/person	13,259.9	14,039.9
Explanatory variables	Land circulation area (*cland*)	Area of land transfer inflow plus the area of land transfer outflow (hm^2)	1.3814	6.8296
Characteristics of the household head	Gender (*gen*)	1 means male; 0 means female	0.9172	0.2756
	Age (*age*)	Age of the household head	51.83	11.50
	Education level (*edu*)	no elementary school as 1, elementary school as 2, junior high school as 3, senior high school as 4, vocational/technical school as 5, technical secondary school as 6, junior college as 7, university undergraduate as 8, and postgraduate and above as 9.	2.736	0.9729
	State of health (*hea*)	1 = very good, 2 = good, 3 = fair, 4 = bad, 5 = very bad	2.20	0.9482
	Being a member of the CPC (*pol*)	1 means a member of CPC 0 means not a member of CPC	0.110	0.3129
	Being a village cadre (*cad*)	1 means yes; 0 means no	0.0492	0.2163
	Participation in a profes-sional cooperative eco-nomic organization (*org*)	1 representative to participation 0 representative not to participation	0.0341	0.1815
Family characteristics	Family size (*pop*)	Total family population	3.80	1.489
	Number of families non-agricultural labor (*lab*)	The total number of households migrant workers and non-agricultural production and operation population (person)	0.9818	2.44
	The net value of fixed as-sets of agricultural opera-tions (*agr*)	The estimated net value of agricultural operating fixed assets at the end of the year	6189.0	84,159.6
	Family-owned land area (*land*)	The total land area operated at the beginning of 2013	5.574	8.868
Region	Central Region	Central region = 1, non-Central region = 0	0.3859	0.4868
	Western Region	Western region = 1, non-Western region = 0	0.2638	0.4407
	East Region	Eastern region = 1, non-Eastern region = 0	0.3503	0.4771

4.4. Benchmark Regression Results

Table 2 shows the estimated results of the RIF regression benchmark for the impact of land circulation on the income gap of rural households. The per capita annual disposable income of rural households was used as the explained variable, and land circulation was used as the explanatory variable. Columns (1)–(3) report the empirical results of the central, western, and eastern regions, respectively. Column (4) shows the empirical results of the national data without a fixed province effect, while Column (5) shows the empirical results of the national data after fixing the province effect. Comparing the results from Columns (1)–(6), it can be seen that the area of land circulation further widens the income gap of rural households.

From Column (5), after controlling for the provincial factors that do not affect the income gap of rural households over time, the larger the land circulation area in the country as a whole, the bigger the income gap of rural households. Moreover, comparing Columns (4) and (5), the coefficient increased from 0.0014 to 0.016 after controlling for the province. The specific impact mechanism will be discussed in detail in the next section. Among the control variables, the variables of whether to participate in a professional cooperative economic organization, family size, the size of the non-agricultural labor force, and the area of land operated by the family have a significant effect on the income gap of rural households. Among them, the variable of participation in a professional cooperative economic organization can widen the income gap of rural households. Meanwhile, the other variables can narrow the income gap of rural households. Families participating in professional cooperative economic organizations can obtain certain advantages in agricultural production, so this factor can widen the income gap of rural households. In the context of the one-child policy, the family population is limited; a large family size and a large non-agricultural labor force can narrow the income gap among rural households. The contracted land area can also reduce the income gap between rural households; this is because large-scale agricultural production reduces production costs. However, a rural household with less land typically chooses non-agricultural management to realize the optimal allocation of limited land resources, which increases the overall income level while also reducing the income inequality of rural households.

The national data on rural households were divided into three regions. Comparing Columns (1)–(3), it can be seen that the most significant effect of land circulation on the income gap of rural households was in the central region. This is because Henan and Hunan, the major grain-growing provinces, are in the central region. Before the emergence of large-scale migrant workers, the main economic source of income for a rural household was agricultural production. Cities in the eastern region are relatively more developed, relying mainly on a non-agricultural economy to drive employment. In particular, the eastern coastal cities have benefited from the reform and opening-up policy and no longer rely on agricultural production. Meanwhile, the population in the western region is relatively small, and the level of land rent is generally low; therefore, the ratio of land rent to the total income of rural households is relatively low. Most rural households are unable to achieve an increase in income in this way, so the effect of land circulation on income inequality among rural households was shown to be non-significant.

Table 2. Baseline estimation results of the impact of land circulation on the income gap of rural households.

	Central Region (1)	Western Region (2)	East Region (3)	Nationwide (4)	Nationwide (5)
Land circulation area *cland*	0.0032 * (1.65)	−0.00002 (−0.01)	0.0011 (1.38)	0.0014 * (1.70)	0.0160 ** (1.89)
Gender *gen*	−0.032 (−1.25)	−0.0113 (−0.39)	−0.018 (−0.71)	−0.0091 (−0.58)	−0.0145 (−0.91)
Age (Logarithm) *age*	0.0491 ** (1.85)	0.0038 (0.11)	0.03224 (0.69)	0.0243 (1.08)	0.0279 (1.18)
Education level *edu*	−0.008 (−1.44)	0.0041 (0.39)	0.0103 (1.16)	0.0053 (1.00)	0.0084 (1.57)
State of health *hea*	0.0089 (1.45)	−0.0045 (−0.54)	0.0017 (0.22)	0.0038 (0.92)	0.0035 (0.80)
Being a member of the CPC *pol*	0.0246 (1.18)	0.006 (0.25)	−0.0228 (−1.13)	0.003 (0.24)	0.0053 (0.42)
Being a village cadre *cad*	−0.0112 (−0.47)	0.059 (1.21)	0.0439 (1.16)	0.025 (1.16)	0.023 (1.07)
Participation in a professional cooperative economic organization *org*	−0.0446 * (−1.88)	−0.0754 *** (−2.88)	0.1428 ** (2.02)	0.0958 ** (2.04)	0.1037 ** (2.36)
Family size *pop*	−0.012 *** (−2.63)	−0.0139 ** (−2.23)	−0.0132 ** (−2.41)	−0.0084 ** (−2.53)	−0.009 ** (−2.35)
Number of family non-agricultural labor *lab*	−0.0037 * (−1.89)	−0.0036 ** (−2.52)	−0.0003 (−0.16)	−0.0028 *** (−3.10)	−0.0023 *** (−2.58)
Net value of fixed assets of agricultural operations *agr* (Logarithm)	0.0028 * (1.73)	−0.002 (−1.08)	0.0016 (0.58)	0.0012 (0.93)	0.0004 (0.27)
Family-owned land area land (Logarithm)	−0.008 (−0.89)	−0.0165 ** (−1.77)	−0.0042 (−0.43)	−0.014 *** (−2.57)	−0.0196 *** (−3.23)
Intercept term	0.2783 *** (2.56)	0.5084 *** (3.17)	0.3049 (1.52)	0.3527 *** (3.62)	0.3485 *** (3.45)
Province dummy variable	No	No	No	No	Yes
F test	Pass	Pass	Pass	Pass	Pass
Observations	3963	2705	3594	10262	10262

Note: *t*-values in parentheses, ***, **, and * represent 1%, 5%, and 10% significance levels, respectively.

4.5. Robustness Test

The RIF regression program operates with robust standard errors by default; this can effectively weaken the endogenous problems caused by the omission of variables, etc., and avoid heteroscedasticity from interfering with the estimation results. In addition, the per capita net income of rural households and the behavior of land outflow and land inflow do not exist exactly in the same year. Therefore, the possibility of endogenous factors between land circulation and the net income of rural households was relatively slight. This paper uses a series of tests to confirm the robustness of the conclusions, such as changing the income gap measurement indicators, replacing explanatory variables and explained variables. The results are shown in Table 3.

Table 3. Robustness test results.

Explanatory Variables	Measuring Inequality with Variance	Inequality Measured by 80–20 Quantile Distance Values	Replacement of Explanatory Variables	Replacement of Explanatory Variables
	(1)	(2)	(3)	(4)
Land circulation area (*cland*)	0.0035 *** (2.04)	0.0223 *** (3.2)	—	0.007 * (1.71)
Area of land transferred to (*tland*)	—	—	0.0014 * (1.66)	—
Control variable	Yes	Yes	Yes	Yes
Intercept term	0.3102 (1.38)	0.2408 (0.23)	0.3483 *** (3.44)	0.3814 *** (3.3)
Province dummy variables	Yes	Yes	Yes	Yes
R^2	0.0176	0.0191	0.0131	0.0124
Observations	10,262	10,262	10,262	10,262

Note: *t*-values are in parentheses, and *** and * represent 1% and 10% significance levels, respectively.

Inequality is measured by variance. Variance is a widely used indicator in the issue of inequality, as well as the Gini coefficient. To verify the robustness of the empirical results, this article uses the variance of the logarithm of household disposable income per capita to replace the Gini coefficient. The regression results are shown in column (1) of Table 3. The core explanatory variable land circulation is positive at the 1% significance level, which is consistent with the result of the benchmark regression.

The income gap is measured using 80–20th quantile values. Quantile distance can better test the income gap between the highest and the lowest income group. This article uses the 80–20 quartiles to replace the Gini coefficient to test the robustness of the empirical results. The regression results are shown in column (2) of Table 3. *cland* is significantly positive at the 1% level, the coefficient of *cland* in the corresponding regression is 0.0223. This shows that when the area of land transfer to all rural households in the sample increases by 1 unit, the difference between the 80th quantile and the 10th quantile of the per capita disposable income of rural households will increase by 0.0223, an increase of 2.2%. After the replacement of the inequality measurement indicators, they are consistent with the benchmark results, indicating that the empirical conclusions of this article have not changed due to different income gap indicators.

Replacement of explanatory variables. The area of land circulation is composed of the area of land transfer outflow and inflow. After rural households have transferred inflow land, they can expand the scale of agricultural production and increase their income, which in theory can better reflect the further widening of the income gap. Therefore, this article replaces the explanatory variable with the area of land transferred inflow by rural households. The regression results are shown in Column (3) in Table 3. The estimated results are positive at the 10% significance level, which is consistent with the benchmark regression results.

Replacement of the explained variable. Compared with income, consumption is more stable and more reliable. It can better reflect the living conditions of rural households and is a more accurate indicator of inequality. Therefore, this article replaces the explained variable with the Gini coefficient of per capita consumption of rural households. The regression results are shown in Column (4) in Table 3. The core explanatory variables are significantly positive at the 10% level, which is consistent with the benchmark regression results, indicating that the conclusions of this article are not affected by the metrics of the explained variables.

4.6. Analysis of the Influence Mechanisms

The previous section mainly confirmed that land circulation increases the income gap among rural households. This part mainly focuses on analyzing the mechanism of the effect of land circulation. The government is continuing to further lift the restrictions on non-agricultural employment, is continuously improving the non-agricultural employment market, and is reducing the gap of registered residence between urban and rural areas, thereby providing more employment opportunities for young migrants workers. With the rapid economic development in developed cities, there is a greater labor shortage. The income from migrant workers and non-agricultural business income is significantly higher than that from agricultural production. More young and middle-aged laborers from families with a low income choose to give up their farmland and switch to non-agricultural operations or go out to work, which increases family income. Table 4 shows the regression results of the dynamic relationship between labor mobility and land transfer out on the per capita disposable income of rural households by region. The labor mobility indicator here was the ratio of non-agricultural labor over the total household population, and the per capita disposable income of rural households was processed in logarithm.

Column (1) in Table 4 shows the OLS estimation. Based on the results of this OLS estimation, it can be seen that the coefficients of the interaction terms between labor mobility and land outflow are significantly positive, and passed the 1% significance level test for both the national region and the central, western, and eastern regions. This shows that the interaction term of labor mobility and land outflow increases the income of rural households. Moreover, the dynamic relationship in the eastern region is significantly higher than that of the other two regions. This is because there are more developed cities and more non-agricultural employment opportunities in the eastern region. The estimation result of OLS was used to compare the result of the quantile regression. The quantile regression results of the different quantile points are given in Columns (2)–(6), respectively, which are the 10th, 30th, 50th, 70th, and 90th points. From the results of the quantile regression, as the degree of dynamic relationship increased, the per capita disposable income of rural households also increased. From the national data, it can be seen that the interaction term has the greatest promotion effect on middle-income families. However, the promotion effect of high-income households is significantly higher than that of low-income households, which means that the benefits of "rich people" from land circulation are significantly greater than those of "poor people", which naturally further increases the income gap. In addition, these estimated coefficients are significant in the different quantiles, however the coefficients are not identical, indicating that rural households with different income levels do not benefit equally from land circulation. Combining the results of quantile regressions by regions, especially the western and eastern regions, the dynamic relationship between high-income households with labor mobility and land outflow has increased. The promotion effect on the per capita income of rural households is much higher than that of low-income households, with the coefficient even exceeding 0.1 in the highest quintile. Therefore, land circulation not only causes inequality in the allocation of agricultural land resources in villages, but also further aggravates this inequality through labor mobility, forming the "Matthew effect" where the rich get richer, leading to further widening of the income gap among rural households.

Table 4. Regression results of the dynamic relationship between labor mobility and land transfer outflow on the per capita disposable income of rural households by region.

Region	Variable	Ols (1)	Q10 (2)	Q30 (3)	Q50 (4)	Q70 (5)	Q90 (6)
Nationwide	Proportion of non-agricultural labor force * land outflow	0.0686 *** (5.69)	0.0486 ** (2.12)	0.0779 *** (5.11)	0.0758 *** (5.21)	0.0588 *** (3.82)	0.065 *** (2.95)
	Other variables	Already controlled					
	Constant term	9.541 *** (61.66)	9.147 *** (31.1)	10.031 *** (48.33)	10.288 *** (56.13)	10.459 *** (55.09)	10.506 *** (44.99)
Central Region	Proportion of non-agricultural labor force * land outflow	0.0593 *** (3.72)	0.0642 *** (3.08)	0.0914 *** (4.63)	0.0538 *** (3.27)	0.0515 *** (2.89)	−0.0004 (−0.02)
	Other variables	Already controlled					
	Constant term	9.97 *** (40.32)	8.843 *** (19.09)	10.31 *** (30.3)	10.382 *** (36.29)	10.543 *** (37.29)	10.491 *** (30.81)
Western Region	Proportion of non-agricultural labor force * land outflow	0.060 ** (2.23)	0.0402 (1.10)	0.03182 (0.68)	0.0896 * (1.76)	0.1003 ** (1.98)	0.0582 (1.2)
	Other variables	Already controlled					
	Constant term	9.497 *** (34.12)	7.973 *** (15.78)	9.071 *** (25.3)	9.749 *** (27.74)	10.367 *** (28.61)	10.323 *** (19.99)
Eastern Region	Proportion of non-agricultural labor force * land outflow	0.107 *** (4.52)	0.0818 ** (2.37)	0.0665 ** (2.15)	0.1037 *** (3.60)	0.1127 *** (3.63)	0.1288 *** (2.79)
	Other variables	Already controlled					
	Constant term	10.157 *** (38.24)	10.739 *** (19.72)	10.156 *** (27.36)	10.445 *** (34.72)	10.354 *** (35.35)	10.42 *** (27.78)

Note: t-values in parentheses, ***, **, and * represent 1%, 5%, and 10% significance levels, respectively. The control variables are not listed due to space limitations.

5. Conclusions and Recommendations

The main purpose of China's current rural revitalization strategy is to promote rural development, increase farmers' income, and narrow the income gap of rural households. Under the background of "separating ownership rights; contract rights; and the right to use contracted rural land," the scale of land circulation is getting larger and larger, and stable land property rights are the foundation of rural development. Based on the 2013 China Household Income Project (CHIP), this article used the re-centered influence function (RIF) regression method to empirically test the impact of rural land circulation on the income gap of rural households in China in three regions: the central, eastern, and western regions. The final result were as follows.

First, as a result of the implementation of the "separating ownership rights; contract rights; and the right to use contracted rural land" by the government, the legitimate rights and interests of land operators are now protected. Land circulation has improved agricultural production efficiency and has liberated labor productivity. Surplus peasants are moving to cities to work and earn an income. Coupled with the acceleration of urbanization, the overall income of rural households has improved. For example, rural households with a low-income lack funds for agricultural production and have limited access to land, but have a surplus of labor. Thus, young and middle-aged people of low-income rural households choose to transfer out their land, and then go out for work or go into non-agricultural employment, thereby greatly increasing the total family income. Rural households with a high-income can increase their income by transferring in their land, expanding the scale of agricultural production and reducing the marginal costs. Therefore, the continuous increase in the scale of land circulation has greatly increased the income of rural households as a whole.

Second, land circulation widens the income inequality of rural households. The inequality of market opportunities leads to different amounts of land circulation by different income groups, resulting in income inequality within groups. In addition, the differences in the abilities and factor endowments between groups also leads to different income returns in land circulation among different income groups. From the central, western, and eastern regions, the region where land circulation has the most significant impact on the income gap of rural households is in the central region. There are several large agricultural provinces in the central region, where farmers are more dependent on land. The adjustment of income distribution is also more sensitive to changes in the land system.

Third, from the perspective of labor mobility, the impact mechanism shows that the interaction term of land outflow and the proportion of non-agricultural labor have a more significant effect on the income growth effect of high-income rural households. The "rich" gain from land circulation to a significantly greater extent than the "poor". On the one hand, the total value of agricultural output is limited, and the average income of each farmer is very small. The surplus of rural laborers can only choose to work in cities. However, farmers who originally owned land resources have expanded their production scale through land circulation. The increase in income of rural household migrant workers is smaller than that of rural households with large-scale production, which has led to a gradual widening in the income gap between rural households. On the other hand, land circulation not only causes inequality in the allocation of arable land resources, but also further contributes to the widening of the income gap among rural households through the "Matthew Effect" of the rich getting richer.

Land circulation has liberated rural surplus productivity, increased the scale of land production, and enabled many young adults to engage in non-agricultural operations, allowing effective allocation of land and human resources. Although land circulation has led to the widening of income inequality for rural households, it is unnecessary to give up land circulation. Instead, we should continue to promote land circulation and improve the land system. Based on the previous analysis and the above conclusions, this article proposes the following policy recommendations.

First, the government should further improve the land circulation market and the efficiency of land circulation, reduce the transaction cost of land circulation, and clarify land use rights. This requires establishing an information platform for the transfer market with transparent and open information to solve the problem of information asymmetry. At the same time, they should reduce the restrictions of non-market factors in land circulation so that farmers can become the real beneficiaries of land circulation.

Second, we should provide certain policy preferences to low-income farmers who do not have comparative advantages in land circulation, and we should strengthen the investment and technical training of farmers so that they can improve their competitiveness in non-agricultural employment. The overall education level of the labor force flowing into the cities and towns from rural areas is relatively low, and most of them can only perform simple labor with low wages. Therefore, the government needs to increase the technical training and improve the cultural quality of farmers. In this way, although rural households with a low-income lack the means of production, they can increase their income and stability by relying on their technology and cultural literacy to work outside the home, thereby increasing the income of the family. At the same time, the government should also improve the welfare protection measures for rural non-agricultural operations, so that low-income farmers can engage in non-agricultural operations without worries and can further liberate productivity. For example, the government grants a quota of low-interest loans to farmers who switch from agricultural production to self-employment, and provides a series of policy supports such as tax cuts.

Third, the government should reform the household registration system and open up the social security system. A certain percentage of migrant workers cannot enjoy complete social security due to the restrictions of the household registration system. Thus, they do not dare to transfer out all of their contracted lands, instead only letting their relatives and friends cultivate or even abandon them. At the same time, household registration restrictions have resulted in farmers lacking a sense of belonging to a city, so land reform must also be coordinated with urban sector reforms. Cities and towns have further improved their medical and educational systems, allowing residents who work in said places to enjoy the same welfare protection as urban residents. In particular, the schooling problem of enrolling the children of migrant workers needs to be solved urgently, and the five social insurances and one housing fund should be fully implemented as soon as possible for migrant workers. The government should eliminate the worries that migrant workers have regarding having nowhere to stay in the city and should increase their sense of belonging to said city, so that land resources can be better allocated.

Finally, the value of total agricultural output on the technical level should be increased. Due to the scissors gap between the prices of industrial and agricultural products, the increase in grain prices lags behind the increase in the prices of agricultural materials and other industrial products, further shrinking the value of agricultural output. Therefore, the government needs to increase its support for rural production technology, the value of total agricultural output, and the capacity of the land for the strong rural labor force.

China is a largely agricultural country, and land policy involves all aspects. With the progress of urbanization and industrialization, the proportion of non-agricultural employment among farmers has gradually increased, as has the scale of land circulation. The study of rural household income inequality from the perspective of land circulation is only one aspect. After a significant increase in higher education, a new generation of farmers began to experiment with more income options. Next, we should study whether the combination of land circulation and education system affects agricultural productivity and the income gap of rural households. In addition, we can also explore the improvement of the land circulation market from the perspective of land circulation prices. Uncertainty in the boundaries of land property rights affects farmers' expectations of land use and also restricts potential land transferees; land right confirmation can eliminate this type of institutional risk. Limited by the availability and applicability of the data, this article only used the 2013 CHIP data. These data only span some provinces, so the research in this article has certain limitations regarding depth and breadth. In future research, we will continue to focus on the inequality of rural household income, and strive to find more suitable data and methods to further improve this research.

Author Contributions: Conceptualization, C.H. and L.C.; methodology, C.H. and L.C.; software, C.H. and L.C.; formal analysis, C.H. and L.C.; resources, C.H. and L.C.; data curation, C.H. and L.C.; writing—original draft preparation, C.H. and L.C.; writing—review and editing, C.H., L.C.; C.H. and L.C.; supervision, C.H. and L.C..; project administration, C.H. and L.C.; funding acquisition, C.H., and L.C. All authors have read and agreed to the published version of the manuscript.

Funding: This research was funded by the youth Project of Philosophy and Social Science Foundation in Hunan Province (No.20YBQ048); Science and Technology Project of Jiangxi Provincial Department of Education (No.GJJ209923, GJJ171069).

Institutional Review Board Statement: Not applicable.

Informed Consent Statement: Not applicable.

Data Availability Statement: The data supporting the findings of the article is available in the 2013 China Household Income Project (CHIP). http://www.ciidbnu.org/chip/chips.asp?year=2013, accessed on 26 January 2021.

Acknowledgments: We would like to express our gratitude to all those who helped us during the writing of this article. Our deepest gratitude goes first and foremost to Guoan-Xiao and Muhua Liu, who are from the Hunan University of Science and Technology in China, for their constant encouragement and guidance. We also greatly appreciate James Johnston who is from the University of the West of Scotland in the UK, for his language polishing and constant encouragement.

Conflicts of Interest: The authors declare no conflict of interest.

References

1. Zhejiang University China Rural Family Research and Innovation Team. *China Rural Family Development Report (2018)*; Zhejiang University Press: Hangzhou, China, 2019. (In Chinese)
2. Department of Policy and Reform, Ministry of Agriculture and Rural Affairs. *Statistical Annual Report on China's Rural Policies and Reforms (2019)*; China Agriculture Press: Beijing, China, 2020. (In Chinese)
3. Yifu, L.J. Rural reforms and agricultural growth in China. *Am. Econ. Rev.* **1992**, *82*, 34–51. [CrossRef]
4. Tian, C.H.; Chen, H.H.; Jia, S.H. The Impact of the Agricultural Land Market on the Fragmentation of Cultivated Land—Theories and Experiences from Jiangsu, Zhejiang and Shandong. *Economics* **2005**, 769–784. (In Chinese) [CrossRef]
5. Smith, A. *The Wealth of Nations. Books I–III*; Penguin Books: London, UK, 1999.
6. David, R. *Political Economics and Taxation Principles. Chinese Translation*; The Commercial Press: Beijing, China, 1962; p. 57. (In Chinese)
7. Coase, R.H. The Problem of Social Cost. *J. Law Econ.* **2013**, *56*, 837–877. Available online: https://www.jstor.org/stable/10.1086/674872 (accessed on 26 January 2021). [CrossRef]
8. Theodore, W.S. *Transforming Traditional Agriculture. Chinese Translation*; The Commercial Press: Beijing, China, 2013; p. 108.
9. Chari, A.V.; Liu, E.M.; Wang, S.-Y. NBER Working Paper Series Property Rights, Land Misallocation and Agricultural Efficiency in China. *NBER Work. Pap.* **2017**, 1–58. Available online: http://www.nber.org/papers/w24099.ack (accessed on 26 January 2021).
10. Xu, Q.; Tian, S.C.; Xu, Z.G. Farmland system, land fragmentation and farmers' income inequality. *Econ. Res.* **2008**, *83–92*, 105. (In Chinese)
11. Xu, H.Z.; Guo, Y.Y. Co-integration analysis of the relationship between farmers' non-agricultural income and rural land transfer: Taking Nanjing City, Jiangsu Province as an example. *China Popul. Resour. Environ.* **2011**, *21*, 61–66. (In Chinese)
12. Jin, S.; Jayne, T.S. Land rental markets in Kenya: Implications for efficiency, Equity, Household income, and Poverty. *Land Econ.* **2013**, *89*, 246–271. [CrossRef]
13. Jin, S.; Deininger, K. Land rental markets in the process of rural structural transformation: Productivity and equity impacts from China. Journal of Comparative Economics. *Assoc. Comp. Econ. Stud.* **2009**, *37*, 629–646. [CrossRef]
14. Ravallion, M.; Van De Walle, D. Land in Transition: Reform and Poverty in Rural Vietnam. *Co-Publ. Palgrave Macmillan World Bank* **2008**, 1–203. [CrossRef]
15. Xue, F.G.; Qiao, G.H.; Su, R.N. Evaluation of the effect of land circulation on farmers' income-based on the analysis of the DID model. *China Rural Obs.* **2011**, *36–42*, 86. (In Chinese)
16. Li, Z. Rural land transfer and farmers' income: A study based on the data of the follow-up survey in Shaoyang, Hunan. *Econ. Geogr.* **2013**, 144–149. (In Chinese) [CrossRef]

17. Han, X.; Zhang, A.L.; Zhu, Q.X.; Wan, K. Study on land circulation and farmers' income growth and farmers' optimal management scale—Taking Hubei and Jiangxi mountain and hilly Areas as examples. *Res. Agric. Mod.* **2015**, *36*, 368–373. (In Chinese) [CrossRef]
18. Chen, F.; Zhai, W.J. A Study on the Inducements of Farmland Circulation and Its Welfare Effects from the Perspective of Farmers' Behavior. *Econ. Res.* **2015**, *50*, 163–177. (In Chinese)
19. Wang, X.Y.; Wang, Y.H.; Zhang, Z. Survey on the impact of land circulation on farmers' income in Shandong province. *Res. World* **2015**, 30–32. (In Chinese) [CrossRef]
20. Carter, M.R.; Zimmerman, F.J. The dynamic cost and persistence of asset inequality in an agrarian economy. *J. Dev. Econ.* **2000**, *63*, 265–302. [CrossRef]
21. Leng, Z.H.; Fu, C.J.; Xu, X.P. Family income structure, income gap and land circulation: A micro-analysis based on the data of the Chinese Family Tracking Survey (CFPS). *Econ. Rev.* **2015**, 111–128. (In Chinese) [CrossRef]
22. Tian, X.H.; Chen, L. "Class Land Rights": An analysis framework of rural land rights allocation. *Manag. World* **2013**, 69–88. (In Chinese) [CrossRef]
23. Lin, L.F.; Wang, J. Comments on the reform of agricultural land property rights in transition and developing countries and its market effects. *State Econ.* **2010**, *12*, 121–125. (In Chinese)
24. Carter, M.R.; Olinto, P. Getting institutions "right" for whom credit constraints and the impact of property rights on the quantity and composition of investment. *Am. J. Agric. Econ.* **2003**, *85*, 173–186. [CrossRef]
25. Möllers, J.; Meyer, W. The effects of migration on poverty and inequality in rural Kosovo. *IZA J. Labor Dev.* **2014**, *3*, 1–18. [CrossRef]
26. Guo, J.P.; Qu, S.; Xia, Y.; Lv, K.Y. Income distribution effect of rural land circulation. *China Popul. Resour. Environ.* **2018**, *28*, 160–169. (In Chinese)
27. Zhu, J.J.; Hu, J.L. A Study on the impact of farmland circulation on the income distribution of Chinese farmers—Based on the data of China's health and pension tracking survey. *J. Nanjing Agric. Univ. (Soc. Sci. Ed.)* **2015**, *15*, 75–83, 124. (In Chinese)
28. Xiao, L.D.; Zhang, B. Land Circulation and the Widening of Income Gap among Farmers—Based on the Investigation and Analysis of 725 Farmers in 39 Villages in Jiangsu. *Financ. Econ.* **2017**, 10–18. (In Chinese) [CrossRef]
29. Li, C.M.; Sun, B.W.; Dong, Z.Y. The heterogeneity of rural households, the transfer of farmland management rights and the distribution of rural income: An empirical study based on the Chinese Family Tracking Survey Data (CFPS). *Rural Econ.* **2019**, *8*, 26–33. (In Chinese)
30. Shi, C.L. Heterogeneity analysis of the income effect of Farmers' land circulation. *Learn. Pract.* **2019**, 37–46. (In Chinese) [CrossRef]
31. Deininger, K.; Jin, S. The potential of land rental markets in the process of economic development: Evidence from China. *J. Dev. Econ.* **2005**, *78*, 241–270. [CrossRef]
32. Wan, G.H.; Zhoug, Z.Y.; Lu, Q. Rural income inequality in China: Regression Decomposition Using Farm Household Data. *China Rural Econ.* **2005**, *5*, 4–11. (In Chinese)
33. Zhang, Q.F. Retreat from equality or advance towards efficiency? Land markets and inequality in rural Zhejiang. *China Q.* **2008**, *195*, 535–557. [CrossRef]
34. Adamopoulos, T.; Brandt, L.; Leight, J.; Restuccia, D. Misallocation, selection and productivity: A quantitative analysis with panel data from China. *NBER Work. Pap.* **2017**, *2*, 1–24.
35. Shi, C.L.; Luan, J.; Zhu, J.F.; Chen, Y.M. The impact of land circulation on farmer household income growth and income gap: An empirical analysis based on survey data of farmer households in 8 provinces. *Econ. Rev.* **2017**, 152–166. (In Chinese) [CrossRef]
36. Yang, Z.; Ma, X.L.; Zhu, P.X.; Ma, D. Research on land circulation and farmers' income changes. *Chin. Popul. Resour. Environ.* **2017**, *27*, 111–120. (In Chinese)
37. Liang, Y.; Zhang, Y.J.; Bi, W.T. The impacts of labor mobility and farmland transfer on farmers' income. *Res. Agric. Mod.* **2021**, *42*. [CrossRef]
38. Sergio, F.; Nicole, M.F.; Thomas, L. Unconditional quantile regressions. *NBER Tech. Work. Pap.* **2007**, *339*, 1–53.
39. Koenker, R.; Bassett, G. Quantile regressions. *Econometrica* **1978**, *46*, 33–50. [CrossRef]

land

MDPI

Article

Environmental Regulations, the Industrial Structure, and High-Quality Regional Economic Development: Evidence from China

Lingming Chen [1], Wenzhong Ye [1], Congjia Huo [1,*] and Kieran James [2]

1 School of Business, Hunan University of Science and Technology, Xiangtan 411201, China; lingming1016@mail.hnust.edu.cn (L.C.); 1300030@hnust.edu.cn (W.Y.)

2 School of Business and Creative industries, University of the West of Scotland (UWS), Paisley PA1 1DD, UK; Kieran.James@uws.ac.uk

* Correspondence: huocongjia@mail.hnust.edu.cn

Received: 19 November 2020; Accepted: 11 December 2020; Published: 14 December 2020

check for updates

Abstract: Environmental regulation is an important means of restraining enterprises and protecting the environment. Rationalization of environmental regulatory policies can promote high-quality regional economic development. The optimization and upgrading of the industrial structure has an intermediary effect on the impact of environmental regulations on the high-quality development of the regional economy. After collating and analyzing previous research, this article proposes to classify 30 Chinese provinces into regions with higher than the national average HDI (human development index) and lower than the national average HDI based on the average HDI of Chinese provinces. We explore the mediating effect of industrial structure on environmental regulation and high-quality regional economic development. The model passed the full-sample robustness test and the robustness test with GDP as the replacement variable. The empirical results show that environmental regulations of different intensities have different effects on the quality of regional economic development. The effect of environmental regulations on development quality is mainly mediated through the transformation and upgrading of the industrial structure. Enterprises need reasonable incentives from environmental regulations to transform and upgrade. The mediating effect of the industrial structure on environmental regulations is greater in regions with below-average HDI values than in regions with above-average HDI values, which shows that the industrial structure is the mechanism underlying the effect of environmental regulations on the quality of regional economic development. This result proves that adjusting environmental regulatory policies can effectively promote the upgrading of industrial structure, thereby promoting high-quality regional economic development. Based on this, the article puts forward several policy recommendations.

Keywords: environmental regulation; industrial structure; regional economy; high-quality development; HDI zone

---

1. Introduction

In the report of the 19th National Congress of the Communist Party of China (CPC), General Secretary Jinping-Xi pointed out that "China's economy has shifted from a stage of high-speed growth to a stage of high-quality development." The traditional extensive development model has been gradually abandoned in the "new era", and maintaining high quality is being used as the foundation of and key guiding ideology for economic development. After more than 40 years of being open, China's economy has grown at a high pace and has entered the stage of high-quality development. The optimization and upgrading of the industrial structure is one of the main manifestations of high-quality development. Therefore, one of the goals of high-quality regional economic development is to promote the optimization and upgrading of the industrial structure. It is a key path for coordinating environmental regulations and high-quality regional economic development. While increasing the proportion of knowledge-intensive and technology-intensive industries, it is also necessary to increase the organic unity between rationality and industrial structure upgrades to promote green and low-carbon development within the economy. Environmental regulation can significantly promote economic growth, which is the claim of the famous Porter hypothesis. Environmental regulations can also hinder economic growth if they are not cost-effective. That is, environmental regulations increase the expenditure of enterprises on environmental protection and can crowd out investments in production so that the innovativeness and profitability of new technologies and new products are reduced. Tongbin-Zhang (2017) [1] combined Chinese provincial-level panel data from 1995 to 2013, and concluded that strong environmental regulations promote economic development through an "innovation compensation" effect, while weak environmental regulations are ineffective. The mechanism underlying the effect of environmental regulation on industrial restructuring involves many factors. However, there is little in the literature that provides persuasive empirical evidence based on actual conditions in China while incorporating the cost effect, the pollution refuge hypothesis, and the Porter hypothesis into a unified analytical framework.

The structure of this article is as follows: Section 1 is the introduction; Section 2 describes the literature review; Section 3 provides the theoretical mechanisms and discusses the classification of environmental regulations and the analysis of the environmental regulation mechanism; Section 4 introduces the empirical model and results and discusses the data sources, variable selection, and data processing; Section 5 presents the analysis of the results, including those from the HDI partition analysis, and the robustness test; and Section 6 presents the research summary and recommendations.

2. Literature Review

Along with scientific and technological innovations and FDI (foreign direct investment), environmental regulations can also affect the high-quality development of the regional economy through industrial restructuring. The greater the extent of industrial structure optimization and upgrading, the more positive the effect of environmental regulation on environmental quality. Environmental regulations also impact the industrial structure [2,3]. There are two possible mechanisms underlying the promotion of the optimization and upgrading of the industrial structure. First, environmental regulations can act as a backward-looking mechanism driving the upgrading of the industrial structure. The implementation of environmental regulations imposes a compulsory "deep cleaning" on industries and enterprises. Environmental regulation achieves the optimization and upgrading of the industrial structure and optimal allocation of resources through survival of the fittest, which in turn improves the high-quality development of the regional economy. Second, environmental regulations form implicit green barriers to entry.

(1) Environmental regulations

Environmental regulations are a series of related policies or measures adopted by the government to protect the environment [4]. Environmental regulation is an essential part of the formal Chinese environmental administration system. Environmental regulation is an essential part of the formal Chinese environmental administration system [5,6]. Economic growth alone cannot alleviate environmental pollution, so environmental regulation must be added. Under reasonable environmental regulations, green innovation has been improved, and the industrial development model has also changed. Currently, a large portion of the literature has explored the various relationships between environmental regulations and green technology innovation, including linear and nonlinear relationships. For example, Ling Li and Feng Tao (2012) [7] established a relationship between environmental regulation and green total factor productivity using a panel data model, trying to find the inflection point in the optimal environmental regulation strength for different industries with a goal of promoting productivity. Fuxin Jiang showed that there is a dynamic U-shaped relationship between green technological innovation and the decline in environmental pollution. As the strength of environmental regulation rises, the offset effect changes to the compensation effect. However, the above literature does not effectively distinguish between the types of environmental regulations, so it is impossible to discover the heterogeneous effects of different kinds of environmental regulations on the efficiency of green technological innovation.

(2) Economic development

Economic growth can be defined as the increase in the inflation-adjusted market value of the goods and services produced by an economy over time. High-quality development in the new era must coordinate the relationship between development pace and quality. Quality must be put first and given priority over efficiency. The key is to increase total factor productivity and replace old drivers of growth with new drivers. Pearce and Turner (1990) [8] proposed that the relationship between development and the environment gives rise to the concept of sustainable development. Papalia and Bertarelli (2013) [9] established a model based on entropy calculations and found that there was obvious convergence in the economic growth of some countries. Some scholars measured the quality of economic growth, constructed an evaluation index for the modern economic system, adopted the spatial Durbin estimation method, empirically analyzed how the modern economic system improves total factor productivity, and considered how TFP affects the quality of economic development.

Some scholars have empirically found that environmental regulation improved industrial green total factor productivity in Norway using data on industrial enterprises (Telle and Larsson, 2007) [10]. Other scholars have concluded that research and development subsidies and carbon taxes can effectively improve green economic growth through the investigation of environmental regulations and the efficiency of green innovation (Ploeg and Withagen, 2013) [11]. When carbon duty tariffs and support for carbon release reduction reach a sufficient level, they promote innovation in green energy technologies, thereby reducing environmental pollution caused by economic activities.

(3) Environmental regulation and economic growth

The relationship between environmental regulation and economic growth has always been a hot topic in economics research. New classical economic theory holds that environmental regulation can lead enterprises to internalize the social cost of their emissions. Therefore, increasing the intensity of environmental governance increases the cost of pollution control and reduces enterprise performance. Mainstream scholars, such as Porter, hold the opposite view. Such scholars believe that strict and reasonable environmental regulation can stimulate enterprises to conduct technological innovation activities and obtain compensation for product innovations. Some scholars also hold the view that the effect of environmental regulation on enterprise performance is uncertain. They believe that only appropriate and effective environmental regulation can stimulate or force enterprises to innovate. Therefore, the effect of environmental regulation on enterprise performance is uncertain.

China's government found that we have to protect the ecological environment, and made efforts to reduce pollutant emission because it is unsustainable to obtain economic benefits that sacrifice the environment (Shuai and Fan, 2020) [12]. The environmental regulation efficiency loss of local and neighboring cities inhibits inclusive growth at both the national and regional levels (Ge, T., Qiu, W., Li, J., and Hao, X., 2020) [13]. The Inclusive Wealth Index provides important insights into long-term economic growth and human well-being (Managi, S. and Kumar, P., 2018) [14]. Other research scholars explored the determinants of inclusive growth, including technological innovation (George et al., 2012) [15], economic growth (Vellala et al., 2014) [16], resource mobilization (Oyinlola et al., 2020) [17], and trade openness (Jalles and Mello, 2019) [18]. Rajapaksa, D., Islam, M. and Managi, S. (2017) [19] analyzed the impact of natural disasters on inclusive growth, trade openness, FDI, and GDP growth rate, which are other important determinants of natural capital.

Presently, there have been many research results on the relationship between environmental regulation, economic growth, and industrial structure. Most scholars believe that traditional industries mainly achieve the purpose of upgrading the industrial structure and improving the quality of economic development by the introduction and integration of green innovative technologies and resources.

However, previous literature focused more on the threshold effect of environmental regulation, and seldom incorporated environmental regulation and industrial structure into the study of high-quality economic development, ignoring the role of industrial structure upgrading. China has stringent requirements for high-quality economic development. The research on the impact of environmental regulations on the quality of regional economic development is particularly important. Therefore, this article uses data from 2006 to 2017, based on the mediating effect of industrial structure, to study the impact of environmental regulations on the high-quality development of regional economies. The article also innovatively uses HDI divisions to compare the intermediary effects of the industrial structure between regions above the human development index and regions below the human development index. This complements the important role of industrial structure in high-quality economic development. Compared with the existing literature, the main contributions of this article are as follows: (1) we explain the mechanism underlying the effect of environmental regulations on high-quality economic development from a theoretical perspective; (2) through HDI zoning, we study the differential impact of environmental regulations on the quality of regional economic development; and (3) we provide relevant policy suggestions on how to improve the quality of regional economic development and how to win the battle against pollution in the contemporary Chinese context.

3. Theoretical Mechanism

Countries around the world have been calling for the development of a low-carbon economy. In this context, Grossman and Kreuger [20] pioneered the environmental Kuznets curve (EKC) hypothesis, which focuses on the relationship between environmental quality and economic growth. The environmental Kuznets curve refers to an inverted U curve relationship between environmental quality or pollutant emission levels and per capita income. Research by the World Bank (1992) [21] concluded that the critical point of pollution emissions is a per capita income of $8000. The EKC curve is continuously verified. In the China Sustainable Development Strategy Report 2000, the Chinese government proposed that the EKC curve can be used to scientifically set the critical point of environmental pollution peaks and the length of time to reach this critical point, and form a reasonable environmental regulatory policy based on that. The following paper introduces China's current environmental regulatory policies. Caruso, G., Di Battista., et.al (2020) [22] studied the past policy approach and offered a potential pathway for academics to work with policymakers in moving towards the realization of local growth policies. D'Adamo, et.al (2020) [23] proposed a new indicator, the socioeconomic indicator for the bioeconomy (SEIB), to measure the socioeconomic performance of bioeconomy sectors.

3.1. *Classification of Environmental Regulations*

Environmental regulation tools mainly include three types: command-and-control tools, market incentive tools, and voluntary participation tools [24,25].

3.1.1. Command-and-Control Environmental Regulation

In reality, the most widely used tool is command-and-control environmental regulations. It refers to the government, through legislation or formulating rules and regulations, to clarify the goals and standards of environmental regulations, and at the same time punish enterprises that violate the regulations to achieve the goal of protecting the environment [26]. The Environmental Protection Law of the People's Republic of China promulgated by China and relevant unit regulations are all typical command-and-control tools. This type of tool is simple to operate and efficient, but the implementation cost is high. If one wants to achieve the same effect as market incentive tools, the cost is often several times higher than the latter [27]. The high cost of command and-control tools is due to information asymmetry. The collection of corporate information, corporate supervision, and punishment requires a lot of manpower to execute, so command-and-control tools are weak in market supervision.

3.1.2. Market Incentive Environmental Regulation

Market incentive tools do not constrain the behavior of polluters through norms or regulations; instead, they use market signals to influence the behavioral decisions of enterprises [28], including the imposition of sewage charge, permit systems, and government subsidies [29]. In 1972, the OECD promulgated the "Polluter Pays Principle". Since then, many countries have adopted this form of sewage regulation. The theory behind market incentive tools is Pigou's [30] theorem. The rational polluter determines the amount of pollutant discharge based on the marginal pollution reduction cost equal to the pollutant tax rate, and uses the different pollutant discharge costs of the enterprise to allow polluters to allocate effective resources spontaneously, reducing the cost of information asymmetry caused by the command-and-control tool. At the same time, companies can obtain additional profits by using green technology for production. Therefore, this tool can provide incentives for green technology for pollution reduction, allowing polluters to promote pollution reduction technology and choose a more green and effective production model.

3.1.3. Voluntary Participation in Environmental Regulation

Voluntary participation tools refer to environmental protection proposals put forward by industry associations, enterprises themselves, or other subjects, and necessitate higher requirements for polluters. This category mainly includes environmental certification, environmental audits, eco-labels, environmental agreements, etc. There are four ways to implement voluntary tools based on strong environmental awareness. The first is the administrative form, which exerts pressure on the government through non-governmental environmental protection organizations to supervise the work of the government. The second is the economic form, raising private funds to finance some environmental protection projects. The third is the legal form, which promotes environmental legislation and provides environmental legal assistance to some organizations or institutions. The fourth is the way of education to popularize the importance of protecting the environment.

Command-and-control, market incentives, and voluntary participation environmental tools are the trinity of the government, market, and society to protect the ecological environment. In reality, environmental policies not only select one tool, but in order to satisfy the principles of efficiency, feasibility, and equity, two or even three types of environmental regulation tools are generally used together. The government should implement the best environmental regulation according to the actual situations [31].

China's environmental regulation policies have gone from monetary incentives to non-monetary incentives, from emphasizing government intervention to strengthening market incentives, and the ultimate goal is actually the active participation of the public. Nowadays, in addition to the participation of enterprises and governments, the public also plays an important role in environmental regulation. With the continuous improvement of public awareness of environmental protection, environmental regulations, and policies in the form of environmental certification, environmental hearings and public participation have received increasing attention. The environment and health are linked, and individuals are more concerned about the environment than the government and enterprises. For example, the smog of Beijing in China once attracted the attention of all of the people in China. They were willing and able to force companies to improve their technology and reduce pollution. At this time, we need the proper guidance from the government, such as requiring companies to indicate whether they use high-pollution emission technology on the product packaging, whether the packaging bag is biodegradable, and so on. The modes of public participation in environmental governance are diverse. For example, from the perspective of participating subjects, there are forms based on the participation of individual citizens, participation based on non-profit organizations, and participation based on media. It is precisely the diversity of public participation that makes up for the shortcomings of monetary-based environmental regulatory policies. Such a wide range of roles can affect the various subjects, processes, and specific instruments of environmental governance. The heightened public awareness of environmental protection has been transformed into practical actions in many aspects, prompting the government and enterprises to take effective environmental protection measures to respond. Under the trinity of government, enterprises, and the public environmental regulation, the adjustment of the industrial structure is the core intermediary factor of environmental regulation that affects the high-quality economic development. Next, the specific mechanism of industrial structure will be explained in detail.

3.2. *Analysis of the Mechanisms by Which Environmental Regulations Drive Industrial Upgrading*

In a low-level industrial structure, resources are tilted towards the primary and secondary industries, and a capital investment model that focuses on scale, speed, and materials is implemented. This extensive economic growth model has low resource utilization efficiency and serious environmental pollution, and hinders high-quality economic development. Environmental regulations restrict the behavior of polluters, and can lead to adjustments in the industrial structure through different channels to promote high-quality economic development. Many factors affect the upgrading of the industrial structure due to environmental regulation, and there is spatial heterogeneity in these factors [32]. The higher the intensity of environmental regulation, the better the upgrading of the industrial structure relatedly, the more optimized the industrial structure, and the more positive the effect of environmental regulation [33]. This article mainly discusses three theoretical mechanisms by which environmental regulation drives the upgrading of the industrial structure.

3.2.1. Crowding Out Mechanism

Technological innovation is the largest driving force behind the upgrading of the industrial structure. The speed of industrial structure upgrades depends on the intensity and speed of technological innovation. Only continuous technological innovation can develop the industrial structure from a low level to high level, and environmental regulations force the optimization and upgrading of the industrial structure through technological progress. Environmental regulations increase the price of production factors that generate pollution, and heavily polluting companies will automatically withdraw from the market. The Porter hypothesis states that environmental regulations promote technological progress in enterprises, and polluters must seek more efficient and green production technologies in order to reduce their production costs. Increasing environmental regulatory standards has eliminated low-end production technologies, and manufacturers using low-end technologies have been crowded out of the market. Environmental regulations have increased the production costs of heavily polluting industries, while clean technology industries are minimally affected and the most concentrated in the tertiary industry. The development of the service industry drives the industrial structure upgrades, helps optimize resource allocation, and improves the quality of economic development.

3.2.2. Green Barriers to Entry

For potential entrants in an industry, investment in environmental regulation is a sunk cost on non-production factors. Eventually, this increase in necessary investment capital and investment risk forms an entry barrier in that industry. In addition, the reduction in environmental pollution increases marginally. The larger the scale of the enterprise, the greater the cost of pollution. To improve the efficiency of environmental protection, the government proposes higher environmental standards for companies that are new entrants to the industry. Potential entrants are already burdened with higher risks and higher production costs. Coupled with the cost of environmental regulations, new entrants are in an extremely disadvantaged position, which increases the barriers to entry into the industry. Environmental regulations have formed invisible barriers to entry in some industries, and these barriers are aimed at promoting green production technologies. That is, the higher the pollution intensity of an industry, the higher the invisible barriers to entry are, resulting in resources being concentrated in clean technology industries. High-polluting industries are concentrated in the secondary industry, and clean technology industries are concentrated in the tertiary industry, and so environmental regulations promote the upgrading of the industrial structure and have an impact on high-quality economic development.

3.2.3. Openness to Trade

Openness to trade refers to the ratio of domestic prices to international prices for the same product. The closer the ratio is to 1, the more open the product. The pollution sanctuary hypothesis argues that firms in countries with stricter environmental regulations and a higher cost for discharging pollution will move to other regions if they are in industries that are more pollution intensive. The transfer of polluting industries from developed countries to developing countries has led to an increase in openness to trade among products from industries with high levels of pollution intensity. To pursue high returns, companies will expand their scale and increase their emissions to encourage the government to increase the intensity of its environmental regulations. As a result, high-intensity environmental regulations force some companies out of the market, surviving companies increase the price of their products and reduce relative international competitiveness, and domestic high-polluting industries shrink. In the context of international integration, the substitution effect due to environmental regulations and the expansion effect due to international trade are mutually exclusive, which prevents developing countries from becoming pollution refuges and leads industrial structure upgrades that improve high-quality economic development.

Intense environmental regulation promotes industrial structure upgrading. Regional production models within a reasonable industrial structure tend to intensify along with rapid scientific and technological progress, high production efficiency, and an increase in the quality of production factors and the capacity for optimal combination. The upgrading of the industrial structure has reduced the scale of industries with high energy consumption and intensive pollution, and resources have been tilted towards clean technology industries, further promoting the innovative development of clean technology. The upgrading of the industrial structure is a key path through which the coordination of sustainable economic development and ecological environment protection is promoted. This continuous adjustment to the industrial structure has brought about high-quality economic development.

Based on the analysis of the mechanism of action, this article proposes the following hypotheses:

Hypothesis 1 (H1). *When the pollution refuge effect caused by environmental regulation is greater than the competition effect of the local government scale, environmental regulation policies can promote the optimization and upgrading of the industrial structure.*

Hypothesis 2 (H2). *Environmental regulations force technological innovation, and high-quality economic development is promoted through advanced and rationalized industrial structure.*

Hypothesis 3 (H3). *In regions with different economic development speeds, environmental regulations have different impacts on the upgrading of the industrial structure.*

4. Measurement Model Construction, Data Sources, and Variable Selection

4.1. Measurement Model Construction

This article mainly focuses on the impact of environmental regulations on the high-quality development of the regional economy. To that end, the following regression model is constructed:

$$hqd_{i,t} = \alpha + \beta + er_{i,t} + \sum_{k=1}^{k}\left(\gamma + x^{j}_{i,t}\right) + u_i + \lambda_t + \varepsilon_{i,t} \tag{1}$$

In the above equation, $hqd_{i,t}$ and $er_{i,t}$ represent the economic development quality and environmental regulation, respectively, in province i and in year t, and $x^{j}_{i,t}$ represents the j other explanatory variables, namely industrial structure, human capital level, trade openness, and the amount of investment in fixed assets; α is a constant, β and γ are the coefficients of each variable, and u_i,λ_t and $e_{i,t}$ represent unobservable individual effects, unobservable time effects, and specific errors, respectively.

To study the mechanism underlying the effect of environmental regulation on the high-quality development of the regional economy, cross-terms are introduced, and the following regression model is constructed:

$$hqd_{i,t} = \alpha + \beta \times er_{i,t} + \sum_{k=1}^{k}\left(\gamma \times x^{j}_{i,t}\right) + \sum_{m=1}^{M} \left(\delta \times er_{i,t} \times Z_{i,t}\right) + u_i + \lambda_t + \varepsilon_{i,t} \tag{2}$$

In the above equation, $Z_{i,t}$ represents the intermediary variable industrial structure, and δ is the coefficient of that variable.

4.2. Data Sources

Based on the above theoretical analysis, this article uses panel data from 30 provinces (including municipalities directly under the central government and autonomous regions) in China from 2006 to 2015, and uses a fixed effects (FE) model to analyze them. This choice is based on two considerations. One is that the Hausman test rejects the random effects (RE) model, and the other is that unobservable factors in each province may have different effects on the quality of economic development. In China, regional endowments affect the quality of economic development. The data in this article come from the China Statistical Yearbook, China Environmental Statistics Yearbook, China Industrial Economic Statistics Yearbook, China Science and Technology Statistical Yearbook, the website of the National Bureau of Statistics of China, and the China Economic and Social Development Statistics Database from previous years. The HDI data for different regions are from the China Human Development Report Special Edition (2017), which is based on cooperation between the Development Research Center of the State Council of China and the United Nations Development Program. This paper uses the national average HDI of 0.752 as a boundary to divide the 31 provinces in China into regions with an HDI higher than the national average HDI (nine provinces) and lower than the national average HDI (22 provinces). The regions with a higher than average HDI include Beijing, Shanghai, Tianjin, Jiangsu, Zhejiang, Guangdong, Liaoning, Inner Mongolia, and Shandong. The regions with a lower than average HDI include Jilin, Chongqing, Fujian, Hubei, Shaanxi, Hunan, Shanxi, Hainan, Heilongjiang, Ningxia, Hebei, Xinjiang, Henan, Jiangxi, Guangxi, Anhui, Sichuan, Gansu, Qinghai, Guizhou, Yunnan, and Tibet (since there are too many missing data on Tibet, this study does not include Tibet). For individual missing data, the mean interpolation method is used to fill in the gap.

4.3. Variable Selection and Processing

This paper uses panel data from 30 provinces in China from 2006 to 2017, for a total of 1000 observations. The human development index (HDI), which covers life expectancy at birth, education level, and quality of life, is used to designate zones that are considered when studying the effects of environmental regulations on the regional economic development quality.

Explained variable: high-quality development in the regional economy (hqd). Baoping Ren (2018) [34] defines the quality of economic development as the effectiveness of, adequacy of, coordination in, level of innovation in, sustainability of, sharing of, and stability of economic development. Hong Zhang (2015) [35] constructed an indicator system to measure economic development based on the five dimensions of effectiveness, coordination, sharing, innovation, and sustainability. Thus, the quality of economic development needs to be judged from a multidimensional perspective. In consideration of the comparability, stability, and sustainability of the development of the regional economy, this article determines six secondary indicators from these three dimensions: the economic development level, the stability of economic development, and the sustainability of economic development. These indicators are used as proxy indicators for high-quality regional economic development. The details are shown in Table 1.

Table 1. Selection of indicators for the high-quality development of the regional economy.

Explained Variable	Level 1 Indicators	Level 2 Indicators	Unit	Indicator Attributes
High-quality development of regional economy (*hqd*)	Level of economic development (*le*)	Logarithm of GDP per capita (*y1*)	/	Positive index
	Stability of economic development (*st*)	CPI (*y2*)	%	Reverse indicator
		Urban registered unemployment rate (*y3*)	%	Reverse indicator
	Sustainability of economic development (*su*)	Green coverage rate of built-up area (*y4*)	%	Positive index
		Per capita public green area (*y5*)	%	Positive index
		Pollution intensity (*y6*)	M^2/person	Reverse indicator

The comprehensive indicator of the economic development quality is different from the sustainable development index. The methods of evaluating the process of sustainable development in economic theory mainly include green GDP accounting [36], the monetary value accounting of natural resource depletion, the four-capital model, and the definition of strong sustainability and weak sustainability [37], etc. The International Institute for Sustainable Development (IISD) has given out the seven principles of the sustainable development indicator system: availability, comprehensibility, measurability, significance, availability, comparability, and universality of indicators [38]. The sustainable development indicator system proposed by China in 1999 is complicated to calculate. It uses 45 indexes, covering 208 indicators [39], and focuses more on humanities and society. The comprehensive indicators of the economic development quality used in this article are mainly to evaluate the current quality of economic development in China. The quality of economic development includes the increase of quantity and the improvement of quality, which is the organic unity of quantity and quality. The level of economic development is used to measure the level of output. The higher the output level, the more prosperous the region, the higher the living standard of residents, and the higher the quality of economic development accordingly. Sustainability development, GDP, natural resources, and foreign direct investment have been added to the Section 2.

(1) Level of economic development (*le*): The article uses the logarithm of per capita GDP to measure economic development, and per capita GDP is deflated to values for the year 2000.
(2) Stability of economic development (*st*): The stability of economic development means that the speed of economic development fluctuates within a moderate range so that resources can be fully utilized. According to the AS-AD model, if the economy grows too quickly and aggregate demand is excessively high, it will usually lead to inflation; if the economy grows too slowly and aggregate demand is insufficient, it will usually lead to unemployment. Therefore, the more stable the economic development, the higher the quality of economic development. This article uses two secondary indicators to measure stability, the CPI and the urban-registered unemployment rate.
(3) Sustainability of economic development (*su*): The sustainability of economic development is the main factor that distinguishes the quality of economic development from the pattern of economic growth. Economic development inevitably leads to environmental pollution. If economic development cannot be sustained, it will inevitably cause serious damage to the ecological environment, and these negative consequences will, in turn, hinder the quality of economic development. This article uses three secondary indicators to measure this, which are the green coverage of completed areas, per capita public area of green land, and pollution intensity. Pollution

intensity is calculated by the entropy weighting method using three indicators: per capita volume of industrial waste water discharged (su_1), per capita volume of sulfur dioxide emissions by the industry (su_2), and per capita emission of industrial soot and dust (su_3). The following describes the specific steps followed in the entropy weighting method to calculate pollution intensity.

First, we sort the 30 provinces into 1, 2, 3 ... 30, that is, i = 1, 2, 3 ... 30, respectively. Then the per capita volume of the industrial waste water discharged of the i-th province is su_{i1}, the per capita volume of sulfur dioxide emissions by the industry is su_{i2}, and the per capita emission of industrial soot and dust is su_{i3}, where j = 1, 2, 3.

The first step is to calculate the proportion of the i-th indicator value in the j-th industry [40,41]:

$$P_{ij} = r_{ij} / \sum_{i=1}^{m} su_{ij} \tag{3}$$

The second step is to calculate the entropy value e_j of the j-th indicator [42]:

$$e_j = -k / \sum_{i=1}^{m} P_{ij} \cdot \ln P_{ij}, k = 1/\ln m \tag{4}$$

The third step is to calculate the entropy weight w_j of the j-th indicator [43,44]:

$$w_j = \left(1 - e_j\right) / \sum_{j=1}^{n} \left(1 - e_j\right) \tag{5}$$

where w_j is the final weight coefficient for each indicator, and the weight coefficient obtained is substituted into $y_i = \sum^{m} w_i x'_{ij}$. Then we can get the pollution intensity of each province.

The three emissions have different weights in different years. An area may have both high pollution and high resource intensity, or it may have both high technology and high pollution intensity. Under different environmental regulatory intensities, the innovation efficiency of pollution-intensive or technology-intensive economic zones may be different [45].

Since the above six indicators have different dimensions, this article draws on the method of Bo-Shi (2018) [46] to measure the quality of economic development. We first adopted the range standardization method to process the original data underlying each indicator into nondimensional indicator values, referred to the United Nations Human Development Index evaluation system, and finally used the DEA-SBM method to calculate the quality of regional economic development. Table 2 shows the high-quality development of the regional economy of different provinces in different years. The range of the indicator is 0-1. The closer the value is to 1, the better the quality of economic development. From the values of the quality of regional economic development given in Table 2, the quality of regional economic development of more developed provinces in China is significantly higher than that of other provinces. For example, Beijing, Zhejiang, Guangdong and other provinces and cities reached the highest value 1 in 2017. The quality of regional economic development in the central and western regions is lower than that in the eastern regions. The quality of regional economic development is poor which is limited by geographical location and resource endowments, such as Qinghai, Ningxia, Gansu and other provinces. However, in general, China has experienced ten years of development, and the quality of economic development in all provinces has been steadily improving, especially in Chongqing. These are basically consistent with the actual situation. This article uses the indicator data in Table 3 as the empirical explained variable *hqd*.

Table 2. The high-quality development of regional economic in China's provinces from 2006 to 2017.

	2006	2007	2008	2010	2011	2012	2013	2014	2015	2016	2017
Beijing	0.68	0.63	0.67	0.89	0.86	0.91	1.00	0.95	0.97	1.00	1.00
Tianjin	0.23	0.23	0.24	0.26	0.28	0.30	0.30	0.32	0.35	0.39	0.44
Hebei	0.22	0.24	0.26	0.32	0.32	0.33	0.35	0.37	0.40	0.44	0.47
Shanxi	0.20	0.22	0.25	0.27	0.31	0.31	0.32	0.34	0.37	0.40	0.49
Inner Mongolia	0.24	0.26	0.27	0.33	0.34	0.39	0.39	0.43	0.45	0.49	0.51
Liaoning	0.31	0.36	0.40	0.46	0.45	0.50	0.52	0.52	0.53	0.51	0.56
Jilin	0.26	0.27	0.29	0.33	0.33	0.33	0.33	0.38	0.39	0.43	0.44
Heilongjiang	0.29	1.00	0.37	0.41	0.41	0.41	0.42	0.43	0.44	0.47	0.49
Shanghai	0.27	0.28	0.31	0.57	0.56	0.59	0.57	0.57	0.57	0.61	0.69
Jiangsu	0.43	0.48	0.50	0.55	0.57	0.59	0.62	0.64	0.65	0.67	1.00
Zhejiang	0.23	0.25	0.27	0.30	0.38	0.44	0.46	0.49	0.51	0.56	0.58
Anhui	0.35	0.33	0.35	0.40	0.42	0.44	0.46	0.48	0.50	0.53	0.57
Fujian	0.14	0.14	0.18	0.23	0.21	0.29	0.30	0.32	0.35	0.41	0.42
Jiangxi	0.19	0.22	0.24	0.30	0.32	0.34	0.34	0.36	0.34	0.34	0.47
Shandong	0.41	0.43	0.43	0.47	0.49	0.53	0.56	0.57	0.57	0.59	0.62
Henan	0.22	0.24	0.27	0.28	0.30	0.34	0.35	0.38	0.39	0.50	0.53
Hubei	0.23	0.24	0.27	0.30	0.30	0.36	0.38	0.42	0.46	0.52	0.53
Hunan	0.19	0.21	0.23	0.25	0.26	0.27	0.29	0.32	0.35	0.41	0.46
Guangdong	0.57	0.62	0.72	0.82	0.81	0.81	0.86	0.86	0.89	0.98	1.00
Guangxi	0.23	0.21	0.21	0.25	0.32	0.32	0.37	0.42	0.47	0.52	0.58
Hainan	0.17	0.83	0.86	1.00	0.87	0.72	0.45	0.33	0.43	0.64	1.00
Chongqing	0.11	0.17	0.21	0.32	0.38	0.42	0.19	0.42	0.43	0.47	0.51
Sichuan	0.25	0.25	0.26	0.35	0.39	0.43	0.17	0.43	0.44	0.49	0.53
Guizhou	0.32	0.33	0.34	0.34	0.32	0.34	0.18	0.34	0.36	0.44	0.46
Yunnan	0.15	0.18	0.21	0.29	0.27	0.31	0.31	0.33	0.33	0.35	0.44
Shanxi	0.20	0.20	0.20	0.24	0.27	0.32	0.35	0.36	0.44	0.46	0.49
Gansu	0.16	0.19	0.21	0.23	0.23	0.27	0.30	0.32	0.34	0.39	0.41
Qinghai	0.08	0.08	0.08	0.08	0.10	0.10	0.12	0.13	0.14	0.15	0.16
Ningxia	0.14	0.18	0.20	0.23	0.25	0.28	0.30	0.32	0.32	0.37	0.39
Xinjiang	0.24	0.33	0.33	0.37	0.39	0.40	0.41	0.44	0.47	0.57	0.59

Data source: Calculated based on data from the China Statistical Yearbook, China Environmental Statistical Yearbook, China Environmental Yearbook, China Industrial Economic Statistical Yearbook, and China Science and Technology Statistical Yearbook. Among them, per capita GDP, CPI, urban registered unemployment rate, green coverage rate in built-up areas, and per capita public green area are derived from the yearbook, and the pollution intensity is calculated by the entropy weight method.

Table 3. Descriptive statistical analysis.

Variables	Sample Size	Mean	Standard Deviation	Min. Value	Max. Value
hqd	360	0.3996	0.193	0.0759	1
ln*er*	360	5.004	0.9609	1.808	7.258
is	360	0.4255	0.0916	0.286	0.806
fa	360	9.04	0.9555	5.966	10.919
edu	360	8.696	0.9039	6.6	11.6
ln*imex*	360	5.792	1.6135	1.504	9.458

Explanatory variable: environmental regulation (*er*). There are no uniform standards in the academic community for measuring the intensity of environmental regulations. Levinson (1996) [47] measured the intensity of environmental regulations based on pollutant emissions. This article adopted the same method as Yaobin Liu (2020) [48], measuring the intensity of environmental regulations based on the logarithm of the total investment in regional industrial pollution control.

Intermediary variable: industrial structure (*is*). The upgrading of the industrial structure is conducive to improvements in the quality of regional economic development, but the dilemma of corporate governance can lead to the simplification of the industrial structure, thereby inhibiting improvements in the quality of economic development. Considering that a change in the proportion of the tertiary industry can reflect the optimization and upgrading of the industrial structure, this article uses the proportion of output from the tertiary industry to GDP to measure the optimization and upgrading of the industrial structure.

Other control variables: (1) Human capital (*edu*). With an increase in educational attainment, awareness of environmental protection should be gradually strengthened. An increase in the average number of years of education per capita will fundamentally improve the quality of workers' human capital and simultaneously boost the effect of labor production, thus promoting the development of the regional economy. This article draws on the calculation for average years of education. Human capital is measured as the average years of education of the labor force in each province [49,50]. The formula for years of education is as follows: years of education = proportion of elementary school students × 6 + proportion of junior secondary school students × 9 + proportion of senior secondary school students (including regular senior secondary school and secondary vocational school) × 12 + proportion of university students (including college students, undergraduates, and graduate students) × 16, where the proportion of the number of students in each grade = the number of students educated in that grade/the total population in a given year. (2) Trade openness (*imex*). Trade openness refers to the degree to which a country or a region's import and export trade (including trade in goods and services) is liberalized. This article uses the total number of goods in and out of warehouses by region and the total import and export of goods according to the domestic destination and source of goods to measure trade openness. (3) Amount of investment in fixed assets (*fa*). The amount of investment in fixed assets refers to the funds of total investment in fixed assets in the whole country by region. It represents the workload of the construction industry and the acquisition of fixed assets in currency. It is a comprehensive indicator that reflects the scale, speed, proportional relationship, and direction of the use of fixed asset investments.

To analyze the reliability of the above variables, Table 3 shows the descriptive statistical results. The mean value of the explained variable (high-quality development of the regional economy) is 0.3996, the maximum value is 1, and the minimum value is 0.0759, which indicates that the quality of economic development varies greatly among different provinces. The standard deviation of the logarithmically transformed core explanatory variable (environmental regulations) is 0.9609, the maximum value is 7.258, and the minimum value is 1.808, indicating that the intensity of environmental regulations varies greatly in different provinces. The standard deviations of the other explanatory variables are all large, indicating that different provinces have large differences in industrial structure, fixed asset investment, human capital levels, and openness to trade.

5. Analysis of Results

5.1. HDI Partition Analysis and Characteristics

The human development index (HDI) is a comprehensive system of indicators proposed by the United Nations Development Programme (UNDP) to measure the level of economic and social development across countries. On the basis of the HDI for various regions in China in 2017, we used ArcGIS to draw a map. As shown in Figure 1, the darker the color, the greater the HDI. Figure 1 reports the 2017 HDI rankings for each region in mainland China. These HDI values are the comprehensive result of environmental regulations and economic development over a total of 12 years from 2006 to 2017. We find that environmental regulations and the quality of regional economic development are relatively high in regions with an HDI above the national average HDI, and the distribution of HDI values across above average regions is more homogeneous. This may be due to a better economic foundation in regions above the national average HDI. The competition between localities indicates a benign competition model in which the environment is first and the economy is second. Environmental protection and high-quality economic development can occur together in a win-win situation. The environmental regulations and quality of regional economic development are relatively low in regions below the national average HDI, and there are large deviations in HDI values across regions. This may be due to the relative underdevelopment in regions below the national average HDI, and there is wide variation in development between regions. The competition between localities indicates a bottom-to-bottom competition model in which the economy is first and the environment is second. For example, Hebei Province and Henan Province have economies that are mainly driven by polluting industries, which have a great impact on the environment, so the intensity of environmental regulations is also greater. On the other hand, due to its geographical features, Tibet has fewer polluting enterprises, a better ecological environment, and less investment in environmental governance, so the level of intensity of its environmental regulations is relatively low.

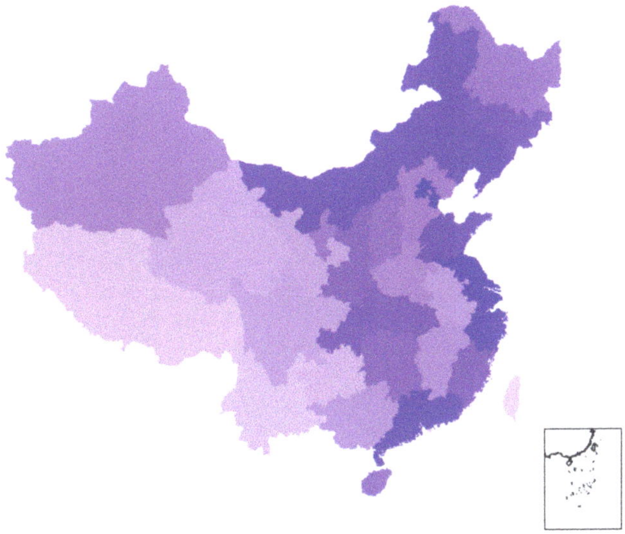

Figure 1. HDI ranking of Chinese provinces in 2017.

5.1.1. Comparison Based on the HDI Partition

Before processing the panel data, the fundamental question is whether to use a fixed effects model or a random effects model. First, we set the null hypothesis. According to the results of Hausmann's test, all of the data reject the null hypothesis, that is, the mean, reject random effects, and accept fixed effects. In addition, the development of different regions in China is uneven. Due to the different factor endowments in different regions, including the influence of other unobservable factors, the quality of economic development varies greatly, which also shows that the fixed effects model is more suitable. Therefore, we used the fixed effects model in the empirical evidence. Table 4 shows the impact of environmental regulations on the quality of economic development based on different HDI zones. Columns 1–2 present the empirical results when the cross-term conditions are not included, and columns 3–4 present the empirical results for the impact of environmental regulations on the quality of economic development when the cross-term conditions of the industrial structure are introduced. This paper proceeds to conduct a detailed analysis of the empirical results.

From the empirical results, we will find that the same intensity of environmental regulation is more effective in regions lower than the national average HDI. Combined with the environmental Kuznets curve (EKC), this shows that the current environmental regulation intensity has not reached the critical point and needs to be further increased.

Table 4. Estimation results for the effect of environmental regulations on the quality of economic development based on HDI zones.

Independent Variables	Dependent Variable (*hqd*)			
	No Cross Term Introduced	No Cross Term Introduced	Introduce Cross Terms	Introduce Cross Terms
	Provinces with Higher than the National Average HDI	Provinces with Lower than the National Average HDI	Provinces with Higher than the National Average HDI	Provinces with Lower than the National Average HDI
ln*er*	0.0171 (0.89)	0.0182 (0.39)	−0.1301 ** (−2.17)	−0.2125 ** (−2.56)
is	1.112 *** (5.83)	0.4583 *** (3.56)	−0.5109 (−0.88)	−2.393 * (−1.74)
fa	0.328 ** (1.98)	−0.0413 (1.36)	0.0594 ** (1.99)	0.053 *** (2.68)
edu	0.0267 (0.51)	−0.0345 (−0.77)	0.0182 *** (3.33)	−0.0185 (−0.48)
ln*imex*	0.0796 (1.32)	0.0715 *** (4.16)	0.0859 (1.37)	0.0616 (0.8)
ln*er*×*is*	-	-	0.2798 *** (2.95)	0.5647 ** (2.14)
Constant Term	−1.256 *** (−6.12)	−0.3697 (−1.03)	−0.622 ** (−2.21)	0.6063 (0.87)
Sample Size	108	252	108	252
Province	9	21	9	21
Method	FE	FE	FE	FE
Adj R-sq	0.8754	0.3501	0.8841	0.3671

Note: *, **, and *** indicate significance levels of 10%, 5%, and 1%, respectively, with standard errors in parentheses.

5.1.2. Analysis of the Results for Provinces with HDI Values above the National Average

The adjusted goodness-of-fit (Adj R-sq) is a correlation coefficient that reflects the degree of model fit, and deducts the influence of the number of included terms in the regression equation. From Table 4, it can be seen that the degree of fit is better in regions with higher than the national HDI, with a value of Adj R-sq close to 1. In comparison, the two models have a fair degree of fit in regions with lower than the national average HDI. The first column in Table 4 shows that environmental regulations in regions with an HDI above the national average HDI have a positive impact on the quality of regional economic development, with a coefficient of 0.0171. This shows that in provinces with a higher than average HDI, strengthening the intensity of environmental regulations can improve the quality of regional economic development and that high-quality development and environmental protection can coexist. However, this result is not statistically significant at the 10% level. This shows that when no cross-term conditions are included, the effect of environmental regulations is relatively small. This may be because the impact of environmental regulations on the quality of regional economic development is mainly exerted through intermediary factors, such as the industrial structure and foreign direct investment. The industrial structure has a positive effect on the quality of economic development, with a coefficient of 1.112, and it is significant at the 1% level. This shows that the industrial structure plays a significant role, and that optimizing and upgrading the industrial structure can greatly increase the speed of economic development. Other explanatory variables—the amount of fixed asset investment, the human capital level, and openness to trade—all have a positive impact on the quality of regional economic development. The volume of fixed asset investment is significantly positive at the 5% level, with a coefficient of 0.328, which indicates that the volume of fixed asset investment plays an important role in improving the quality of regional economic development. The level of human capital and trade openness are insignificant at the 10% level. The levels of human capital and trade openness in regions above the national average HDI have little impact on the high-quality development of the regional economy.

The third column in Table 4 shows the impact of environmental regulation in regions with HDI values above the national average on the quality of regional economic development estimated with cross-terms included. The cross-term between environmental regulations and the industrial structure has a negative effect on the quality of economic development in provinces with a higher than average HDI, with a coefficient of −0.1301. This result is not statistically significant at the 10% level. The reason may be that the intensity of environmental regulations is insufficient, which negatively impacts economic development. However, the cross-term between environmental regulations and the industrial structure has a positive impact on the quality of economic development in regions with HDI values above the national average, with a coefficient of 0.2798. This is significant at the 5% level. The reason is that environmental regulations affect the quality of economic development through the industrial structure, mainly due to the crowding-out mechanism and the growth in green barriers to entry. Environmental regulations have improved the efficiency of green innovation in industry, leading to industrial structure upgrades and improvements in the quality of economic development. Moreover, low efficiency and polluting enterprises are forced out of the industrial sector, and resources are concentrated in clean technology industries. Therefore, the higher the intensity of environmental regulation, the more obvious the positive effect of the cross-term between environmental regulation and industrial structure on economic development.

5.1.3. Analysis of the Results for Provinces with HDI Values below the National Average

The second column in Table 4 shows the empirical results of the estimation of the impact of environmental regulations on the quality of economic development in regions with an HDI below the national average, without the inclusion of the cross-terms. The results show that in regions with below average HDI values, environmental regulations have a positive impact on economic development, with a coefficient of 0.0182. This result is insignificant at the 10% level. In regions with below average HDI values, strengthening the intensity of environmental regulations does not improve the quality of regional economic development. This may be because regions with an HDI below the national average are relatively underdeveloped and there are large differences in regional development. In terms of economic development, the governments in these areas have fallen victim to the misconception of pollution first, treatment later. For the sake of economic growth, the government has sacrificed the environment, resulting in the inability of environmental protection and high-quality economic development to coexist. The amount of investment in fixed assets and the level of human capital have little effect on economic development. The industrial structure and openness to trade in regions with below average HDI values have a positive impact on economic development.

The fourth column in Table 4 shows empirical results of the estimation of the impact of environmental regulations on the quality of economic development in regions with HDI values that are below the national average, with cross-terms included. The results show that in regions with below average HDI values, environmental regulations have a negative impact on economic development, with an impact coefficient of −0.2125. This effect is greater in provinces with higher than average HDI values, and is significant at the 5% level. The coefficient on the cross-term with industrial structure is 0.5647, which is also significant at the 5% level. In addition, the cross-term has a positive impact on economic development, which is consistent with, and even greater than, the results for regions with an HDI above the national average.

5.2. Robustness Test

To confirm the reliability of the empirical results obtained above, this article conducts robustness tests both using the full sample and conducting the estimation with a substitute variable.

5.2.1. Full Sample Testing

Table 5 shows the results of the robustness test of the effect of environmental regulations on economic development quality using the full sample. The first column in Table 5 presents the results of the robustness test of the effect of environmental regulations on economic development quality without including the cross-term conditions. The second column in Table 5 presents the results of the robustness test of the effect of environmental regulations on economic development quality including the cross-term conditions. The first column in Table 5 shows that the impact of environmental regulations on the quality of economic development is not significant at the 10% level in the full sample, with a coefficient of 0.021. This shows that environmental regulation does not have a significantly positive effect on the quality of economic development, which is consistent with the empirical results for provinces with higher than average HDI values and regions with below average HDI values. Other explanatory variables, including industrial structure and trade openness, have a positive effect on the quality of economic development. The effect of the industrial structure is significantly positive at the 1% level, with a coefficient of 0.786, and that of human capital is significantly positive at the 1% level, with a coefficient of 0.0738. This is consistent with the results from the first and second columns of Table 4, which are for provinces with higher than average HDI values and provinces with below average HDI values, respectively. These results show that environmental regulation has not played a role in the governance of corporations, leading to insignificant optimization of the industrial structure. After the introduction of the cross-terms in the second column of Table 5, the full-sample empirical results show that environmental regulations have a negative impact on the quality of economic development. The impact of the cross-term with the industrial structure is positive and significant at the 1% level, which is consistent with the empirical results in the third and fourth columns in Table 4. The results listed in Table 5 are consistent with the results in Table 4, indicating that the results in Table 4 are highly robust.

Table 5. Estimated results for the effect of environmental regulation on the quality of economic development (full sample robustness test).

Independent Variables	**Dependent Variable (*hqd*)**	
	No Cross-Term Introduced	**Introduce Cross-Terms**
ln*er*	0.021 (0.72)	−0.1796 *** (−4.32)
is	0.786 *** (5.57)	−1.594 ** (−2.52)
fa	0.0164 (0.71)	0.0546 *** (3.64)
edu	0.0087 (0.26)	0.0007 (0.02)
ln*imex*	0.0738 *** (4.00)	0.067 *** (3.98)
ln*er*×*is*	-	0.4356 *** (4.03)

Table 5. *Cont.*

Independent Variables	Dependent Variable (*hqd*)	
	No Cross-Term Introduced	Introduce Cross-Terms
Constant Term	−0.6911 *** (−2.74)	0.1533 (0.4)
Sample Size	360	360
Province	30	30
Method	FE	FE

Note: **, and *** indicate the significance levels of 10%, 5%, and 1%, respectively, with standard errors in parentheses.

5.2.2. Substitute Variable Test (GDP)

This article draws on Shiyi-Chen (2018) [51], who adopted labor productivity (GDP) to measure the quality of economic development. This provides an estimate of the effect of environmental regulations on a substitute for the explained variable. Table 6 reports the robustness test results from the estimation of the impact of environmental regulations on the quality of economic development as measured by the substitute variable. Columns 1–2 of Table 6 present the robustness test for the effect of environmental regulations on the quality of economic development when no cross-term is introduced and the estimation is conducted at the HDI zone level. Columns 3–4 of Table 6 present the robustness test results for the effect of environmental regulations on the quality of economic development when the cross-terms are introduced and the estimation is conducted at the HDI zone level.

Table 6. Estimation results for the effect of environmental regulations on the quality of economic development (robustness test with a substitute variable).

Independent Variables	Replace Dependent Variable (*hqd*)			
	No Cross-Term Introduced	No Cross-Term Introduced	Introduce Cross-Terms	Introduce Cross-Terms
	Provinces with Higher than the National Average HDI	Provinces with Lower than the National Average HDI	Provinces with Higher than the National Average HDI	Provinces with Lower than the National Average HDI
ln*er*	−1.912 (−0.86)	2.72.4 (0.22)	−2.3776 ** (−2.24)	−1.0116 ** (−2.24)
is	8.5171 ** (3.30)	2.2220 *** (2.67)	−1.55776 (−1.19)	−1.06194 ** (−1.97)
fa	1.0911 ** (2.20)	3.484 ** (2.00)	1.4871 *** (3.77)	4.008 *** (2.64)
edu	6.565 (1.01)	−1.854 (−0.88)	5.311 (0.98)	−1.133 (−0.54)
ln*imex*	7.136 (1.05)	5.953 *** (4.39)	8.067 (1.60)	5.510 *** (3.83)
ln*er*x*is*	-	-	4.1542 * (1.88)	2.5427 ** (2.34)
Constant Term	−2.19494 *** (−4.47)	1.3546 *** (−3.19)	−1.25338 ** (−2.42)	7.88.1 (0.04)
Sample Size	108	252	108	252
Province	9	21	9	21
Method	FE	FE	FE	FE

Note: *, **, and *** indicate significance levels of 10%, 5%, and 1%, respectively, with standard errors in parentheses.

The first column of Table 6 shows that environmental regulations in regions with above average HDI have a negative impact on the high-quality development of the regional economy, but this result is not significant at the 10% level. After introducing the cross-term with industrial structure, the results in the second column of Table 6 show that at the 5% level, the impact of environmental regulations on the quality of economic development is negative, with a coefficient of −2.3776. This is consistent with the results in Table 4, indicating that the model is robust.

The second column of Table 6 shows that environmental regulations in regions with below average HDI have a positive impact on the quality of regional economic development, but the effect is not significant. After introducing the cross-term with industrial structure, environmental regulation does have a significant impact, with a coefficient of −1.0116. In columns 3–4 of Table 6, the impact of the cross-term with industrial structure on the quality of economic development is significant and positive, with a coefficient of 4.1542 for provinces with HDI values lower than the national average and with a coefficient of 2.5427 for provinces with HDI values higher than the national average. This shows that the mechanism of technological innovation induced by environmental regulations promotes industrial upgrades, which in turn promotes high-quality regional economic development. This is consistent with the empirical results in columns 3−4 of Table 4. The results presented in Table 6 are consistent with the estimated results presented in Table 4, indicating that the results in Table 4 are highly robust.

6. Conclusions and Policy Recommendations

Based on theory and the institution of environmental regulations, this paper constructs a comprehensive index of regional economic development quality using the three dimensions of the economic development level, the stability of development, and the sustainability of development. We used panel data from 30 provinces as our research sample, classified them into regions with HDI values above the national average and regions with HDI values below the national average, and conducted a comparative study of their regional differences. We draw the following conclusions.

(1) The same intensity of environmental regulations in regions with different HDI levels has different effects on the quality of economic development. The development of different provinces in China is extremely imbalanced. Limited by the influence of natural endowments and strategic positions, the economic development of the central and western regions is not as high as that of the eastern region. However, environmental regulations in provinces with above average HDI values and below average HDI values show the same trend in the quality of economic development. Moreover, the mediating effect of the industrial structure on environmental regulations is greater in areas with below average HDI values.

(2) Environmental regulations in provinces with above average HDI values have a negative impact on the quality of regional economic development. The cross-term between environmental regulations and the industrial structure has a significantly positive impact on the quality of regional economic development. The mechanism underlying the effect of environmental regulations on the quality of economic development is mainly the transformation and upgrading of the industrial structure. Reasonable environmental regulations can motivate enterprises to transform and upgrade. Highly polluting enterprises are eliminated through the survival of the fittest mechanism, and social resources are transferred to clean technology industries. This promotes the transfer of resources from the primary and secondary industries to the tertiary industry, optimizes the industrial structure, and promotes high-quality economic development.

(3) Environmental regulations in provinces with below average HDI values have a positive impact on the quality of regional economic development, but the effect is not significant at the 10% level. This shows that environmental regulations play a minor role in determining economic development quality. After the introduction of the cross-terms, environmental regulations and industrial structure are found to have a negative impact on the quality of regional economic development. Cross-terms between environmental regulation and industrial structure have a significantly positive impact on the quality of regional economic development. Moreover,

the mediating effect of the industrial structure on environmental regulations is greater in low HDI regions than in high HDI regions. This shows that the mechanism underlying the effect of environmental regulations on high-quality economic development is the industrial structure. In addition, the current intensity of environmental regulations is insufficient, and the intensity of environmental regulations should be increased.

In the process of economic development, with the endless emergence of environmental problems, the upgrading of the industrial structure has gradually become an effective means of coordinating environmental protection and economic growth. This article discusses the intermediary role of industrial upgrading in the impact of environmental regulations on the quality of economic development, and proposes the following recommendations and policies based on this discussion.

(1) Increase the intensity of environmental regulations by industry. The interaction between environmental regulation adjustment and industrial structure upgrading can promote high-quality economic development. However, the development of different industries is uneven, and the direction of industrial structure adjustment is also different. For example, the manufacturing industry needs to be adjusted from an extensive labor force development to a technology-intensive development. Therefore, policymakers are required to refine environmental regulations by industry and adopt environmental regulations that are most suitable for the development of the industry. In addition, some high-polluting industries should formulate more stringent environmental regulations and policies to help the industry cross the inflection point of the U-shaped curve as soon as possible to promote high-quality economic development by optimization and upgrading of the industrial structure.
(2) Local governments need to grasp the positive effects of environmental regulations on the upgrading of industrial structure, and guide healthy competition in environmental regulations among regions. Regional development in China is uneven, especially the economic development between the eastern and western parts of the country. Local governments must take into account the actual local conditions when formulating environmental regulations and policies. Regions above the national average HDI should sum up their experience, and regions below the national average HDI should strengthen environmental regulations. The local government should adjust measures to local conditions, tilt resources to green development industries, vigorously introduce technical talents, and gradually adjust the structure of pollution-intensive industries.
(3) Actively play the role of green technological innovation in promoting the upgrading of industrial structure. Talent is the foundation of innovation. The Chinese government should vigorously cultivate high-level green innovative talents, continuously invest in green innovative technologies, and accelerate the advancement and rationalization of the industrial structure.
(4) Improve public awareness of environmental protection. Environmental protection is not an individual's business. It requires the economic participation of all sectors of society, and ordinary people are also an important part of that. As consumers, we resolutely refuse to purchase goods produced by highly polluting technologies. This can fundamentally increase the enthusiasm of enterprises and enhance their environmental responsibility. At the same time, the government also encourages institutional and technological innovation and strengthens the role of enterprises as the main body and main force of technological innovation.

This article only studies the mediating effect of the industrial structure. The mechanism of environmental regulation in the high-quality development of regional economy is complex, and the path is extensive. We only studied the mediating effect of industrial structure, which has certain limitations. In addition to upgrading the industrial structure, the technology spillover effect of FDI may also affect the high-quality development of the regional economy. In the future, we can study the effects of environmental regulation on the high-quality development of the regional economy when FDI, technological innovation, and other factors are used as intermediary variables at the same time.

In addition, China's digital economy has developed rapidly in recent years. For the next research direction, we can also consider the digital economy as an explained variable or mediating variable in the study of high-quality development of the regional economy.

Author Contributions: Conceptualization, L.C. and C.H.; methodology, W.Y. and L.C.; software, K.J. and C.H.; formal analysis, W.Y. and K.J.; resources, L.C.; data curation, L.C.; writing—original draft preparation, L.C. and C.H.; writing—review and editing, L.C., C.H., W.Y. and K.J.; supervision, W.Y. and K.J.; project administration, W.Y.; funding acquisition, L.C. All authors have read and agreed to the published version of the manuscript.

Funding: This research was funded by Science and Technology Project of Jiangxi Provincial Department of Education, grant number is GJJ191661.

Acknowledgments: We would like to express our gratitude to all those who helped us during the writing of this article. Our deepest gratitude goes first and foremost to Guoan-Xiao, for his constant encouragement and guidance.

Conflicts of Interest: The authors declare no conflict of interest.

References

1. Tongbin, Z.; Qi, Z.; Qingquan, F. Research on Corporate Governance Motivation and Public Participation Externalities under Government Environmental Regulation. *China Popul. Resour. Environ.* **2017**, *27*, 36–43. (In Chinese)
2. Burton, D.M.; Gomez, I.A.; Love, H.A. Environmental regulation cost and industry structure changes. *Land Econ.* **2011**, *3*, 545–557. [CrossRef]
3. Chong, Z.; Qin, C.; Ye, X. Environmental regulation, economic network and sustainable growth of urban agglomerations in China. *Sustainability* **2016**, *8*, 467. [CrossRef]
4. Eiadat, Y.; Kelly, A.; Roche, F.; Eyadat, H. Green and competitive? An empirical test of the mediating role of environmental innovation strategy. *J. World Bus.* **2008**, *43*, 131–145. [CrossRef]
5. Jahanshahi, A.A.; Brem, A. Antecedents of corporate environmental commitments: Te role of customers. *Int. J. Environ. Res. Public Health* **2018**, *15*, 1191. [CrossRef]
6. Gray, W.B.; Shadbegian, R.J. Plant vintage, technology, and environmental regulation. *J. Environ. Econ. Manag.* **2013**, *46*, 384–402. [CrossRef]
7. Ling, L.I.; Feng, T.A.O. Selection of Optimal Environmental Regulation Intensity for Chinese Manufacturing Industry—Based on the Green TFP Perspective. *China Ind. Econ.* **2012**, *5*, 70–82. (In Chinese) [CrossRef]
8. Pearce, D.W.; Turner, R.K. *Economics of Natural Resources and the Environment*; JHU Press: Baltimore, MD, USA, 1990.
9. Papalia, R.B.; Bertarelli, S. Nonlinearities in economic growth and club convergence. *Empir. Econ.* **2013**, *44*, 1171–1202. [CrossRef]
10. Telle, K.; Larsson, J.D.O. Environmental regulations hamper productivity growth? How accounting for improvements of plants' environmental performance can change the conclusion. *Ecol. Econ.* **2007**, *61*, 438–445. [CrossRef]
11. Van der Ploeg, R.; Withagen, C. Green growth, green paradox and the global economic crisis. *Environ. Innov. Soc. Transit.* **2013**, *6*, 116–119. [CrossRef]
12. Shuai, S.; Fan, Z. Modeling the role of environmental regulations in regional green economy efficiency of China: Empirical evidence from super efficiency. *J. Environ. Manag.* **2020**, *261*, 110227. [CrossRef] [PubMed]
13. Ge, T.; Li, J.; Sha, R.; Hao, X. Environmental regulations, financial constraints and export green-sophistication: Evidence from China's enterprises. *J. Clean. Product.* **2020**, *251*, 119671. [CrossRef]
14. Managi, S.; Kumar, P. *Inclusive Wealth Report 2018*; Taylor & Francis: London, UK, 2018.
15. George, G.; McGahan, A.M.; Prabhu, J. Innovation for inclusive growth: Towards a theoretical framework and a research agenda. *J. Manag. Stud.* **2012**, *49*, 661–683. [CrossRef]
16. Vellala, P.S.; Madala, M.K.; Chhattopadhyay, U. A theoretical model for inclusive economic growth in Indian context. *Int. J. Hum. Soc. Sci.* **2014**, *4*, 229–235.
17. Oyinlola, M.A.; Adedeji, A.A.; Bolarinwa, M.O.; Olabisi, N. Governance, domestic resource mobilization, and inclusive growth in sub-Saharan Africa. *Econ. Anal. Pol.* **2020**, *65*, 68–88. [CrossRef]
18. Jalles, J.T.; Mello, L.D. Cross-country evidence on the determinants of inclusive growth episodes. *Rev. Dev. Econ.* **2019**, *23*, 1818–1839. [CrossRef]

19. Rajapaksa, D.; Islam, M.; Managi, S. Natural capital depletion: The impact of natural disasters on inclusive growth. *Econ. Disasters Clim. Chang.* **2017**, *1*, 233–244. [CrossRef]
20. Grossman, G.M.; Kreuger, A.B. Economic growth and the environment. *Q. J. Econ.* **1995**, *110*, 353–377. [CrossRef]
21. World Bank. *World Development Report 1992: Development and Environment*; China Financial and Economic Press: Beijing, China, 1992.
22. Caruso, G.; Di Battista, T.; Gattone, S.A. A Micro-level Analysis of Regional Economic Activity through a PCA Approach. In *Decision Economics: Complexity of Decisions and Decisions for Complexity. DECON 2019. Advances in Intelligent Systems and Computing, 1009*; Bucciarelli, E., Chen, S.H., Corchado, J., Eds.; Springer: Cham, Switzerland, 2020. [CrossRef]
23. D'Adamo, I.; Falcone, P.M.; Morone, P. A New Socio-economic Indicator to Measure the Performance of Bioeconomy Sectors in Europe. *Ecol. Econ.* **2020**, *176*, 106724. [CrossRef]
24. Weitzman, M.L. Prices vs. quantities. *Rev. Econ. Stud.* **1974**, *41*, 477–491. [CrossRef]
25. Xie, R.; Yuan, Y.; Huang, J. Different types of environmental regulations and heterogeneous influence on "green" productivity: Evidence from China. *Ecol. Econ.* **2017**, *132*, 104–112. [CrossRef]
26. Hahn, R.W. The political economy of environmental regulation: Towards a unifying framework. *Public Choice* **1990**, *65*, 21–47. [CrossRef]
27. Tietenberg, T.H.; Lewis, L. *Environmental and Natural Resource Economics*; Routledge: London, UK; New York, NY, USA, 2018.
28. Hahn, R.W. Market power and transferable property rights. *Q. J. Econ.* **1984**, *99*, 753–765. [CrossRef]
29. OECD. *Guidelines for the Application of Environmental Economic Instruments*; China Environmental Science Press: Beijing, China, 1994; pp. 12–13.
30. Pigou, A.C. *The Economics of Welfare*; MacMillan: London, UK, 1920.
31. Stavins, R.N. Market-Based Environmental Policies. *Public Policies Environ. Prot.* **2007**, *2*, 159–173.
32. Levinson, A. Environmental regulations and industry location: International and domestic evidence. *Fair Trade Harmon. Prerequisites Free Trade* **1996**, *1*, 429–457.
33. Antweiler, W.; Copeland, B.R.; Taylor, M.S. Is free trade good for the environment? *American Econ. Rev.* **2001**, *91*, 877–908. [CrossRef]
34. Baoping, R.; Fengan, W. Judgment criteria, determinants and realization methods of China's high-quality development in the new era. *Reform* **2018**, *4*, 5–16. (In Chinese)
35. Hong, Z. *Research on the Quality Measurement of the Economic Development of the Yangtze River Economic Zone*; China University of Geosciences: Wuhan, China, 2015. (In Chinese)
36. Xiurui, X.; Gang, W.; Xiaoan, Z.; Jianguo, S.; Deyong, S. Research advances in assessment of green GDP indicator. *Chinese J. Ecol.* **2007**, *26*, 1107–1113. (In Chinese)
37. Taosheng, S.; Jinbin, W. On the integration between ecology and economics based on sustainable development. *Econ. Surv.* **2001**, *18*, 13–15. (In Chinese)
38. Anderson, V. *Alternative Economic Indicators*; Rutledge: London, UK, 1991.
39. Study Group on Sustainable Development, Chinese Academy of Sciences. *1999 Report on China's Sustainable Development Strategy*; Science Press: Beijing, China, 1999. (In Chinese)
40. Gorgij, A.D.; Kisi, O.; Moghaddam, A.A.; Taghipour, A. Groundwater quality ranking for drinking purposes, using the entropy method and the spatial autocorrelation index. *Environ. Earth Sci.* **2017**, *76*, 269. [CrossRef]
41. Li, X.G.; Wei, X.; Huang, Q. Comprehensive entropy weight observability-controllability risk analysis and its application to water resource decision-making. *Water SA* **2012**, *38*, 573–579. [CrossRef]
42. Dong, G.; Shen, J.; Jia, Y.; Sun, F. Comprehensive evaluation of water resource security: Case study from Luoyang City, China. *Water* **2018**, *10*, 1106. [CrossRef]
43. Liu, L.; Zhou, J.; An, X.; Zhang, Y.; Yang, L. Using fuzzy theory and information entropy for water quality assessment in Three Gorges region, China. *Expert Syst. Appl.* **2010**, *37*, 2517–2521. [CrossRef]
44. Amiri, V.; Rezaei, M.; Sohrabi, N. Groundwater quality assessment using entropy weighted water quality index (EWQI) in Lenjanat, Iran. *Environ. Earth Sci.* **2014**, *72*, 3479–3490. [CrossRef]
45. Bartzokas, A.; Yarime, M. *Technology Trends in Pollution-Intensive Industries: A Review of Sectoral Trends*; No 1997-06; UNU-INTECH Discussion Paper Series; United Nations University-INTECH, USE: Tokyo, Japan, 1997.

46. Shi, B.; Ren, B. Measurement and analysis of the high-quality development of China's inter-provincial economy. *Econ. Issues* **2018**, *4*, 1–6. (In Chinese)
47. Levinson, A. Environmental regulation and manufacturers' location choices: Evidence from the census of manufactures. *J. Public Econ.* **1996**, *62*, 5–29. [CrossRef]
48. Liu, Y.; Xiong, Y. The Impact of Environmental Regulation on the Quality of Regional Economic Development—Based on the Comparison of HDI Regions. *Econ. Survey* **2020**, *37*, 1–10.
49. Li, T. Environmental Regulation, Service Industry Development and the Industrial Structure Adjustment in China. *Econ. Manag.* **2013**, *35*, 1–10. (In Chinese)
50. Zeng, Q.; Zeng, X.; Liu, J. The Impact of Environmental Regulation Tools on Environmental Quality from the Perspective of Industrial Structure. *Econ. Survey* **2018**, *35*, 94–100. (In Chinese)
51. Chen, S.; Chen, D. Haze pollution, government governance and high-quality economic development. *Econ. Res.* **2018**, *2*, 20–34.

Publisher's Note: MDPI stays neutral with regard to jurisdictional claims in published maps and institutional affiliations.

MDPI
St. Alban-Anlage 66
4052 Basel
Switzerland
Tel. +41 61 683 77 34
Fax +41 61 302 89 18
www.mdpi.com

Land Editorial Office
E-mail: land@mdpi.com
www.mdpi.com/journal/land

www.ingramcontent.com/pod-product-compliance
Lightning Source LLC
Chambersburg PA
CBHW061552210926
51850CB00009B/2082
* 9 7 8 3 0 3 6 5 2 9 9 1 2 *